Ecological Systems of the Geobiosphere

Heinrich Walter   Siegmar-W. Breckle

Ecological Systems of the Geobiosphere

1 *Ecological Principles in Global Perspective*

Translated by Sheila Gruber

With 130 Figures

Springer-Verlag Berlin Heidelberg GmbH

Professor Dr. Heinrich Walter
Universität Hohenheim
Postfach 700562
7000 Stuttgart 70 / FRG

Professor Dr. Siegmar-W. Breckle
Universität Bielefeld
Postfach 8640
4800 Bielefeld 1 / FRG

Translator:
Sheila Gruber
Nibelungenstraße 32
5300 Bonn 2 / FRG

Title of the Original Edition:
Walter/Breckle, Ökologie der Erde, Band 1
© Gustav Fischer Verlag Stuttgart 1983
ISBN 3-437-20297-9

Library of Congress Cataloging in Publication Data. Walter, Heinrich, 1898–. Ecological principles in global perspective. (Ecological systems of the geobiosphere; v. 1)
Translation of: Ökologische Grundlagen in globaler Sicht. 1. Ecology—Collected works.
I. Title. II. Breckle, Siegmar-W. III. Series: Walter, Heinrich, 1898–. Ökologie der Erde. English;
v. 1. QH540.3.W3513 vol. 1 551.5 s [551.5] 84-23663 [QH541]

ISBN 978-3-662-02439-3   ISBN 978-3-662-02437-9 (eBook)
DOI 10.1007/978-3-662-02437-9

© Springer-Verlag Berlin Heidelberg 1985

Originally published by Springer-Verlag Berlin Heidelberg New York in 1985.
Softcover reprint of the hardcover 1st edition 1985

Typesetting and printing:
Petersche Druckerei GmbH & Co. Offset KG, Rothenburg ob der Tauber.

2131/3130-543210

Preface

Des Menschen Werk auf Erden kann vergehen,
Doch Leben stets im Wandel wird bestehen.
Heinrich Walter

The importance of ecology for the fate of mankind is receiving ever wider recognition. A syncretic-holistic approach to ecology was recently given unexpected support by the well-known atomic physicist and pupil of Heisenberg, Fritjov Capra. In his book *The Turning Point*, published in 1982 in the U.S.A., Capra comments critically, from the viewpoint of the latest findings of subatomic physics, on the mechanical-analytical approach which still predominates in the biological sciences, and adds some philosophical reflections. The following quotations are important ecologically and may be of interest to biologists in general: "It is now becoming apparent that overemphasis on the scientific method and on rational, analytic thinking has led to attitudes that are profoundly anti-ecological. In truth, the understanding of ecosystems is hindered by the very nature of the rational mind. Rational thinking is linear, whereas ecological awareness arises from an intuition of non-linear systems....

The Cartesian view of the universe as a mechanical system provided a 'scientific' sanction for the manipulation and exploitation of nature that has become typical of Western culture....

The problem is that scientists, encouraged by their success in treating living organisms as machines, tend to believe that they are nothing but machines....

Modern physics can show the other sciences that scientific thinking does not necessarily have to be reductionist and mechanistic, that holistic and ecological views are also scientifically sound....

Scientists will not need to be reluctant to adopt a holistic framework, as they often are today, for fear of being unscientific. Modern physics can show them that such a framework is not only scientific but is in agreement with the most advanced scientific theories of physical reality."

We, however, have never been troubled by the fears alluded to here (vide Walter 1982). Indeed, it has always been our aim to present a picture of the terrestrial biosphere as an entity, the separate parts of which are linked in a dynamic way as a result of constant changes in their living components.

This three-volume work is an attempt to encourage such an integrative or holistic approach, by treating the many results of analytical-ecophysiological research not in isolation, but as parts of the whole.

One of the problems facing biological science today is the overabundance of detailed information amassed by all sorts of specialists. This makes it very difficult for the student to visualize the whole. There is no lack of symposia at which ecologists address themselves to some particular, frequently very limited aspect of a narrowly defined problem; nor is there any shortage of well-documented reviews reporting the latest investigations into various special questions; almost no attempt, however, has been made at a synthesis; that is, a presentation of relationships on a grand scale, with the whole as the starting point.

Many valuable scientific "building bricks" form but a pile of rubble if they are not, figuratively speaking, formed into a meaningful structure or an artistically valuable mosaic picture. It is only when the structure or pattern is recognizable in outline that it is possible to identify the gaps, and thus undertake purposeful investigations to discover the missing "bricks". Otherwise it will be a matter of pure luck whether, amongst the accumulation of new facts, any truly useful "building bricks" turn up. This is particularly relevant in ecology, as research is now predominantly physiological and biochemical, mainly analytical, and concerned especially with processes which can be described and explained in quantitative physicochemical terms. To this end, short-term laboratory investigations are made in conditions which are unlike those in the natural environment and, in keeping with the American motto "publish or perish", the results are immediately submitted for publication.

The ecologist has, however, to deal with processes which continue throughout the period of vegetational growth and take place under constantly changing conditions. Here not only the quantitative but also the qualitative aspects are important, as, for example, in growth processes, which lead to qualitative changes; in the developmental phases from germination, through vegetational growth to the reproductive phase; in competitive relationships which are very difficult to quantify; in the interaction of all the environmental factors, and so on. Only when they are seen as parts of a whole is it possible to understand these separate aspects; not the other way round.

The global approach we have attempted here must be seen as a first step towards a holistic treatment of the subject. It is based largely on worldwide personal experience, together with an assessment of already available "building bricks". The hope is that it may stimulate further ecological research as did Schimper's *Pflanzengeographie auf physiologischer Grundlage*[1] almost a century ago.

The new approach represents for the senior author the culmination of more than 60 years of ecological research, for the co-author the start of a comprehensive and integrative treatment of the subject matter. If, in the process, we go beyond the bounds of phytocoenology itself, to take account of related fields of research, this should be seen as a stimulus to research workers in these fields to attempt to close the large gaps which still exist in our knowledge, and to direct their research to the ecosystem as a whole. Only when global ecological principles have been grasped will it be possible to take the correct steps to avert the dangers which threaten our environment. There is no time to be lost.

The German original *Ökologie der Erde. Band 1. Ökologische Grundlagen aus globaler Sicht*, published in 1983, is, in part, a revised edition of *Vegetation der Erde*, first published in 1962 and now out of print. In this new work the previous subdivision of the geobiosphere into vegetational zones has been replaced by an ecological subdivision (Walter 1976), because the vegetation, albeit a very important living component of ecological systems, is only one of many, and should not be considered in isolation from the rest. The most important component, which is entirely independent of all the others, is the climate.

This volume is the first in a series of three. It deals with ecological problems of a general nature, knowledge which will be assumed in

---

1 English edition: Schimper, A. F. W.: Plant geography upon a physiological basis. Clarendon Press, Oxford, 1903

Volumes 2 and 3. As far as possible, the problems are treated on a global basis, to avoid the distortion that arises through viewing everything on the basis of relationships in the northern hemisphere, the part of the earth most thoroughly investigated. Volumes 2 and 3 deal with "special ecological relationships" in the different climatic zones of the earth, Volume 2 with tropical and subtropical zones and Volume 3 with temperate and arctic zones.

Since this is intended for use as a university textbook, the aim is not to bring together as much factual information as possible, but rather to attempt a synthesis and present an overall view of the most important ecological problems of the geobiosphere globally seen; that is, of ecological events on a macro-scale. No attempt has been made to describe the processes involved at the molecular or biochemical level, while models for which there is as yet no well-founded global basis are not considered. We will be grateful for any criticisms or important information which has been overlooked.

We wish especially to thank Mrs. Sheila Gruber for the exceptionally careful translation of the German edition and the Springer-Verlag for the publication of the English version of Volume 1. We hope that the translation and publication of Volume 2 will soon be completed.

December 1984   Heinrich Walter
                                                            Siegmar-W. Breckle

Contents

1 Introduction

1.1 A Survey of the Scope of Ecological Research

Ecology is, in broad terms, a field of study concerned with the relationship of living organisms to one another and to their non-living environment. It is thus a *biological science* which, however, encroaches on other, related fields. Where no living organisms exist, as on the moon, no ecological research can be undertaken. It is necessary, however, to differentiate more clearly between ecology and other branches of biology.

Within biology, in its broadest sense, three basic units can be distinguished—the *cell*, the *organism* and the *community*. The study of plant and animal cells, which are fundamentally alike both in their submicroscopic structure and in their basic metabolism, is the concern of *electron microscopy*, *molecular biology*, *biochemistry* and also *genetics*.

Evolution from single-celled to many-celled organisms and finally to the most highly developed forms of life took two directions, leading to the plants on the one hand and to the animals on the other. These groups of living organisms differ so markedly in their anatomy, morphology and physiology that biological science came to be separated into the two distinct disciplines of *botany* and *zoology*.

In recent times it has become clear that at the highest level of complexity of organization, that of the community, or biocoenose, plant communities (phytocoenoses) are inseparably inter-linked with the communities of animals which are dependent on them, and that living organisms, together with various environmental factors such as climate and soil, form complex, open systems described as *ecosystems*. *Ecology* is the study of such ecosystems, which in turn form

part of larger groupings, or biomes, all of which together make up the biosphere of our earth. Here we will consider ecology from a global viewpoint.

In ecological research a broadly based, integrated approach is essential. Such research has to be conducted in the field under varying environmental conditions, and must take into account the effects of competitive interactions as well. Such an *integrated treatment, leading to a final synthesis of the many different aspects, is a fundamental feature of ecology*. "Autecology" or "ecophysiology", which is more analytical, is based increasingly on laboratory investigations; it is concerned with special problems and is essentially physiology seen from an ecological viewpoint (Lange et al. 1981). In any integrated approach such results must certainly be taken into account, but their applicability under natural and ever-varying conditions in the field must be very critically appraised.

An ecosystem is a complex and dynamic structure, built up of *abiotic components* (environmental factors such as climate and soil), and *biotic components*, the *producers* —or autotrophic green plants, the heterotrophic *consumers—*mainly animals, and the *decomposers—*microorganisms and small invertebrates (vide Sect. 4).

An ecosystem is characterized by a cycling of matter. This starts with the organic compounds formed during photosynthesis, and these serve as food for consumers. Plant and animal remains are ultimately completely broken down by decomposers, and the mineralysed nutrients released are again taken up by plants, thus closing the cycle.

Parallel to this is an energy flow, for in photosynthetic assimilation radiant energy from sunlight is converted into chemical energy, which in turn is utilized by consumers and decomposers (cf. Sect. 4.1).

There are two main types of ecosystem; these reflect the nature of the environment and differ markedly in their structure, namely, terrestrial ecosystems on land and aquatic ecosystems in water.

Terrestrial ecosystems are readily distinguished one from another, since each is associated with a particular phytocoenose, or better, a biogeocoenose; aquatic ecosystems in large bodies of water are, however, usually not clearly demarcated.

Different ecosystems are interrelated in some particular way so as to form clearly recognizable larger units; for these the term biome (either terrestrial or aquatic) has come to be used. All the biomes together make up the largest ecological unit, the biosphere. Here, too, a distinction is made between the *geobiosphere* on land and the *hydrobiosphere* of waters and rivers.

We will limit our attention to the geobiosphere; the hydrobiosphere is the concern of the special field of hydrobiology, which includes oceanography and limnology.

While in this work the various terrestrial ecological units described above will be examined down to the level of the small ecosystem, this first volume will be limited to questions of a more general nature. It should be seen as an introduction to Volumes 2 and 3.

The geobiosphere includes the layer of the atmosphere nearest to the earth's surface as far as plants extend into it (maximally 100 m) and also the upper layer of the lithosphere or the soil (known also as the pedosphere) that is penetrated by roots. Only in these layers of soil and air do animals have any permanent habitats, even though they may temporarily, in flying or being blown passively by the wind in a dormant condition, enter higher layers of the atmosphere. Organisms living in caves may be found deeper below the earth's surface.

The hydrobiosphere, by contrast, extends to the greatest depths of the oceans.

The following will make clear how marked is the difference in structure between the hydrobiosphere and the geobiosphere.

In *aquatic systems*, if we exclude the narrow litoral zone, the producers are autotrophic algae, which are usually either unicellular or occur in colonies; they are suspended in the water and form part of the plankton. They multiply rapidly by cell division. Since they require light for photosynthesis, they are found only in the upper, illuminated water layers. They serve as food for animal organisms in the micro- and macroplankton; these in turn are eaten by larger animals and this chain extends to fishes, aquatic mammals and those birds which take their prey out of the water. All dead organic matter from these organisms is converted to inorganic compounds as a result of the activities of decomposers both in the water and in the mud layer beneath. *The available phytomass (total dry weight of producers) in water is small, whereas the primary production, that is, the total amount of dry organic matter produced in the course of a year is, as a result of rapid multiplication of the algae, relatively large.* Since this primary production serves as food for animals, and to a large extent becomes incorporated into their body structures (secondary production), the zoomass (dry weight of consumers) is very large compared with the phytomass; it is often more than 15 times larger.

The situation is quite different in the *terrestrial* ecosystems of the geobiosphere. Here the producers are higher plants, in comparison with which the lower plants play an unimportant role. In higher plants the biomass of the photosynthetic tissue is only a small part of the total phytomass, which includes organs like buds and roots which have no chlorophyll. In woody plants the dead wood in the centre of trunk and branches remains part of the phytomass until the tree dies. *For this reason the standing phytomass is always very large, particularly in wooded areas. In comparison with this, the annual production of organic dry matter, that is, primary production, is relatively small,* particularly since the cells involved in photosynthesis no longer divide and are not active all through the year. Only a small part of the phytomass, generally much less than 1%, serves as food for animals living aboveground; their zoomass is thus also small, being on average less than 1% and often as little as 0.1% of the phytomass. The following figures show the average relationship between phytomass and primary production for all terrestrial and aquatic ecosystems of the biosphere:

	Phyto-mass		Primary production
Terrestrial ecosystem	14	:	1
Aquatic ecosystem	1	:	350

The hydrosphere, which includes the oceans, covers an area of $361.1 \times 10^6 km^2$, or 71% of the surface of the globe. It is twice the area of the geobiosphere which, with all the continents, has a surface area of $148.9 \times 10^6 km^2$, or only 29% of the earth's surface (for more precise data see Walter 1979, p 255).

1.2 The Relationship Between the Geo- and Hydrobiospheres

The geo- and hydrobiospheres are not entirely independent parts of the biosphere, but are interrelated by virtue of a constant exchange of water. This water exchange also results, of course, in exchange of a certain amount of other matter. Some of the water vapour formed over the sea as a result of evaporation is carried by air movements to the geobiosphere, where it is deposited as rain; it is thence carried back to the sea by way of rivers and groundwater streams. The water vapour carries with it latent heat to the geobiosphere; thus not only the water balance, but also the temperature and hence the general climate of land masses are influenced by the global exchange of water. Baumgartner and Reichel (1975) provide the following data on the worldwide water balance.

The total water content of the earth, as a fluid, a gas or in solid form, is 1384120000 km³, or in round figures, $1.4 \times 10^9 km^3$. The salt water of the oceans accounts for 97.4% of this; 2% is in the form of ice (polar ice, sea ice and glaciers) and only 0.6% as freshwater on land (0.58% as groundwater or moisture in the soil and 0.02% in lakes and rivers). Water vapour in the atmosphere,

making up 0.001% of the total, is relatively insignificant. On average, 425000 km³ ($\triangleq$ 1176mm rainfall) of water evaporates annually from the surface of the sea; of this 385000 km³ falls back as rain into the oceans. The difference of 40000 km³ ($\triangleq$ 110 mm related to ocean surface) reaches the land as rain from the sea. The average annual evaporation from the land surface is 71000 km³ ($\triangleq$ 480mm) of water, while the total amount of rain is 111000 km³ ($\triangleq$ 748 mm). The difference of 40000 km³ ($\triangleq$ 266 mm related to land surface) flows back to the sea, thus equalizing the loss of water from the sea by evaporation. In any one year, the amount of water circulated over the entire earth is, in round figures, 500000 km³ ($\triangleq$ 1000mm), yet the water turnover of the land masses accounts for only 18% of this.

The distribution of land between the northern and southern hemispheres is asymmetrical, with the land area north of the equator being approximately twice that south of the equator. The flow-off to the sea is 25000 km³ in the northern hemisphere, but only 15000 km³ in the southern hemisphere. The balance flows as ocean currents from north to south across the equator. To equalize this, the same amount of water, in the form of water vapour, must flow in the opposite direction across the equator to the northern hemisphere; as a result 600×10^6 kcal (2520×10^6kJ) of latent heat is transferred each year. This additional quantity of heat received by the Earth's surface in the northern hemisphere contributes to the somewhat warmer climate of the northern landmasses compared with those of the southern hemisphere, a factor which is of importance in our further considerations. Furthermore, there is an exchange of heat between sea and land effected by wind.

The hydrobiosphere is of importance also in stabilizing the CO_2 content of the atmosphere. The quantity of CO_2 dissolved in the sea is approximately six times that in the atmosphere. In addition, in warmer parts of the oceans, weakly soluble $CaCO_3$ is precipitated, often as a result of the activities of microorganisms.

There is a balance between the amount of CO_2 dissolved in the sea and the partial pressure of CO_2 in the air. If the latter rises, more CO_2 dissolves in the water; if the partial pressure falls, CO_2 is released into the atmosphere from the oceans, which thus act as a buffer. A further exchange between the geo- and hydrobiospheres involves substances dissolved or suspended in river waters. In recent times, as a result of pollution of rivers by effluents, the quantity of various substances reaching the sea has risen alarmingly. Valuable nutrient elements, like phosphorous, are being continuously lost from the geobiosphere in this way. Restoration of such substances to the geobiosphere can take place only through the utilization of marine animals as food for man, or if, in the course of geological time, a continental shelf were to be lifted and become part of the land mass. This is especially important in the exchange of salt between sea and land, which will be discussed in the next section. Although the geo- and hydrobiospheres interact in various ways, their ecological characteristics are, as we have seen, so fundamentally different that they must be treated separately, although this is not usually the case in textbooks of general ecology. Contrary to normal practice, we will deal here exclusively with the geobiosphere.

Semiaquatic ecosystems, characterized by temporary flooding or by a permanently high goundwater level, link aquatic and terrestrial ecosystems. Where in this case the producers are predominantly mosses, as on moors, or higher plants, as in mangrove swamps, we will treat these together with terrestrial ecosystems.

1.3 Salt Exchange Between the Hydrobiosphere and the Geobiosphere

Salt plays an ecologically very important role in the relationship between the hydrobiosphere and the geobiosphere. Among readily soluble salts, sodium chloride (NaCl) is the major constituent of sea salt and it is also widely distributed in arid areas on land.

Apart from NaCl, there are very few chlorine-containing minerals. Thus the occurrence of chlorine is more or less identical with that of NaCl. According to Rubey

(1951), the distribution of chlorine is as follows:

Chlorine in seawater of the hydrobiosphere	276×10^{14} t
Chlorine in marine sediments on land	30×10^{14} t
Chlorine in mineral rock on land	5×10^{14} t

The chlorine content of marine sediments is derived from the NaCl of seawater included in them at the time of sedimentation. When such rock formations on land weather, the NaCl is washed out by rain and returned to the sea by rivers. In desert areas where there is no drainage, the scanty rain washes the salt into depressions; here the water evaporates and salt accumulates, forming either saline soils or salt pans. These are mostly found in deserts where marine deposits occur (Jurassic, Cretaceous, Tertiary) as, for example, in the northern Sahara. Where, however, the bedrock is crystalline or volcanic, salt accumulation is less marked.

It may therefore be said that almost *all deposits of salt on land are of marine origin*, even in the deserts of central Asia.

Salt also reaches the land from the sea by way of the atmosphere. According to Korte (1980), this involves an annual total of 10^8 t of NaCl. On coasts with heavy surf, a considerable quantity of seawater is flung up as spray. Spray is formed also in the open sea as a result of the action of wind and waves. The water evaporates from the minute droplets of spray, and the resulting fine salt-dust is blown far inland by sea winds. This salt falls to the ground; in wet climates it is constantly washed into rivers by rain and is carried back to the sea. One may thus speak of a *salt cycle* or of *cyclic salt*. If the climate of a coastal area is dry, this cycle is, however, interrupted, and in an area with no river drainage, salt accumulates, being washed by the scanty rainfall only from higher ground into the hollows. Here, too, salt pans and salt lakes arise, with associated saline soils. In the dry season, the surface of a salt pan may dry up and become covered with a salt crust; from such pans large quantities of salt may be blown inland. In this way, many salt pans and salt lakes have been formed several kilometers inland from the coast in the arid south-west of Australia. In this region the estimated amount of salt-dust deposited annually with rain 300 km inland from the coastline is 40 kg ha^{-1}; at a distance of 600 km approximately 10 kg ha^{-1} (Teakle 1937). In many other arid areas, for example Iran (Breckle 1981), cyclic salt contributes a large part to salt deposits in closed basins, since rainwater always contains a certain amount of salt, of the order of 5 to 20 ppm.

Along the coastal strip of the Namib Desert (South West Africa), fine sea spray mixed with fog may be carried up to 50 km inland, resulting in salination of the outer margin of the Namib (Walter 1936 and see Vol. 2).

The salt marshes of the "Putrid Sea" (Sivash), between the Crimea and the mainland, dry out in summer. Salt from the surface is carried to the southernmost part of the steppe and gives rise to the slightly saline "solonetz soil" of this region.

Saline soils also arise in areas once covered by the sea, for instance, in the Caspian lowland (Walter and Box 1983). Subsequent redistribution of the salt occurs in spring; higher ground is desalinated by water from the melting snow, while the salt content of the more low-lying terrain is increased. Saline soils are also created when large freshwater lakes dry out, since there are always traces of NaCl, even in fresh water. Lake Bonneville in Utah (USA), a large freshwater lake which once had a surface area of 31 800 km^2 and a depth of 300 m, dried up in the Postglacial period, forming the far smaller Great Salt Lake, with an average depth of only 5 m, near Salt Lake City. There are today large areas of mineral salt deposited around the lake (Flowers and Evans 1966, see also Walter 1971/72). Similarly, the Dead Sea of Palestine was formed from the earlier Lisan Sea. The kawirs of Iran, however, seem to have arisen in a quite different way (Gabriel 1957, Krinsley 1970, Scharlau 1958, Breckle 1981). It is believed that in this area there were no natural basins where lakes could form, but that over a long period of geological time, areas of tectonic sinking became filled with large quantities of stones, sand and clay carried by water running off the slopes. Water which spread out over the wide, flat surface of the clay evaporated, leaving accumulations of salt. These areas of salty clay are called *takyres;* the kawirs are circular formations of such takyres.

Saline soil is found also on all sea coasts, particularly in the tidal zone. In a wet climate this is usually only a narrow strip along the beach, because salt is washed out by rain, and groundwater streams of freshwater, which are permanently at the same level as the sea, flow into the sea. On very flat coasts, where the climate is dry and the rate of evaporation potentially high, seawater can penetrate into the soil far inland with resulting salination of wide areas (e.g. the Rann of Kutch, western India).

In secondarily saline soil the salt is, of course, also of marine origin. Soils of this type arise in areas where artificial irrigation causes the groundwater level to rise to such an extent that water reaches the upper surface of the soil by capillary action. As this water evaporates on the surface, the dissolved salts accumulate, forming a salt crust which renders the soil unfit for further cultivation. Such an upward movement of groundwater must at all costs be avoided. The golden rule is: no irrigation without drainage, otherwise the irrigated field becomes like a sink without an outlet. All river water used in irrigation contains NaCl, even if only in traces, and this salt, too, originates from the sedimentary rock in the area where the river arises. Such secondarily saline soils are widespread in Mesopotamia; there irrigation was practiced even in Babylonian times and led gradually to the formation of deserts. Even in modern irrigation schemes, salination of the soil may occur as a result of carelessness. In many countries, even in Long Valley, California, the productivity of large areas of agricultural land has been reduced or entirely destroyed in this way.

2 The Geobiosphere in the Geological Past

2.1 The Origin of Continents, Continental Drift and Plate Tectonics

Historical thinking is generally alien to an ecologist; he attempts to explain existing conditions in a purely causal manner as the result of the interaction of factors which now prevail. The geobiosphere as we know it today is, however, the result of a very long process of development. The past has left many traces which can be understood only if conditions prevailing at earlier times are borne in mind (Palaeo-ecology). It is necessary, therefore, to discuss briefly certain aspects of this history, but mentioning only those features which contribute to an understanding of the geobiosphere of today. The earlier "permanence theory" was based on the assumption that the past distribution of oceans and continents was the same as it is now. The only deviation from this view which was considered acceptable in order to explain the undeniable past exchange of animals and plants between the continents, was a possible rising or sinking in the region of the continental shelves, with a consequent formation of certain land bridges. Such assumptions are, however, inadequate to explain various problems of biological distribution.

Thus it was akin to salvation for biogeographers when, in 1915, Wegener proposed his theory of continental drift and polar shift (Wegener 1936). These views were subsequently elaborated by Köppen and Wegener (1924), taking prehistoric climates into account.

Irmscher (1922, 1929) and his pupil Studt (1926) showed that this theory offered a plausible explanation of the present distribution of certain plant families, but that it is essential to bear in mind both continental drift and also polar shift.

In contrast to the reaction of biologists, geophysicists and geologists initially completely rejected Wegener's theory. The reason for this was that Wegener's idea that the continents float on a heavy, viscous layer, the sima, was not in accord with geophysical concepts.

A complete change in attitudes has taken place during the past two decades since facts about palaeomagnetism have become known, and more especially since results became available from studies of deep-sea cores collected since 1968 by the US *Glomar Challenger* in all the oceans of the world. The new findings of palaeomagnetism have made it possible to determine the previous relationships both of the continents to one another and to the poles. The position of the magnetic poles has changed considerably in the course of geological time.

Deep-sea boring has also provided data which have made it possible to explain the mechanism of continental drift in terms of plate tectonics and sea-floor spreading from mid-oceanic ridges. This explanation of the way in which continental platforms are pushed apart is in accordance with geophysical ideas.

While many of the details are uncertain and still under discussion, and several problems have not yet been satisfactorily solved, continental drift is today no longer called in question. Thenius (1977) gives a short summary and review of the present position and provides much support for the theory from the field of zoogeography.

According to seismological investigations, the continental platforms have an average thickness of 30–35 km and extend deep into the outer crust of the earth. They are geologically old and consist of particularly light, mainly granitic rock. By contrast, the earth's crust on the sea floor is on average only 6 km thick and beneath the thin, upper sedimentary layer consists of particularly heavy basalts, together with a third, compact, deep layer (gabbro or peridotite).

In contrast to the continental platforms, the ocean floor has proved to be relatively young. Palaeontological data and palaeontological datings of deep-sea sediments have shown that the oldest, central part of the Atlantic, between North America and West Africa, arose only in the Jurassic. In the Pacific, too, the oldest deposits, which lie east of the Mariana islands, are of Jurassic origin, while the rest of the floor of the Pacific is Cretaceous and Tertiary.

Today the Pacific ocean is of significantly greater size than the Atlantic; this is a result of the greater rate of spreading of the ocean floor of the former, 5–20 cm annually, compared with only 1–10 cm in the Atlantic. The Indian ocean is even younger, but little is known of its detailed history. The oldest sediments near the west coast of India are of Cretacean origin. The Indian continental plate, drifting from the south, reached the Asian continent only in the late Tertiary.

The ocean floor expands outwards on both sides of the mid-oceanic ridges. The ridges have a central trough where molten rock from below wells up and is extruded; the further this material is from the ridge, the cooler it becomes, and in cooling it shrinks; this leads to the formation of a slope from the trough towards the sea floor on each side of the ridge.

A sensational discovery of ecological interest has been made in the region of the mid-oceanic ridge near the Galapagos islands (Corliss and Ballard 1977). Here, at a depth of 2.5 km, the temperature of the water on the sea floor is near freezing point, but increases to 17°C above the ridge. A rich fauna was found on rocks of hardened lava at several places on an otherwise lifeless sea floor. This included mussels, 30–40 cm in size, 45 cm long tube worms (Pogonophora), sea lilies, crustaceans and cuttle-fish, the whole forming veritable deep-sea oases of a most unusual ecosystem. Seawater penetrating through cracks in the cooled lava on the sea floor comes into contact with warmer rock, and emerges again as hot springs, forming oases of warmth. At these depths the water is enriched with metallic salts, which are precipitated as oxides at those places where the hot water emerges. At such hydrothermal vents, water samples were found to contain the isotopes radon 222 and helium 3, both of which arise deep in the interior of the earth. In addition,

the thermal water smells strongly of hydrogen sulphide; this is formed in the cracks of the lava as a result of reduction of sulphates contained in the sea water. This hydrogen sulphide is used in chemosynthesis by autotrophic sulphur bacteria, which are thus the producers of the ecosystem of these oases, making it possible for the animals to live there. The source of the energy which flows through this ecosystem is thus not sunlight, but energy liberated by the oxidation of H_2S. At one point the sea floor was covered with dead mussels; clearly here a hydrothermal vent had ceased to flow.

For a certain period of geological time, lasting until the Permo-carboniferous, all the continents were joined, forming the *Pangaea*. By the beginning of the Triassic, there were two large land masses: *Laurasia*, corresponding to North America, Europe and Asia, including Indonesia, but not the rest of southern Asia, and *Gondwana*, consisting of the platforms forming what is now South America, Africa and Madagascar, the Indian subcontinent, Australia, New Zealand and the Antarctic. These two land masses were at times separated by a sea, the *Tethys* sea, sometimes described as the original Mediterranean (Fig. 1). Wegener assumed that the original land surface was comparable in size to that of today; it could, therefore, not have covered the entire surface of the globe. Two thirds of the surface must have been covered by the original ocean, for which no detailed data are available, and the fate of which is unknown.

Carey (1968) has pointed out that if the radius of the earth were reduced by ⅓ to ½, Pangaea as a whole would have covered its surface completely. It would then not be necessary to postulate the existence of any separate ocean at that time; the whole of the solid surface of the earth would originally have been covered by water. Only when large, deep troughs formed between the continents was the water drained off into the ever-widening deep-sea basins, giving rise in this way to the first dry land.

Carey proposed, in his expansion theory, that the earth was originally a completely cold planet, which gradually increased in volume as a result of radioactive processes, while at the same time the temperature of the earth's core rose enormously until its present condition was reached. This hypothesis is supported by the fact that the Pacific coast of North America can be perfectly fitted into that of eastern Asia, provided the assumption is made that the earth once had a smaller radius. This theory has, however, not found general

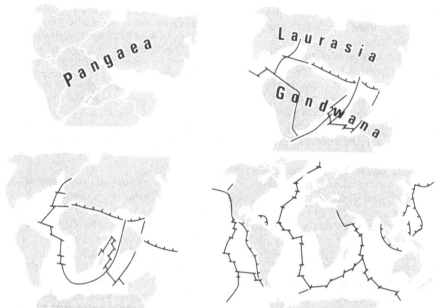

Fig. 1. The probable development of the continents. *Upper left* the Pangaea stage in the Permo-Carboniferous; *upper right* Laurasia and Gondwana separated by the Tethys Sea at the beginning of the Cretaceous; *below left* the separate continents at the beginning of the Tertiary; *below right* their position today. *Lines* show the position of the central oceanic ridges. (Alexander Atlas, Ernst Klett Verlag, Stuttgart)

acceptance, since it is impossible to explain an expansion of the required order of magnitude.

Subsequent changes in the two land masses are illustrated in Fig. 1. Laurasia remained intact for a long time. In the north there was a connection across Greenland until the Eocene, 50 million years ago. Iceland separated about 20 million years ago, or rather it represented a largely new development from the activity of mighty shield volcanoes.

The southerly Gondwana land mass behaved differently. The formation of the south Atlantic, which led to the separation of South America from Africa, is now believed to have occurred only by the mid-Cretaceous. Even before the Australian plate separated from the Antarctic, Madagascar broke away from the African land mass, as did also the western part of the Indian plate. The western Indian plate was pushed northwards against the Asian continent, resulting, in the Tertiary, in the formation of the Himalaya mountain range, which continues to rise even today.

About the same time, the opening of the Tethys sea to the east closed (Whyte 1976) and the large arid continental area of central and southern Asia arose. The drift northward of the Australian plate over many degrees of latitude in the direction of southern Asia began in the Eocene and led to the formation of the young mountains of northern New Guinea. Early in the Tertiary, the connection between the southern tip of South America and Antarctica was severed, following which the Antarctic continent moved slowly to its present position at the South Pole and became ice-covered. Besides such changes in position of the terrestrial plate, there was, during the Tertiary, extensive formation of mountain ranges. The alpine mountain formation extended from the Atlas mountains in North Africa, over the Sierra Nevada System in Spain and the Pyrenees; through the whole of Europe over the Alps, Carpathians and the Illyrian-Balkan mountains to western Asia and the Caucasus; to central Asia and the Himalayan range, as well as further to Japan, including Kamtchatka in the north and New Guinea and New Zealand in the south. In the same way, the Andes arose in America as a result of subduction, that is, the submergence of oceanic crust beneath the lighter continental platform. The formation of the Alps in Europe was caused by the northward movement of the great African shield. This gave rise to a shearing movement which caused Spain to turn, broadening the Bay of Biscay, and resulted also in the folding of the Pyrenees. The most recent fracture is the east African rift system which extends into the Red Sea and the Jordanian rift valley with the Dead Sea. The Red Sea split off the Arabian peninsula and forced it northeastwards. By about the late Tertiary, formation of the earth's surface as we know it had largely been completed, yet even today, movements and breaks continue undiminished. The most important feature of the Pleistocene was the occurrence of many glacial and inter-glacial periods; the exact cause of these is not yet known, but they have left clear traces right up to the present time.

2.2 The Colonization of the Land by Living Organisms

The history of the continents having been briefly outlined, we turn now to their settlement by plants and animals. Apart from autotrophic bacteria, the first settlers could only have been autotrophic green plants which by producing organic compounds, created conditions suitable for heterotrophic consumers and decomposers (cf. Sect. 1.1).

The appearance of the first land plants is placed at the end of the Upper Silurian or the beginning of the Lower Devonian, 400–370 million years ago. These were the small psilophytes, precursors of the pteridophytes (vascular cryptogams).

A statement like this made in textbooks, however, conveys the impression that before this time there were no living things on land, but only in the sea: this certainly cannot be correct.

A source of water is a prerequisite for life. From the beginning, water was available on land, since the cycling of water between sea and land (cf. Sect. 1.2) would result in rain falling on land. There must have been rivers and basins with freshwater, and the products of the weathering of rocks were wetted by the rain. The equatorial areas must have been particularly warm and moist, and the interior of the continents particularly dry. We may thus assume that the first living things so establish themselves on the land surfaces were prokaryonts; primitive autotrophic cyanophytes (blue-green algae), and the heterotrophic schizomycetes (bacteria). These organisms were already capable of forming simple ecosystems. With further biological evolution, eukaryotic algae, fungi and protozoans appeared. All of these organisms could easily be distributed through the air in a dormant phase. Since these microorganisms have left no fossil record, their original distribution on land cannot be determined. By contrast, even in the Cambrian there were marine algae of complex structure with a calcareous skeleton, as, for example, the Dasycladaceae; these gave rise to marine rock formations. The fossil forms of these algae include 120 genera, but only 10 genera are extant. This applies also to several marine animal groups. Diatoms appeared first in the Tertiary.

These very early ecosystems could have formed only a thin skin on the upper surface of rocks. They are still found today in places unsuitable for colonization by higher organisms (vide Sect. 5.2.1). One can only really speak of an actual covering of vegetation, suitable as food for larger animals, after the establishment on land of vascular plants (cormophytes), the first representatives of which were the psilophytes. It is not known when the first mosses (bryophytes) appeared; although they do show some relationship, they are not ancestral to the pteridophytes.

The psilophytes themselves must have evolved from the green algae (Chlorophyceae), as the following features are unique to these two groups: (1) the same chloroplast pigments, (2) the same assimilate, namely, the polysaccharide starch, and (3) the same compound as cell wall material, namely, cellulose. Of the various orders of the Chlorophyceae, the most probable ancestors of the Psilophyta are the chaetophorales, and, in particular, those forms resembling the extant *Fritschiella tuberosa*. This soil alga occurs in Africa and India and has a relatively strongly differentiated thallus. From creeping rhizoid cells in the soil, threads lead to the soil surface, and there form upright branching bundles; the outer cell walls of these latter exhibit a sort of cutinization which protects them against desiccation. There is, however, a wide gap between algae such as these and the psilophytes with proper tissues and vascular bundles with tracheids. Intermediate forms have not yet been found.

The chaetophorales are freshwater algae and only such algae, not marine algae, can be considered as possible ancestors of land plants (Walter and Stadelmann 1968). The reasons for this are physiological; saline soils are particularly unsuitable for cormophytes and were only colonized very late by a few specially adapted angiosperms. These salt-adapted plants form an ecological group known as the halophytes (Sect. 5.2.5). Further, *the fact that neither sodium nor chlorine is essential to plants,* other than those which live in the sea, speaks against the psilophytes having arisen directly from marine algae. Indeed, quite the opposite is the case, for, the halophytes apart, *sodium chloride is a poison to all terrestrial plants.*

It is significant that amongst the mosses or vascular cryptogams there are no true halophytes. No moss can grow in seawater and only very few tolerate slightly brackish water; the mosses *Pottia heimii, Amblyostegium compactum* and *Schistidium maritimum,* for example, do occur in the Elbe estuary.

Of the pteridophytes, *Asplenium maritimum* is found from Scotland to northern Spain in the coastal strip within reach of sea spray. The leaves are not infrequently wetted with droplets of seawater, but in these areas the rainfall is good and the soil must therefore be largely free of salt. Likewise, the mangrove fern *Acrostichum aureum* is not a halophyte. In East Africa it grows only in the brackish water mangroves at river mouths, where the salt content of the soil is very low or where the soil has never been saline (Walter and Steiner 1936). Similarly, *Acrostichum* is found on the rainy coast of Brazil, together with non-halophytic plants such as *Crinum,* above the high-tide zone and immediately

below the slope of the bank on which palms grow; in other words, it occurs where freshwater flows out into the sea. Soil tests in this zone showed that 100 ml of soil solution from the surface contained 0.65–0.68 g NaCl and at a depth of 15 cm, 0.54–0.84 g; the chloride content of a mangrove zone was, however, 2.5 g (Lamberti 1969).

Halophytic gymnosperms are likewise unknown. The remarkable *Welwitschia mirabilis,* which grows in the Namib desert in South West Africa, avoids saline soils, although 20–25% of the osmotic potential of cell sap is due to chloride (Walter 1936b). The halophytic angiosperms will be dealt with in more detail in Section 5.2.5.

The fact that fossils of the first cormophytes are found in sediments of earlier sea coast does not prove they were not descended from freshwater algae. From an ecological viewpoint, the most suitable places for transition from an aquatic to a terrestrial life were the oceanic lagoons on the coasts in the humid tropics. These were separated from the sea only by a sand bar but contain freshwater with a rich algal flora. There is a steady flow of groundwater from the land into these lagoons, so that the soil of the low-lying banks is always moist and salt-free. Today such places are covered with a luxuriant bog vegetation of halophobes. Schaarschmidt (1974) believed that he had found the fossilized remains of a succulent halophyte in mud deposits on a prehistoric sea coast of Lower Devonian dating near the Moselle river; he was, however, not certain whether it belonged to the psilophytes. Judging by the photographs, it could well be the jointed thallus of a phaeophyte with some conducting structures, and had simply been washed up onto the beach. Brown algae cannot be considered as possible ancestors of the cormophytes because of their chemical composition.

Once the higher plants had conquered the land, the preconditions had been created for its colonization by true land animals, culminating in the giant dinosaurs of the Jurassic, the mammals in the Tertiary, and finally man. But unlike land plants, some terrestrial animals are of marine origin. In contrast to land plants, sodium and chloride are essential ions even for mammals and man.

The transition from life in water to living on land was easier for vertebrates than for the cormophytes. The gills were replaced by internal breathing organs (lungs), the fins by four appendages with which locomotion on land was possible. With their compact bodies and relatively small surface area, animals are less subject to water loss than are plants through transpiration; not only are plants immobile, but the large upper surface of leaves is exposed to the effects of sunlight. The water requirement of animals is thus low and can be met by feeding on food with a high water content, or supplemented by direct uptake of water from the many scattered sources available on land. Being mobile, animals can visit these at any time. Water is thus not as important in determining animal distribution as it is in determining plant distribution. While there is a danger of salt accumulation within plants exposed to saline water, drinking such water is far less dangerous to animals, which can void salt in the urine. Therefore, in arid areas, brackish water can be supplied to animals but cannot be used to irrigate plants.

2.3 Floristic Realms and Faunistic Regions

Both higher plants and animals could spread over the whole land surface. So long as Pangaea existed, this was easy and, since connections were established from time to time between Laurentia and Gondwana, an exchange of plants and animals between them was not precluded. Climatic differences would lead to differentiation of the fauna and flora. Thus, in the Permo-carboniferous, part of Gondwana around the South Pole was covered with ice. Traces of this glaciation have been found by geologists in South Africa, Australia and western India, which at that time had not separated from one another. On the other hand, during the Carboniferous, tree-like pteridophytes formed moist forests on the equator and these gave rise to the coal seams. In the Triassic and Jurassic, the pteridophytes were replaced by the gymnosperms. Forests of *Taxodium*, *Sequoia* and *Metasequoia* contributed to the formation of brown coal during the Tertiary.

Shortly before Gondwana broke into separate continents, the angiosperms appeared. They were thus able to spread over all the continents and, during the Tertiary, replaced the gymnosperms. Today the

angiosperms predominate; it is only in the climatically unfavourable areas of the boreal and on soils poor in nutrients, such as marshy habitats, that gymnosperms, mostly conifers, dominate.

The most highly developed group of animals, the mammals, appeared later than the angiosperms. The primitive marsupials are the only members of this group to have reached Australia, while the isolated islands of New Zealand have no indigenous mammals. It was only later that two species of bat were able to fly there.

Evolution of the flora and fauna subsequently took different directions as a result of isolation; the earlier the separation, the greater the differences. This led to the formation of six major floristic realms and, corresponding to these, a number of faunal regions (Fig. 2). The floristic realms are differentiated on the basis of large taxonomic groups, such as families, but also the orders of the angiosperms. These six floristic realms are the Holarctic, the Palaeotropical, the Neotropical, the Australian, the Cape (in South Africa) and the Antarctic.

The Holarctic Realm

This covers the whole of the non-tropical land masses of the northern hemisphere; these separated only at a very late stage into the New and Old Worlds. It is the main area of distribution of such families as the Salicaceae, Juglandaceae, Betulaceae, Fagaceae, Ranunculaceae, Cruciferae, Caryophyllaceae, Saxifragaceae, Rosaceae and Aceraceae, of particular taxa among the Cyperaceae, especially *Carex*, as well as the Gramineae and other families.

The glaciations during the Pleistocene, the most recent of past geological periods, brought about a certain amount of differentiation within the Holarctic. In Europe the glaciations resulted in a large part of the Pliocene flora dying out, thereby impoverishing the flora in general. Many of these families, like the Magnoliaceae, Hamamelidaceae, Styracaceae and others, still occur today in North America and Asia, but not in Europe. This applies also to a very large number of genera, all of which grow well in European gardens and include, amongst others, *Morus*, *Mahonia*, *Dicentra*, *Astilbe*,

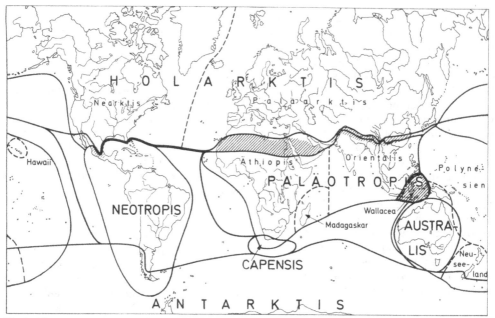

Fig.2. Floristic realms of the earth *(capital letters)* and faunistic regions *(small letters)*; Australia is identical with Notogaea. *Cross-hatched areas* are intermediate between faunistic regions. Further explanations in text. (After Walter and Ziswiler 1984)

Deutzia, Hydrangea, Physocarpus, Wistaria, Ampelopsis and *Phlox*, etc.

The Palaeotropical Realm

In the tropical zone two floristic realms can be distinguished; this is because Africa separated at an early stage from South America. The Palaeotropical realm is thus the tropical part of the Old World and is characterized by such families as the Pandanaceae, Zingiberaceae, Anonaceae, Myristicaceae, Sterculiaceae, Dipterocarpaceae, Combretaceae, most of the Araliaceae, Moraceae (including 1000 *Ficus* species and 40 of *Artocarpus*), Euphorbiaceae, many of which are stem succulents, as well as the genera *Aloë, Sansevieria* and *Dracaena* amongst others.

This floristic realm is divided into three subrealms: the African subrealm, including the very distinct flora of Madagascar (Rauh 1973), the Indo-Malaysian subrealm, including New Guinea, and the Polynesian subrealm.

The Neotropical Realm

This encompasses the tropical areas of the New World; that is, Central America and South America, excluding its southernmost tip. Particularly characteristic families are the Cactaceae, the Bromeliaceae, but also many others, such as the Tropaeolaceae. Amongst the genera, *Agave* and *Yucca* deserve special mention. The family Palmae occurs in both the neotropical and palaeotropical floristic realms, but is represented by different genera in the two areas.

The Australian Realm

The land mass of the southern hemisphere is markedly subdivided and, correspondingly, three very different floristic kingdoms have arisen. In Australia certain unique changes occurred. After separation from the Antarctic, the continent reached the arid, subtropical zone. The original antarctic flora, which included *Nothofagus*, died out almost completely. A floral vacuum must have arisen in which two genera adapted to the new conditions and gave rise to many new species. These were *Eucalyptus* and *Acacia*. The former, with today more than 450 species, forms the tree layer in all the different types of forest from sea level to the alpine tree line. The genus *Acacia* established itself in the dry interior of Australia; it includes approximately 750 species, mostly leafless, the leaves being replaced by broadened petioles or "phyllodes". As well as these there are many genera of Proteaceae which are also represented by many species (*Grevillea* with 250, *Hakea* 140, *Banksia* 50, to mention but a few examples); 75% of all these species are found only in Australia. The vegetation of Australia is thus unique and has very little in common with that of the other continents (Beadle 1981).

The Cape Realm

This, the smallest of the floristic realms, is restricted to the extreme south-west corner of South Africa, the Cape. Nonetheless, it is distinguished by a remarkably large number of species and is quite unique. As well as some endemic families, the Proteaceae are important here also, although represented by a different subfamily which includes *Protea* with 140 species, *Leucodendrum* with 73 and *Leucospermum* with 40. Similarly, the Restionaceae belong to genera different from those found in Australia. What is particularly remarkable is that there are here about 600 species of *Erica,* one genus of the otherwise holarctic Ericaceae. This genus must also have encountered something of a vacuum created by climatic changes. This floristic realm is separated from the rest of Africa by a very effective climatic barrier, a desert strip.

The Antarctic Realm

This includes the southern tip of South America, the subantarctic islands and South Island, New Zealand. All these were once joined to the antarctic continent. Today that continent is almost completely covered with ice and only two flowering plants, *Colobanthus crassifolia* (Caryophyllaceae) and the grass *Deschampsia antarctica,* belong to the natural flora. New Zealand separated from Australia earlier than from Antarctica, so that it has little in common with Australia botanically. Although most of the New Zealand flora shows Melanesian features of the Palaeotropical realm, South Island has many antarctic floral elements, as does also Tasmania, which otherwise belongs to the Australian realm.

The flora of any area constitutes the units from which the plant cover, that is the vegetation, is built up; these units are the producers of the ecosystem. The composition of this vegetation will affect the structure of the ecosystem and the cycling of matter within it. For this reason, an ecologist must take into account the historically determined composition of the vegetation in any area to be studied.

In a similar way the history of the formation of the continents has affected the distribution of animals. Animals could live on land only after the development of terrestrial plants which would provide the necessary source of food. The first land vertebrates, the Ichthyostegaea, date from the late Upper Devonian. Terrestrial arthropods, scorpions, are known as early as the Silurian. The reptiles evolved from the amphibia and from these the birds and mammals arose independently. The oldest mammals are represented by fossils from the Triassic; they were small, probably insignificant members of the fauna. It was only during the Cretaceous that the two main lines, the marsupials and the placental mammals, evolved. By the beginning of the Tertiary all the extant orders of mammal are represented, and during this period they achieved a dominant position among the large land animals.

The zoogeographical regions, based on the distribution of mammals, largely coincide with the floristic realms (Fig. 2). Certain differences arise from the fact that mammals are mobile and are thus able to spread more rapidly across land connections than can plants. Movement of animals across the narrow land bridge between North and South America was far more extensive than that of plants. Both ungulates and carnivores migrated from North to South America, displacing the original marsupial fauna of this region. Similarly, the Pleistocene land link across the Bering Strait between Asia and North America played an insignificant role in the dispersal of plants, but was of great importance for the migration of animals. It was over this land bridge that man migrated from Asia to North and South America.

The mobility of animals accounts also for the fact that there is no faunistic region corresponding to the Cape floristic realm. The desert strip between the South African and Palaeotropical floristic realms which constitutes a barrier for plants has been crossed by mammals.

The following major zoogeographical regions are recognized in the schemes proposed by de Lattin (1967) and Illies (1971): the *Palaearctic* and *Nearctic* which together correspond to the Holarctic floristic realm; the *Ethiopian*, being the area of Africa south of the Sahara, but including the sub-region of Madagascar; this, together with the *Oriental* region, including tropical Asia with India, the Philippines and Formosa, correspond to the Palaeotropical realm, while *Wallacea* is an intermediate zone between Australia and India (Fig. 2); *Neogaea*, South and Central America, is the equivalent of the Neotropical realm and *Notogaea*, Australia with New Guinea as well as New Zealand, Polynesia and Hawaii, corresponds to the Australian realm. The indigenous monotremes and very many marsupials are characteristic of Australia. The only other mammals in Australia—several rodents, bats and the dingo—are late migrants from Asia. Apart from two species of bat, there are no indigenous mammals in New Zealand. *Antarctica* is generally not regarded as a zoogeographical region, as the land fauna consists of only a few invertebrates such as mites and non-flying insects.

3 Ecological Zonation of the Geobiosphere

3.1 The Principles of Subdivision

The nature of an ecosystem is determined first of all by abiotic factors—climate and soil, and secondly by biotic factors—vegetation and fauna. It is therefore mistaken, when proposing ecological subdivisions, to take into account only one of these factors, as is common practice when maps of climate, soil conditions or vegetation are drawn. On the other hand, it is impossible to consider all these factors at the same time. In a stepwise process of subdivision into ever smaller units, consideration must be given to the priority to be accorded to each factor. The first decision to be made is which of these four factors should form the basis on which to establish the first major subdivisions.

The fauna is the least suited for this purpose since zoogeographical regions are based on the very well-known distribution of mammals and birds; these animals are, however, characterized by great mobility and show little tendency to be confined to a single region. The movements of carnivores, ungulates and migratory birds are all examples of this.

Plants are, by contrast, immobile and therefore far more suitable. Schemes for zonation of the Earth on the basis of the nature of the vegetation have indeed frequently been proposed. In these the assumption is made that the life-forms of plants, used as a basis for differentiating one plant "formation" from another, are ecologically determined, and from this it is suggested that such "formations" are a very good reflection of prevailing ecological conditions. This approach is furthermore regarded as having the advantage of eliminating the purely historically determined differences between different floristic realms. It is, however, a mistake to believe that life-forms are simply ecologically determined; historical factors play just as important a role. The evergreen conifer forests, for example, are relics of the earlier dominant gymnosperms, which were pushed back and are now restricted to areas unfavourable to angiosperm forests.

There is, furthermore, no agreement about which life-forms should be used as a basis for a world-wide zonation. In central Europe Raunkiaer's system is generally adopted, but others have been suggested for application to western Asia (Zohary 1952, Orshan 1953), while Australian botanists consider that the Raunkiaer life-forms are quite unsuitable for their purposes. In New Zealand there is a life-form, unknown elsewhere, which is important: this includes the divaricate branching, hemispherical bushes and also juvenile forms of trees with fine, divaricate, that is, dichotomous branching of their annual shoots. In Patagonia, the many examples of convergence amongst the species of broom are very striking, as are also the different types of cushion plants, both here and in the Pamirs.

Ideally, very many life-forms, based on a worldwide consideration of the whole of the plant kingdom, should be taken into account. By no means all of these would be determined by ecological factors. Often they will depend upon taxonomic relations, such as membership of a particular family; for example, the grass form, palms, and so on. Indeed, the likelihood of particular, ecologically determined life-forms occurring within individual families is very small. This applies also to such growth-forms as trees, bushes, low shrubs, grasses, etc. It is almost impossible, for example, to name even one general ecological characteristic of trees, other than their height. The formation described as "evergreen trees" includes both the evergreen tropical trees and the ecologically totally different boreal conifers, while the larch *(Larix)*, which is closely related to the latter group, is placed in the formation "deciduous trees". The Australian casuarines occupy a special position in that externally they resemble *Pinus,* as does also *Acacia aneura* with its needle-shaped phyllodes. Trees can also be halophytes; these include not only the various mangroves, but also *Haloxylon* spp. from the deserts of central Asia. The tree-shaped dendrosenecios of the alpine belt of the East African mountains look like tufted trees, but are really stalked rosette plants. *Bambusa* spp. are the trees amongst the grasses, yet ecologically they are quite different from typical trees. The same considerations apply to the formations of shrubs, dwarf shrubs and so on; they, too, are ecologically very heterogeneous.

This whole system of plant "formations" is a highly simplified and artificial concept, conceived at a desk without detailed knowledge of the diversity of the vegetation of other continents. It bears little relationship to reality.

Even less satisfactory is the use of soil types as the basis for a major subdivision of the geobiosphere; this is because soil profiles cannot be seen and can be examined only by taking random samples. Soil type frequently does not reflect climate, since the profile is influenced by the nature of the mother rock; that is, by the geological substratum. In tropical areas soil profiles are usually extremely old, and show characteristics that were established under quite different climatic conditions from those now prevailing. There are many "fossil" soils. The "terra rossa", for example, was once regarded as the typical profile for an area with a Mediterranean climate, but today it is considered to be a fossil tropical soil. Many other soils are relics of some other climatic regime. Finally may be added the fact that there is still no single, generally accepted system of soil typing.

The soil and the vegetation form, moreover, an inseparable whole. Soil contains living organisms: apart from the root system of higher plants with their associated rhizosphere or mycorrhiza, there are also free-living fungi and bacteria, as well as a very rich soil fauna. Without these organisms, the formation of a soil profile would hardly be possible, while water uptake by plant roots strongly influences soil water content. Furthermore, soil scientists generally examine only the inert "skeleton" of a soil, its structure, texture and chemistry, and then mostly that of agriculturally exploited soil, that is, soil of a secondarily artificial type.

The macroclimate alone is free from historical influence (vide Sect. 4.3), for it is determined by the global circulation of air and this latter changes immediately when the distribution of the continents and oceans changes. It is thus not possible to speak of fossilized climatic factors. Climate is the sole primary factor which influences all the others—the soil, the vegetation and, to a lesser extent, the fauna; it is in turn affected by these factors only at the level of the microclimate.

The macroclimate is therefore particularly suitable as the basis for subdivision of the geobiosphere into larger units. The difficulty is, however, that climate on the whole is not easy to define. Meteorological stations provide average values of various measurable climatic factors. The interaction of these factors at one place at any one time is called the "weather"; considered over a short period (days, weeks) this may be considered as forming the local "weather conditions". These conditions over the year as a whole, based on average values for many years' observation, are used to define the climate. No climatologist can tell us, however, how an integration of the measured meteorological factors should be made. The climate is indeed usually described in a rather complicated way. Such descriptions are not satisfactory for preparing climatic maps and commonly one specific meteorological factor, such as temperature or rainfall, is selected to differentiate one climatic zone from another; or else the border of a climatic zone is regarded as being identical with that of a vegetational zone, giving the impression that climatic and vegetational maps are in remarkably close agreement.

Frequently climatic "indices" are calculated; such, for example, as those for aridity or continentality. These, however, give no picture of the climate as a whole, as

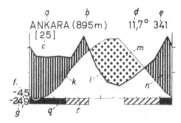

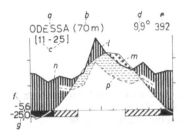

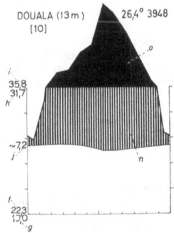

Fig. 3. Typical examples of climatic diagrams with explanations. Temperature in °C; rainfall in mm. The letters and numbers on the diagrams indicate:

a   Meteorological station
b   Altitude in metres above sea level
c   Number of years of observation; where there are two figures, the first refers to temperature measurements, the second to rainfall
d   Mean annual temperature
e   Mean annual rainfall
f   Mean *daily* minimum during the coldest month
g   Absolute minimum (lowest recorded temperature)
h   Mean *daily* maximum during the warmest month
i   Absolute maximum (highest recorded temperature)
j   Mean daily temperature fluctuation
k   Curve of mean monthly temperature (1 unit = 10°C)
l   Curve of mean monthly rainfall (1 unit = 20 mm)
m   Relatively arid period *(dotted)* = drought season
n   Relatively humid season *(vertical bars)*
o   Mean monthly rainfall above 100 mm: *black area* (scale reduced to ¹⁄₁₀) (shown for Douala)
p   Curve for rainfall on a smaller scale (1 unit = 30 mm); above it, horizontally striated area indicates relatively dry period = dry season (shown for Odessa)
q   Months with mean daily minimum below 0°C; *black* below zero line
r   Months with absolute minimum below 0°C; *diagonally striated* below zero line
s   Average duration of period with daily mean temperature above 0°C; number of days in *standard type:* alternatively, average duration of frost-free period; number of days in *italic type* (shown for Hohenheim).

Such data are not available for every station; where no information is available, the appropriate place is left blank in the diagram. Values for *h, i* and *j* are given only for tropical stations with a diurnal climate (for example, Douala)

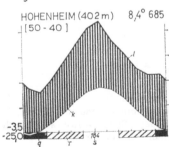

this involves changing weather conditions throughout an entire year. It therefore seemed to us essential to represent the overall, annual climate in the form of "ecological climatic diagrams" (Walter 1955), based on the ombro-therm curves of Gaussen. Ecologists are, however, not interested in events in the outer atmosphere, although knowledge of these is essential for an understanding of the genesis of climatic types; they are interested solely in the climate of the lower atmosphere, as this determines conditions in the geobiosphere. For this reason meteorological data from ground stations are used in

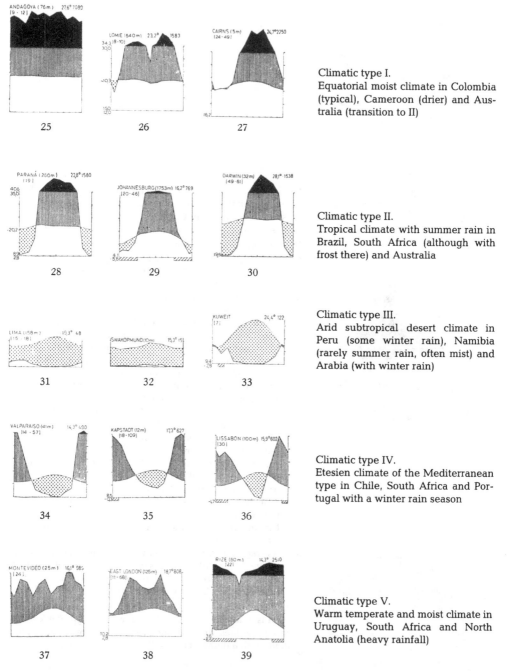

Climatic type I.
Equatorial moist climate in Colombia (typical), Cameroon (drier) and Australia (transition to II)

Climatic type II.
Tropical climate with summer rain in Brazil, South Africa (although with frost there) and Australia

Climatic type III.
Arid subtropical desert climate in Peru (some winter rain), Namibia (rarely summer rain, often mist) and Arabia (with winter rain)

Climatic type IV.
Etesien climate of the Mediterranean type in Chile, South Africa and Portugal with a winter rain season

Climatic type V.
Warm temperate and moist climate in Uruguay, South Africa and North Anatolia (heavy rainfall)

Fig. 4. Examples of climatic diagrams for the climate types (zonobiomes) *I–IX*. Numbers in the middle of the $+10°C$ and the $-10°C$ lines (e.g. in Type VIII) indicate the number of days with a mean daily temperature above $+10°C$ or above $-10°C$. Months begin with January for the northern hemisphere, with July for the southern hemisphere. (Summer season always in the middle of the diagram)

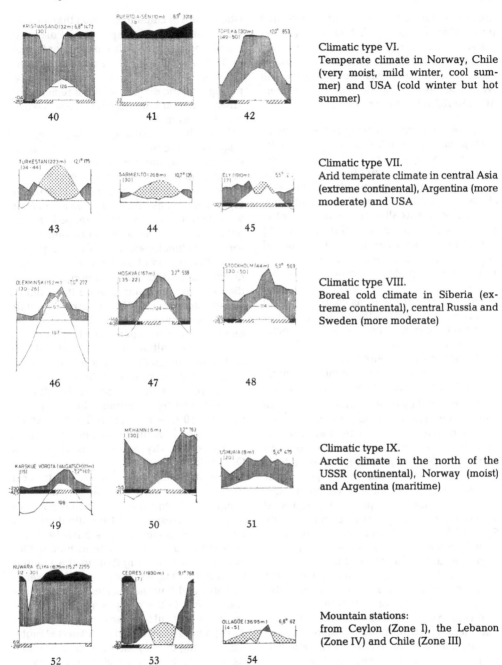

Climatic type VI.
Temperate climate in Norway, Chile (very moist, mild winter, cool summer) and USA (cold winter but hot summer)

Climatic type VII.
Arid temperate climate in central Asia (extreme continental), Argentina (more moderate) and USA

Climatic type VIII.
Boreal cold climate in Siberia (extreme continental), central Russia and Sweden (more moderate)

Climatic type IX.
Arctic climate in the north of the USSR (continental), Norway (moist) and Argentina (maritime)

Mountain stations:
from Ceylon (Zone I), the Lebanon (Zone IV) and Chile (Zone III)

these climatic diagrams in an unmodified form; that is, as published by weather stations, being long-term averages of measurements made under standard conditions. How climatic diagrams are prepared and what conclusions can be drawn from them are described by Walter et al. (1975).

Ecological climatic diagrams are a graphical representation of the climate as a whole, clearly showing its seasonal course. They contain, even if in simplified form, all the information an ecologist needs about macroclimatic conditions. These diagrams are now used by ecologists all over the world. For the reader still unfamiliar with these, an explanation is provided in Fig. 3. Ecological climatic diagrams show at a glance:
1. the annual pattern of temperature and rainfall;
2. the wet and dry seasons characteristic of any area, as well as their intensity, since the rate of evaporation is almost directly related to temperature (vide Sect. 3.7);
3. the occurrence or otherwise of a cold season of the year, and the months in which early and late frost have been recorded.

In addition they give information on mean annual temperature, mean annual rainfall, the mean daily minimum temperature during the coldest month, the absolute minimum recorded temperature, the altitude of the station and the number of years over which observations have been made. Climatic diagrams for equatorial stations at which there is a "diurnal climate" (vide Sect. 3.2.1), show also the mean daily fluctuation in temperature; others, for semi-wet steppe localities, show the time of the dry season, and yet others, for cold climates, the average duration of frost-free periods or else the number of days on which the temperature is above 10°C or above −10°C. Further information has been deliberately omitted because, for ease of reading, a climatic diagram should not be overloaded with data (see also Walter 1963).

Climatic diagrams including these features, and based on data from approximately 8000 meteorological stations, have been compiled in the World Atlas of Climatic Diagrams of Walter and Lieth (1967). These diagrams, when attached to large wall maps of each of the continents, provide a more real-

istic picture of climatic zonation than do conventional climatic maps. Such diagrammaps are to be found also, in smaller format, in Walter et al. (1975), which further provides an introduction to the methods of preparing climatic diagrams and of recognizing areas of similar climate, that is, areas which are "homoclimatic".

From all the available ecological climatic diagrams, nine main types were selected as characterizing the nine most important climatic zones on land. Representative examples for the different zones are shown in Fig. 4. One, for example, arose in connection with climatic diagrams for a steppe in a temperate region. The area concerned had a semi-arid climate, without a drought period, yet with a period of summer dryness which did not show up clearly on a normal climatic diagram, in which one unit on the ordinate represents either 10°C, or 20 mm rainfall. The reason for this was that in the dry summer months, rain comes in the form of heavy cloud bursts; although much rain falls at such times, it has little effect on the vegetation, since it does not soak into the soil but mostly runs off the surface. This difficulty was resolved by including a broken line at the scale 10°C ≡ 30 mm rain. The area above this line and beneath the temperature curve indicates the dry season, as shown for Odessa in Fig. 3.

In warm-temperate, almost subtropical steppe areas, such as the South American pampas, even this reduced scale is not adequate to represent the true situation, and indeed some meteorologists have been led to describe the climate there as humid. Our own thorough investigation in the pampas showed that it is, without any doubt, a semi-arid area and thus natural steppe, not degenerate forest. In the pampas summer rainfall occurs as particularly heavy, almost tropical downpours. These are almost always at night, while during the day the sun shines brightly. As a result, the heavy soil dries out within a few hours in the morning. This example serves to show that it is very important to have long first-hand experience of a particular climatic region and not to assess its characteristics by sitting at a desk and studying monthly averages based on years of observation. This can be clearly illustrated by references to many climatic diagrams for arid parts of Australia: for these the rainfall curve runs almost horizontally and gives the impression of a scanty rainfall spread evenly throughout the year. Observations on the spot show, however, that this is quite misleading: rain falls in frequent, sharp showers, and the time of year in which it falls differs from one year to the next. The rainfall is thus unpredictable. Between rains there are extremely long periods of drought. The average taken over many years gives, however, the same low value for each month of the year. Such cases will be considered in more detail in Volumes 2 and 3.

3.2 Climatic Diagram Types and the Large Ecological Units of the Geobiosphere: Zonobiomes and Zonoecotones

The ecological climatic diagram types shown in Fig. 4 relate to the nine major low-altitude climatic zones of the earth, and indicate, at the same time, the variation which is found within any one type. We can differentiate between the following zones:

I. The equatorial zone, roughly from 10°N to 5–10°S, with a "diurnal climate"; that is, one in which the average daily temperature fluctuation is greater than the difference between the mean temperatures of the warmest and coldest months of the year: there are no annual seasons based on temperature differences. Annual rainfall is typically high, being more than 100 mm a month; the rainfall curve shows two equinoxial maxima (zenith rain).

II. The tropical summer-rainfall region, to the north and south of zone I, extending approximately to 25°–30°N and S, with a marked seasonal temperature difference, heavy rain in summer and extreme drought conditions during the cooler periods of the year; the further from the equator, the longer the period of drought, and the lower the annual rainfall.

III. The subtropical desert zone, lying further towards the poles. Here there are descending air masses that are warmed and dried in the process, so that rain seldom falls. The radiation of the sun is very strong, and there is great heat loss by irradiation at night, with resulting extreme daily fluctuations in temperature: night frosts may occur but are infrequent. The annual rainfall is less than 200 mm and, in extreme desert conditions, below 50 mm.

IV. The intermediate zone with winter rain and a long period of summer drought; this lies roughly between 35° and 40° latitude in the northern and southern hemispheres. This is the typical mediterranean climate.

V–VIII. These are zones of temperate climate, mainly in the northern hemisphere, with cyclonic rains at all seasons of the year; the amount of rain which falls is less, the greater the distance from the coast. We may thus differentiate between wet oceanic and dry continental climatic conditions. The general area is divided into the following zones:

V. A warm temperate climate, with almost no cold winter season, and typically very wet summers.

VI. The typical temperate climate, as pertains in central Europe, with a relatively short cold winter period, warm, even hot summers and an adequate supply of moisture. The oceanic regions have almost no winter cold, but cooler, moister summers.

VII. The arid-temperate climate of continental areas, with a marked difference between summer and winter temperatures and a low rainfall. The following degrees of aridity are distinguished:

VII. the semi-arid steppe climate with a dry summer period, but with drought indicated only slightly,

VIIa, arid semi-desert climate with a clearly marked drought period and short wet season. This is a transitional zone to VII (r III), where the rainfall pattern (r) is similar to that of a desert climate of type III, with drought lasting all through the year, but distinguished by cold winters. This is encountered mainly in central Asia.

VIII. The cold-temperate or boreal climatic zone, with usually cool, wet summers and very cold winters, which last for half the year. This is found, for example, in northern Eurasia. There is no such climatic zone in the southern hemisphere, but in the northern hemisphere it covers a very large circumpolar area. The corresponding subantarctic zone in the southern hemisphere has a quite different climate because of its oceanic situation; here there is little seasonal variation in the climate, which is characterized by a high rate of precipitation and strong winds.

IX. The arctic zone, which, as tundra, is limited almost exclusively to the northern hemisphere; this has a low rainfall evenly distributed over the whole year and, as a result of the low temperatures, the summer is short, wet and cool, despite the fact that there is no night. By contrast, the long winter nights are cold (Fig. 4, Karskije Vorota). The antarctic in the southern hemisphere is an icy desert, without vascular plants.

These climatic diagram types are linked with one another by intermediate types, indicated by their numeration; a diagram which is labelled I–II is intermediate between types I and II. The description I(II) indicates that it is closer to type I, II(I) that it is more like type II. Thus a climatic diagram indicated as V(IV) represents a warm temperate climate with mainly winter rainfall, and so on.

The climates of the higher mountain areas are a special case and seldom correspond to any one of the zonal climatic types; but the climates of mountain formations within any one climatic zone have similar characteristics.

The highest reliably recorded air temperatures are 57.8°C in San Luis (Mexican highland) on 11.8.1933 and 56.7°C in Death Valley (California) on 10.7.1913. More recently, 58°C was recorded in Al Aziziyah (Libya). The coldest place on Earth is said to be Wostok in the Antarctic, with a mean annual temperature of −57.8°C, and an absolute minimum of −91.5°C. In the northern hemisphere, the lowest temperature, recorded in Verkhoyansk and Oimyakon (northeastern Siberia), was −77.8°C. Here the mean annual variation, that is the differ-

ence between the lowest monthly average (−50°C in January) and the highest (+15°C in July) is 65°C (Troll 1973); the greatest recorded difference is 106.7°C. The highest temperature in Europe, 50°C, was recorded in Seville (Spain) (Edelmann 1982).

This separation into different ecological climatic diagram types when displayed on climatic diagram maps makes it possible to appreciate the general principles, and also some special features, on which climatic subdivision of the geobiosphere is based.

If the ecological subdivision of the macroclimate of the earth's surface is compared with the genetic subdivision made by meteorologists on the basis of the general circulation of the atmosphere and its seasonal shifts, as Flohn (1973) has done in a brief review, the two are found to correspond almost completely. Zones I–IV are the same. The temperate zone of the meteorologists, with a rainfall that is usually continuous throughout the year, and which covers a very large area of the northern hemisphere, has been divided by us, for ecological purposes, into zones V to VIII, on the basis of the length of the cold period and the degree of continentality. We have grouped together the subpolar and polar zones of the meteorologists into zone IX, because the extreme polar zone of Antarctica lacks vegetation, apart from the sporadic appearance of some lower plants. This correspondence of our classification on the basis of climatic diagram types with the genetic classification of the meteorologists shows that our subdivisions are entirely natural ones, which treat climate as an entity, taking into consideration the interaction of all the factors involved.

The nine major ecological units, represented by the nine different climatic diagram types, we call zonobiomes—in other words, zoned biomes (abbreviated ZB). We use the Roman numerals I to IX to distinguish them. These can be summed up as follows:

ZB I Equatorial zonobiome, with a diurnal climate (perhumid)

ZB II Tropical zonobiome with summer rainfall (humid-arid)

ZB III Subtropical arid zonobiome (desert climate)

ZB IV Zonobiome with winter rain and summer drought (Mediterranean, arid-humid)

ZB V   Warm-temperate (oceanic) zonobiome

ZB VI   Typical temperate zonobiome with short frost period (nemoral)

ZB VII   Arid-temperate zonobiome with cold winters (continental)

ZB VIII   Cold-temperate zonobiome with cool summers (boreal)

ZB IX   Arctic and antarctic zonobiome

Between the individual zonobiomes there are intermediate zones where there is a smooth transition from one climatic type to another. We describe these ecological spanning areas as *"zonoecotones"* (ZE) and give them the Roman numerals for the neighbouring zonobiomes, thus ZE I–II, ZE II–III, and so on, up to ZE VIII–IX. In this way we avoid the artificially sharp boundaries drawn on all previous climatic maps. Furthermore, the mountainous areas, with their completely divergent vertical climatic zonation, must be excluded from the zonobiomes and zonoecotones; they are indicated as black areas on the maps. We will return later to these *"orobiomes"* (OB).

Figure 5 shows the distribution of the zonobiomes and zonoecotones on a map of the world. The zones are distributed north and south of the equator, but not symmetrically. The size of the land mass between the 40th and 70th degree of latitude is far smaller in the southern than in the northern hemisphere, and the climate therefore more oceanic and cooler (see p 3) than in the northern hemisphere. As a result, the ZB III, IV and VII cover only small areas in the southern hemisphere, while ZB VIII is absent altogether. ZB IX is represented only by the southermost tip of South America and the subantarctic islands, while on the Antarctic continent, as has already been mentioned, apart from two flowering plants, there are only a few species of moss, many lichens, less than 100 algal species (mainly Cyanophyta), a few fungi (*Penicillium*, yeasts) and bacteria.

If a zonobiome covers a very extensive area, climatic conditions differ to some extent in different parts of it. A subdivision into several subzonobiomes (sZB) can be made, making use of climatic diagrams for their delimitation. As an example, we may take Venezuela. This is in a zonobiome I, and includes very many subzonobiomes, dif-

fering in their annual rainfall patterns. The situation is described later in section 10.4.

Zonobiome III is divided into several desert subzonobiomes:

1. with summer rain,
2. with winter rain,
3. with two rainy seasons,
4. with episodic rainfall only,
5. without rain, but often mist.

This will be described in more detail in Volume 2.

Sometimes different parts of a single zonobiome belong to different floristic realms or animal regions, so the producers show marked floristic differences, and the consumers, too, are very different. In this case it is possible to speak of historically determined *"biome groups"* within a zonobiome. In some cases a zono-ecotone connects not two but three zonobiomes. Examples of such triangular zono-ecotones are provided by the pannonic area of Hungary, spanning ZB VI, ZB VII and ZB IV, and also by the frontier area between Pakistan and Afghanistan, where there is a meeting of a typically central Asiatic area, of another of more arid-mediterranean type and of an extension of the Indian monsoon area.

3.3 Orobiome, Pedobiome and Eubiome

Within a zonobiome, very widespread areas frequently stand out clearly from the rest because of some special feature of their climate or their soil. They are ecosystems atypical of their particular zonobiome, and distinguished on the basis of their producers, that is, their vegetation, and consequently, also of their consumers. These are the mountainous areas, or *orobiomes* (OB) already mentioned, and the *pedobiomes* (PB), areas with soil which is not typical of the climatic area. In general, however, a climatically defined zonobiome corresponds closely with vegetational and soil-type zones; this is shown in Table 1. Deviations from this will be dealt with in greater detail in a separate section.

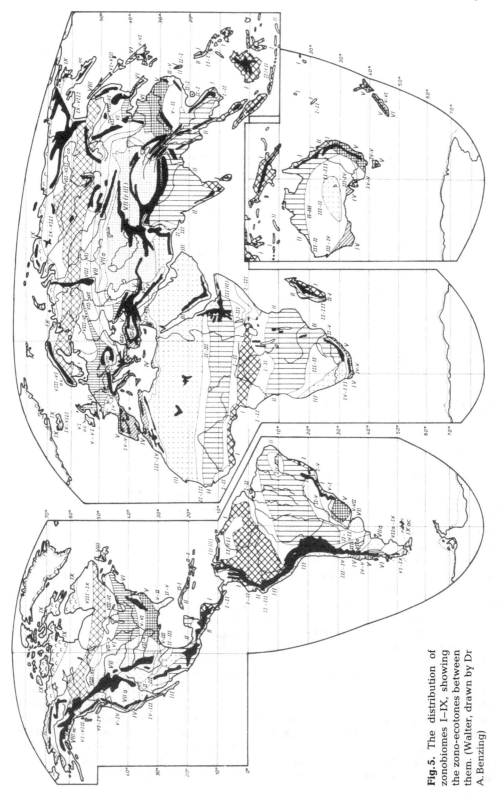

Fig. 5. The distribution of zonobiomes I–IX, showing the zono-ecotones between them. (Walter, drawn by Dr A. Benzing)

Table 1. Relationship of climatically defined zonobiomes to zonal soil types and zonal vegetation

ZB	Zonobiome; climate	Zonal soil type	Zonal vegetation
I	Equatorial; with diurnal climate, usually always humid	Equatorial brown clays, ferrallitic soils—latosols	Evergreen tropical rain forest; almost no seasonal aspects
II	Tropical; with summer rains and cooler drought period (humid-arid)	Red clays or red earths—savanna soils	Tropical deciduous forest or savanna
III	Subtropical; arid desert climate, scanty rainfall	Sierozems or syrozems (raw desert soils), also saline soils	Subtropical desert vegetation; landscape characterized by rock
IV	Mediterranean; with winter rain and summer drought (arid-humid)	Mediterranean brown earths; often fossil terra rossa	Sclerophyllous woody plants; sensitive to prolonged frost
V	Warm temperate; with maximum rainfall in summer or mild maritime climate	Red or yellow forest soils, lightly podzolic	Temperate evergreen forest; somewhat frost-sensitive
VI	Nemoral; cool temperate with a short period of frost	Forest brown earths and grey forest soils, lightly podzolic	Nemoral broadleaf deciduous forests; bare in winter, frost-resistant
VII	Continental; arid-temperate with a cold winter	Chernozem, Castanozem, Burozem to Sierozem	Steppe to desert with cold winters; frost-resistant
VIII	Boreal; cold-temperate with cool summer and long winters	Podzols (raw humus-bleached earths)	Boreal coniferous forests (taiga); very resistant to frost
IX	Polar; arctic and antarctic, with very short summers	Tundra humus soils with heavy solifluction	Treeless tundra vegetation, usually on permafrost soils

3.3.1 Orobiomes

In orobiomes, or mountainous areas, the ecosystems change with climate vertically; this is known as an *altitudinal belt series*. In the basal areas at the foot of a mountain conditions are the same as for the rest of the zonobiome from which the orobiome arises. The altitudinal belts, from the bottom to the top of the mountain, are described as follows: *colline—montane* (lower and upper)—*alpine* (lower and upper)—*nival*. This terminology can be applied to all mountains, irrespective of the types of ecosystems found on the belts, the sharpness of their demarcation from one another, or their actual altitude.

The statement that the series of altitudinal belts corresponds, in a condensed form, to the series of zones from south to north is an unacceptable generalization, based on imprecise observations made in central Europe and North America. A certain similarity may arise from the fact that in a mountain range which runs meridionally, species could only disperse along its length during the Pliocene-Pleistocene period of mountain formation, "orogenesis", since there was a continuous decrease in mean monthly temperature associated with increasing height of the mountains. This happened also in the Alps as the glaciers receded in the late glacial period. Thus certain arctic floral elements (*Dryas, Salix* spp. and others) are found today in the alpine belt and boreal floral elements (*Larix, Pinus cembra, Linnaea*, amongst others) in the high montane belt. Furthermore, orogenesis led to an increase in the rate of mutation and formation of new species, resulting in the development of new endemic species. It is for this reason that one finds in the alpine belt of the northern Alps new species of genera typical of the periglacial steppe (cf. ZB VII), such as

Artemisia, Festuca, Poa, Avena s.l., *Astragalus, Oxytropis, Leontopodium, Anemone* s.l., amongst others. In the southern Alps, as would be expected, the new species are of more mediterranean genera, such as *Crocus, Lilium, Primula* and *Campanula*. Such speciation due to orogenesis led, in isolated mountain areas, to the formation of a distinct flora with many new endemic species.

Climate changes far more rapidly in a vertical than in a horizontal direction; thus in the northern Alps the mean decrease in temperature over 1 m change in altitude is roughly the same as that over a distance of 1 km in the direction of the pole from the foot of the Alps. In comparison with the zonobiomes, altitudinal zonation therefore takes the form of quite narrow belts running around a mountain. Because of this, the ecotones between the belts are not marked. One exception is the subalpine ecotone between the high montane and the lower alpine belts, but then only provided the high montane belt is forested and the alpine treeless. In extremely arid mountains with hot summers, the forest belt is completely absent and the montane steppe belt passes smoothly into the alpine zone.

Climatic change in vertical and horizontal directions is generally quite different or even reversed. On mountains there is usually a gradient of increasing rainfall with altitude, due to rain moving up the mountain; above normal cloud level, however, it falls off sharply. For zonobiomes there is no comparable change in rainfall pattern from the equator towards the poles. Another difference between orobiomes and zonobiomes is the way in which day-length is affected: this does not change with increasing altitude, but does so in a polewards direction; the same applies to the position of the sun at noon. Furthermore, direct radiation of sunlight increases with increasing altitude and diffuse radiation becomes less, so that the contrast between sunny and shady positions becomes more marked; the opposite is true of movement towards the poles.

Ecological climatic diagrams for the upper altitudinal belts of various mountains are shown in Fig. 6. Each orobiome is named after the zonobiome within which the mountains arise. Thus orobiome I includes all the mountains which lie in the equatorial zono-

biome I (e.g. Mount Kilimanjaro), orobiome III those which arise from the hot desert zone ZB III (e.g. the Tibesti mountains), and so on. Orobiomes like these are called *unizonal orobiomes*. *Multizonal orobiomes* are those which extend over several zones; the Urals, for example, stretch from zonobiome IX in the north, to zonobiome VII in the south. The various belts of the altitudinal series must in this case be treated separately as suborobiomes, representing single zonal segments. *Interzonal orobiomes* lie between two zonobiomes and often form a sharp climatic border, as do the Alps, for example, between zonobiomes VI and IV in Europe. In this case the altitudinal series of belts on the northerly and southerly slopes differ markedly from one another and must be described as suborobiome VI and suborobiome IV. It must be borne in mind, however, that intramontane central mountain valleys, with their more continental climate, have their own belt series, forming "central orobiomes". Conditions in the Andes are particularly complicated (Ellenberg 1975).

3.3.2 Pedobiomes

By "pedobiomes" (PB) we distinguish those areas within a zonobiome in which the nature of the soil is determined more by the mother rock than by the climate. Pedologists speak of an "intrazonal soil" if the type of soil is found in one zone only, and of an "azonal soil" if it occurs unaltered in several zones. Such soils support a distinctive vegetation, but in most cases there is no detectable difference between the vegetation on intrazonal and azonal soils within one zonobiome. It is therefore useful to refer to *azonal soils and azonal vegetation, but only if the composition of the soil exerts a greater influence than does the climate.*

Pedobiomes are designated according to soil type: lithobiomes (stony soil), psammobiomes (sandy soil), halobiomes (salty soil), helobiomes (marshes), hydrobiomes (waterlogged soil), peinobiomes (soils poor in nutrients, from *peine*, Greek = hunger) and amphibiomes (soils which are flooded only part of the time, such as river banks, mangroves, etc.).

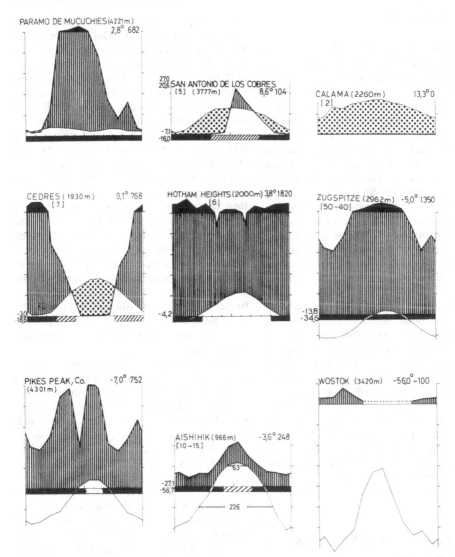

Fig. 6. Examples of mountain climates: orobiomes of various zonobiomes. *Upper row I* Venezuela; *II* Peru; *III* Chile; *middle row IV* Lebanon; *V* Australia; *VI* central Europe; *lower row VII* Colorado, USA; *VIII* Alaska; *IX* Antarctis. These climatic types do not occur anywhere in the lowlands. (Walter et al. 1975)

Pedobiomes often form a mosaic of small areas, and as such they are found in all zonobiomes, but they can extend over vast areas, such as the Sudd marshes on the White Nile (150000 km²); the even more extensive moorlands of western Siberia; the lava fields in Idaho (USA); fluvoglacial sandy plains; the nutrient-poor soil of the Campos Cerrados in central Brazil; the tree-less Nullarbor Plain in a limestone area of Australia, and so on. These will be dealt with in a separate section. In our subdivision of the geobiosphere there are thus three series: the first is climatic, represented by the zonobiomes; the second is orographic and is represented by the orobiomes; the third is pedological, with various pedo-biomes.

3.3.3 Eubiomes

The basic unit for all three series is the relatively large ecological unit, the *eubiome* (or simply biome). By this we imply a unit in which the landscape as a whole has a characteristic appearance, as may be seen in the climatic series of deciduous forests in central Europe; among deserts, in the Sonoran Desert, and also in the orographic series of both the Kilimanjaro massif and the Sierra Nevada in southern Spain; the Alps, on the other hand, are subdivided into several biomes, the Rocky mountains and the Andes into many more. The examples mentioned above in the pedological series are all large biomes; smaller ones are the Sand Hills in Nebraska (USA), the Karst formations of Jurassic limestone and the East African mangroves.

Future ecological monographs on particular countries must identify their eubiomes and describe them in greater detail. This is already being done in the Soviet Union for all area monographs as, for example, those on the Caucasus, the Caspian basin, western Siberia, etc., although in these they use the term "rayon". In our general survey of the geobiosphere we can give only a few examples of eubiomes for their actual number runs into several hundreds; nor are we in a position to give a complete list of them; this is more the task of biogeography and requires detailed knowledge of each locality.

3.4 Biogeocenes and Synusiae; Biogeocene Complexes

We have used the term eubiome to describe a large ecological entity; that is, an area which, although extensive, is readily surveyed. The definition of still larger units has been based on climatic conditions, but also on orography and on certain soil types, with characteristics determined by the mother rock and only to a limited extent by the climate.

To define smaller ecological units it is most practical to use the vegetation.

In an area which is topographically and geographically more or less uniform, one corresponding to a eubiome, genetically and historically determined differences in the flora are unimportant. The macroclimate, too, shows only insignificant deviations. The microclimate, however, plays a more important role, that is, the situation on a slope, peak or hollow with small differences in altitude, the degree of exposure, the inclination and, particularly, differences in soil, ground-water level, and so on. It is precisely to such differences that the vegetation reacts very sharply, so that plant communities, the phytocenoses, change quantitatively and qualitatively. As long as man does not interfere destructively, a plant community can thus be regarded as an indicator of the ecological conditions of a locality in which living organisms are growing, that is of a biotope. The composition of plant communities can be established by conventional methods and these communities can be distinguished from one another, although the more natural the vegetation, the greater the difficulty, since plant cover is in principle continuous.

The ecological unit which supports one particular plant community (phytocenose) is known as a biogeocene (BGC), or according to Sukatchev, a biogeocenose (Sukatchev and Dylis 1964). Associated with each phytocenose is a characteristic fauna, especially of smaller animals. These, the smallest of ecological units, might therefore be described as biocenes; the term biogeocene is, however, preferable, as it takes into account the effects of abiotic factors as well.

A *biogeocene* is the smallest basic unit of ecological systems, but it can be further divided into still smaller partial units—the *synusiae*. These are "working communities" of ecologically similar species and correspond to the "ecological groups" of Ellenberg (1956).

What synusiae are can best be explained by means of an example. Consider a deciduous forest of zonobiome VI with a mild nemoral climate: the tree layer and also the shrub layer can be regarded as large synusiae; further, smaller synusiae are to be found in the herbaceous layer; for example, the spring geophytes with a very short growth period before the forest trees come into leaf, the early summer herbaceous plants of the forest floor, the late summer and evergreen species: all of these are separate synusiae. There are also the synusiae of the lower plants, such as that of lichens on tree trunks, and of moss on the base of tree trunks or on tree stumps (vide Sect. 4.2).

Fungi are not producers; as heterotrophic plants, they must be classed, when parasites, with the consumers, or, when saprophytes, with the decomposers.

Synusiae are not micro-ecosystems, but only partial systems, since they have no independent material cycles or flow of energy; the circulation of substances within them forms part of the cycling within the ecosystem as a whole, and the primary production of the whole ecosystem is the sum of the production of the individual synusiae.

Between the basic unit of the large ecological systems, the eubiome, and the basic unit of ecosystems as such, the biogeocene, there is a need for ecological units of an intermediate size. Such units have not yet been subjected to close ecological study. We wish to describe them provisionally as *biogeocene complexes*, and see them as biogeocenes which form a particular series, related to one another either spatially or in time sequence (succession series). Thus, for example, a biogeocene complex may be made up of a mosaic of several biogeocenes where, within a biome or some topographical entity with a particular relief, there are raised areas each with its own microrelief and, as a result, differing in soil moisture content. Another biogeocene complex is associated with the slopes of very indented valleys, where the soil on the slopes forms a so-called "catena"; that is, there is a "chain" of soils with a steady decrease in size of the soil particles down the gradient. Associated with this there is a change in drainage and retention of nutrient elements. Such a catena is not merely a "chain" of soil types, each forming a biogeocene with its own vegetation and fauna, etc. and happening accidentally to follow one another: it is rather a series of biogeocenes linked by their origin, that is, genetically, and further connected by the exchange of substances which takes place between them. A third example is seen in valley river banks where several biogeocenes have developed in response to the greater period of flooding, the closer they are to the river. In each eubiome of an orobiome every altitudinal

belt will be a particular biogeocene complex; the constituent biogeocenes here differ from one another in degree of exposure or in inclination. Other possible biogeocene complexes include erosion channels, talus slopes and plains watered by springs or seepage.

All these ecological units are real units, for only such can be studied in ecology. They correspond to particular phytocenoses, and not to the abstract units of plant sociology, with its "associations" and "alliances", derived from the evidence of field surveys of characteristic species, species which are always encountered so that the constancy or frequency of their occurrence can be calculated.

3.5 Diagrammatic Representation of the Hierarchy of Ecological Units

The ecological subdivision of the geobiosphere, as part of the biosphere, can be summarized in a diagram, which shows the hierarchical sequence of the ecological units described above (see page 30).

3.6 The Extent to Which Animals Are Zonally Distributed

In this discussion of the subdivision of the geobiosphere, the animal world has not been mentioned. In fact, warm-blooded animals are affected only to a small degree by environmental temperature and are therefore not limited to warm climatic zones. In this connection an interesting experiment was the establishment of the very large nature reserve Askania Nova by Falz-Fein in 1887 in the steppe north of the Crimea (Walter 1974). African game, such as several antelope species, zebras and many others, which never encounter frost in their natural conditions, survived the very cold and often stormy winters on the open steppe and reproduced normally. Ostriches, however, need some protection from the cold in winter.

Lions were found in Iran in earlier times, but were exterminated by man because they are dangerous; they continue to exist, like

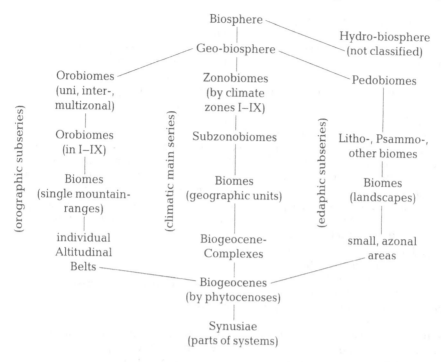

many other species of game animals, only in the sparsely populated areas of Africa. Migratory birds which live in arctic areas during the summer avoid the winter by moving to warmer climates. Animals, through their mobility, are also very independent of rainfall, for in periods of drought they can seek out widely scattered watering holes, or feed on the juicy parts of plants; they may even, to some extent, survive with metabolic water, that is, water formed in the body during the oxidative metabolism of reserve fat, as in camels or fat-tailed sheep. Many rodents in arid areas stay in underground holes during the day, where the air is almost saturated with water; they emerge only at night, when even in the hot desert the air is cool and damp.

Even cold-blooded animals are not entirely without protection against climatic temperature changes, for they can seek out places where there are favourable temperature conditions; during the midday heat, for example, they may move into the shade or underground. Many, like desert rodents, are active above the ground only at night. This helps also to conserve water. Even frogs can exist in dry areas; special adaptations en-

able them to survive 6 or more months of a dry period belowground. The less mobile soil fauna, which includes many arthropods and types of worm, is not exposed to the climate of the zone, but only to the climatic conditions prevailing in the soil itself; here temperature fluctuations are much less marked and the effect of the seasons much reduced, there are almost no diurnal changes and the humidity is usually constantly high. For these reasons the distribution of animals is determined less by climate than by historical—that is, geological and evolutionary— processes.

Chernov (1975) made a thorough investigation in the east European, north Asiatic region, where there is an especially clear zonation of climate and vegetation, of the extent to which animal species are limited in their distribution to a particular climatic zone. This is so only in very few cases. One very clear example of distribution confined to what are in fact subzones is provided by birds of the genus *Calidris* (sand pipers) on the Taimyr peninsular of the Siberian Arctic; this is summarized in Table 2. Further examples are shown on distribution maps; these include *Tipula carinifrons* (an Arc-

Table 2. Distribution of *Calidris* spp. on the Taimyr peninsula in northern Siberia. (Chernov 1975)

Species of the genus *Calidris*	Subzones				
	Dwarf-shrub tundra	Typical tundra		Arctic tundra	
		Southern	Northern	Southern	Northern
Calidris maritima	−	−	−	−	++
C. alba = sanderling	−	−		++	+++
C. canutus	−	−	−	+	++
C. testacea	−	−	+	+++	+++
C. minuta	+	+++	+++	+	−
C. alpina	++	+++	++	+	−
C. melanotos	−	+	+	−	−
C. ruficollis	−	−	+	−	−
C. temminckii	+++	+++	++	−	−

+++ Present in large numbers, ++ dispersed, + only solitary nests found, − absent

tic stream midge), *Bombycilla garrulus* (Boreal), *Eliomys quercinus* (Nemoral s.l.), the rodents: *Sicista subtilis* (steppe zone), *Allactaga saltator* (a large jerboa of the semi-desert zone), *Microtus arvalis* (a vole of the temperate zone, excluding the boreal), and *Marmota bobak* (steppe zone); and also the birds *Calcarius lapponicus* (the Lappland longspur) (arctic), *Aegolius funereus* (Tengmalm's owl) (boreal).

The dwarf hamster *Phodopus sungorus* is also strictly limited to the steppe zone, as are the steppe cockroach *Ectobius duskei*, the grasshopper *Platycleis eversmanni*, which feeds on the tussock grasses of the dry steppe, the rhinoceros beetle *Otiorrhynchus velutinus*, and, of course, other insects which feed on zonally distributed plant species. Limitation of an animal species to a particular climatic zone is, in fact, frequently an indirect result of its association with a plant species; competition between closely related species for a particular niche also plays a role.

The area of distribution, however, often extends over several zones; reindeer *(Rangifer tarandus)*, for example, is found from the Arctis to the Taiga. The roedeer *Capreolus capreolus* formerly inhabited the deciduous forest zone as well as the forested steppe and steppe, just as the aurochs *Bos primigenius* was spread throughout the steppe, but now both have been driven back into the forests by man. Indeed, all over the world,

large game animals, both ungulates and carnivors, have withdrawn into the sparsely populated forests and deserts.

In earlier times, zono-ecotones like the tundra forest and forested steppe, with their greater variety of vegetation and ecological niches, had a particularly rich fauna. Taking the relatively immobile earthworms as an example, the steppe has a large number of species, but only a small number of these, 13%, are zonally limited; this may be compared with values of 75% on the forest steppe, 60% in the mixed forest zone, and 42% in the Taiga zone. For many widely distributed insects, the "law of the relative habitat constancy and the biotope change" (see Sect. 9) can be applied. The distribution of the ladybird *Coccinella septempunctata*, for example, extends from the tundra to the desert; this is possible because the insect selects, in each climatic zone, biotopes where more or less the same conditions prevail. In the case of species of the bark beetle of the genus *Pityogenes*, the limitation to one single zone is more marked the thinner the bark of the twigs which it infests; species which live under the thick bark of tree trunks are polyzonally distributed. This applies also to *Scolytus multistriatus* and *S. scolytus*, which inhabit the lower part of tree trunks and are thus exposed to a constant microclimate. The ant *Lasius fuliginosus*, which is capable of regulating the temperature of its nest, is also widely distributed. This probably applies also to bees. Other species of ant are able to survive in deserts, because they build their nests in damp layers of soil. A polyzonal distribution is also possible for animals limited to a particular pedobiome which, in different zones, provides similar habitat conditions; an example is seen in the shore banks on sea coasts where rotting seaweed has been deposited.

Most of the adaptations of animals are related to their way of life, behaviour and nutrition. Adaptation to climatic conditions is often preadaptive; that is, characteristics already possessed by certain animal groups enabled them, at a later stage, to invade certain habitats in a particular climatic zone; for example, many lemmings and other rodents have the ability to overwinter in a metabolically active state beneath a covering of snow. On the other hand, in contrast to its steppe relatives, the Alaskan ground squirrel *Citellus undulatus* is able to live in the tundra because of its ability to hibernate, an adaptation which probably arose in response to a climate with long drought periods.

In the region of eastern Europe and Siberia, the boundaries of the areas inhabited by many animal species run in a north-south direction; that is, perpendicular to the climatic and vegetational zonation.

In summing up, it can therefore be said that, apart from soil dwellers, the distribution of the fauna, in contrast to that of the vegetation, only seldom displays a close correlation with climatic zones and zonobiomes. Animal ecologists therefore seldom study this question. Chernov (1975) has even questioned the validity of the Bergmann-Allen rule that body size of warm-blooded animals is greater in colder regions for reasons connected with thermoregulation. While this rule does seem to apply to the penguins in the southern hemisphere and to the bears in the northern hemisphere, a wider examination does not, according to Chernov, support the notion.

In discussing the consumers of individual zonobiomes, we will therefore have to make do with a few general observations. Tischler (1976) is the only zoologist who has given lists of animal species for particular zonobiomes; they are to be found in the section on Landschaftsökologie in his *Einführung in die Ökologie.*

3.7 Tropical and Subtropical, Humid and Arid Areas

Lauer (1975) has made a study of the characteristics of the tropics, from the point of view of the temperature and moisture regimes, and mapped them very precisely. In Section 3.2 we outlined a subdivision of the earth into nine climatic zones; here we will first examine their relationship to the conventional tropical and subtropical zones of climatology.

a) The term *tropical* is usually applied to a diurnal climate in which a steady, high temperature is maintained throughout the year, so that there are no seasons based on temperature differences. Such seasons as do occur can be associated only with rainy or dry periods, but in the equatorial zone the climate is often humid throughout the year. "Tropical deserts", which are dry almost the whole year, are exceptional; they are found in parts of Venezuela, of northeastern Brazil and in the interior of East Africa.

It follows that zonobiomes I and II belong to the tropics, although ZB II does have a cooler season.

Subtropical is the term applied to the intermediate zone between a tropical and a temperate climate; in some years, an occasional light frost may occur when there is a clear sky at night and consequently high radiative loss. This is the case in the arid, subtropical zonobiome III. This zonobiome is not found on the eastern sides of the continents, which are very rainy. As one moves from the equator polewards, the mean annual temperature falls steadily, the cooler season becomes ever more marked, until finally, regular frosts occur, while there is still no actual cold season. We have thus a gradual transition from the tropical zonobiome I to the warm, temperate and humid zonobiome V. Between is the humid, subtropical zonobiome II, which is difficult to separate clearly from zonobiomes I and V. Such a delimitation might be based on some particular mean annual temperature, or on the mean or the absolute minimum temperature, but such subjective delimitation would not reflect natural conditions.

The concepts tropical, subtropical and temperate are associated with a particular notion of the natural vegetation and cultivated plants. Typical tropical plant crops are cocoa, hevea-rubber trees, coconut palms, coffee, tea and bananas, amongst others; in the subtropics date palms, citrus fruits (usually with irrigation), olives, figs, pomegranates and others can still be grown, although these crops are also typical of zonobiome IV with its mediterranean climate. The latter zonobiome has, in fact, a subtropical climate in summer, but the winters, with much rain and occasional very cold periods, are more characteristic of a temperate climate. In some years, the crops may be badly damaged by frost. The arid-humid climate of zonobiome IV is thus an intermediate subtropical-temperate type.

The frost line is usually accorded great significance, but is not really critical for the vegetation. It is not zero temperature as such, but the freezing of their living cells which is damaging to frost-sensitive plants, and this occurs only after the temperature has fallen several degrees below 0°C, or after even heavier frost if freezing of the plant tissues occurs only after supercooling. On the other hand, the tropical-subtropical species are damaged by "chilling" even at temperatures above 0°C. They are not immediately killed, but die after a time, and the sensitive tropical species in low-lying situations even at relatively high temperatures (cf. p 89 and Sect. 5.1). A systematic study of this has, however, not yet been made.

We will return to the problem of "hardening" of plants in the temperate zonobiomes VI and VIII in Volume, 3 of this series.

b) The delimitation of *arid* and *humid* areas is even more difficult. Penck (1910) suggested that areas be described as arid if the annual precipitation is less than the potential annual evaporation, and as humid if there is a surplus of precipitation, which is drained back into the sea by way of rivers. The "aridity border" divides these areas and is to be found where precipitation and evaporation (expressed in mm) are of equal magnitude. This cannot be a sharp dividing line, for in years when the rainfall is low, it will be shifted in the direction of the humid area. It can represent only the average position over several years.

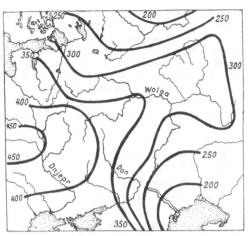

Fig. 7. Actual evaporation (in mm) in eastern Europe. (Kotscharin and Oppokow, from Walter 1960b)

Evaporation is here to be understood not as the actual but the "potential" evaporation, which increases with decreasing saturation of the atmosphere; it increases with rising temperature and increasing saturation deficit, whereas the reverse is true for actual evaporation, since in warm, dry areas the vegetation is sparse and the soil surface usually dry and, as result, very little water can be given off to the layer of air nearest the soil surface. This is illustrated in Fig. 7. In eastern Europe the highest actual evaporation occurs over the extensive marshes of the Pripet basin, while it falls steadily with decreasing rainfall in the direction of the Caspian Sea. The actual evaporation can never be greater than the amount of precipitation, unless additional water reaches an arid area by way of irrigation schemes or rivers.

Potential and actual evaporation are almost equal only in marshes which are wet throughout the year, or where there is an exposed water surface. Meteorologists therefore use a standard tank, the "class A pan" of the U.S. Weather Bureau, to measure the potential evaporation.

This round tank is made of galvanized iron, 25 cm depth, 120 cm diameter and stands on a wooden platform 10 cm above the ground, so that air can circulate beneath it. The tank is filled with water to within 5 cm of the brim, and topped up when the level has fallen by 2.5 cm. Wild's brass cup, still much used in meteorology for measuring evaporation, is not suitable because it is set up inside the hut of the weather station, where it is protected from the radiation of sunlight; it thus gives values which are far too low. Thornthwaite (1948) suggested that, instead of the tank, the loss of water from a short lawn with an optimal water supply, namely, groundwater level constantly at 50 cm depth, should be measured, and that these values should be regarded as potential "evapotranspiration".

Evaporation from a "class A" tank, of course, shows the "oasis effect"; it is greater than the evaporation from a lake, because energy for vaporization is supplied to the tank not only in the form of radiant energy, but also as warmth from the environment or from air warmed by the earth being blown by the wind. This factor is more important in arid than in humid areas. To obtain a value which would approximate the evaporation from a lake, the measurement made with the tank is reduced by multiplying by a factor of 0.8–0.6 and in extreme deserts 0.5. These factors have been established by comparing values for lakes with tank values in the USA (Kohler et al. 1959). It is doubtful whether this reduction is valid, for it is the potential evaporation over the continent that we wish to establish; if this were covered with large lakes, the whole climate would be markedly different. Even the climate of large oases with a plentiful supply of water is significantly cooler than the surrounding desert.

Measurements made with "class A" tanks are available from a few stations only. Penman and Thornthwaite (quoted in Henning and Henning 1976) have therefore worked out formulae which can be used to calculate the potential evaporation using the sort of data published by most meteorological stations. In a slightly modified form, the formula of Penman was used to calculate the so-called Penck aridity border on each of the continents (Henning and Henning 1976). The problems associated with this calculation are described by these authors. The formula postulates, however, a body of water without any heat capacity, and such a body of water does not, of course, exist. The potential evaporation depends always on the nature of the surface from which the evaporation takes place, and it is essential to take this into account if the values are not to be entirely theoretical.

A body of water with a very low heat capacity corresponds to the filter paper of a Piche evaporimeter, kept constantly saturated with water. Meteorologists use white paper; since their measurements are made in the hut of a weather station, away from the effect of radiation, the colour of the paper is irrelevant.

For measurements in the open, green paper is used so as to be comparable with the transpiration from green leaves. In direct sunlight evaporation from green paper is 30% greater than from white paper, as a result of the greater absorption of radiant energy by the green paper (Walter 1960b, pp 160ff). The Piche evaporimeter is indeed suitable only for establishing differences in the rate of evaporation of microclimates in different localities.

Since there can be no such thing as climatological evaporation in abstracto, Henning's map (Henning and Henning 1977, 1977a) showing the position of the aridity border must be re-examined, using features which do allow humid and arid areas to be differentiated. These features are as follows:

1. Endorheic areas where there is no water outlet can be regarded with certainty as arid.
2. Accumulation of salt in the soil and the occurrence of halophytes are only possible in arid areas, if we exclude sea coasts and saline springs.
3. Soil profiles with an effervescent layer (CaCO$_3$ precipitation) are encountered, apart from limestone districts, only in arid areas. Podzolic and ferrallitic soils are typical of humid areas.
4. If the headwaters of rivers lie in arid country, their streams will flow only intermittently.

If the course of Henning's aridity border is examined on the basis of these features in as far as our knowledge of the terrain allows, it becomes clear that in North America the border line between arid and humid areas running from North to South has been placed far too far to the east in the humid zone. In the long-grass prairie there are no hollows without drainage and no signs of saline soils; the soil profiles of the prairie have no effervescent layer, rather, they are moist right down to the water table. This is thus a semi-humid area. Only further west in the mixed prairie, and even more clearly in the short-grass prairie are all the features of an arid area encountered. Henning's aridity border must be moved about 400 km west at the point where it intersects the Canadian border, and at a latitude of 38°N, it must be shifted 1700 km to the west. Likewise, there is no evidence of an arid area in the south-east of the USA extending as far as Florida in the south. On the contrary, from North Carolina to Georgia there are extensive moors and heaths with poor soils and insectivorous plants *(Dionaea, Sarracenia)* and also *Taxodium* marshes. The whole eastern part of North America is humid. In eastern Europe the most northerly black earth has, like the prairie soils, no effervescent layer. This makes its appearance as pseudo-mycelia (mould-like threads) only in the profile of the deep Thick Black Earth. The semi-arid area begins at this point and not more than 400 km further north within the Polesje. There is similarly no evidence of arid areas in central Europe or in the Loire in central France or even in the south-east of England. It is true that small halophilic meadows are found in localized inland places in central Europe, but these occur *only* where springs of brackish water emerge from deep underground salt layers and they are strictly limited to the region of these springs. The whole of western Europe is by its nature a deciduous forest area. Only rocky and sandy areas are relatively dry, as are chalk or gravel slopes with a southerly inclination.

On the other continents, too, the arid areas are much smaller than Henning's maps indicate. In South America our work showed that the border in Argentina is correct, for the pampas is already semi-arid; in southern Brazil, however, there are no arid areas. It is very difficult to assess aridity in areas with a wet and dry season; that is, in the humid-arid zonobiome II, with wet summers and dry, cool winters, or in the arid-humid (mediterranean) zonobiome IV, which has very dry summers and heavy rain with a low rate of evaporation in winter. The winter rainfall is high, and since there are no undrained catchment areas, a large volume of water is transported in the rivers to the sea. As a result, mineral salts which ac-

cumulate in summer are washed out of the soil. The only sign of aridity is the dry river beds in summer; water flows in them only occasionally after a sudden downpour.

What is remarkable in zonobiome IV (in Europe) is that even *Isoetes* grows in the very shallow winter lakes; this genus is typical of nutrient-poor waters of extremely wet areas. The lakes of Sardinia on the basalt plateau Plain of Gesturia form a good example of this: on the unpopulated and inaccessible Table Mountain (520 m above sea level) wild horses graze on the water plants, while their dung accumulates in the pools and lakes. In summer, when the lakes dry up, ants carry all the organic remains to their nests so that only a thin, nutrient-poor layer of soil, containing the dormant remains of *Isoetes* plants is left. As a result, the water which accumulates after rain in winter is poor in nutrients, or oligotrophic, and this is favourable for *Isoetes*. There is also a rapid increase in the numbers of the branchiopod *Lepidurus apus*. The floor of the lake with a thin layer of sediment is characterized by an extremely low phosphorous content and a pH value of 5.3 to 6.3 (Margraf 1981). Around the lake the vegetation is, by contrast, typically mediterranean, with cork trees, *Citrus* species, *Myrtus communis* and *Asphodelus ramosus*.

In the humid-arid climate of zonobiome II the summer rains are so heavy that here too the signs of an arid winter climate are not found. It is only in the border area with zonobiome III that the rainfall decreases, and undrained salt-pans can be formed, as, for example, in the Kalahari, where they occur alongside non-saline chalk pans which have underground drainage.

Apart from the formation of saline soils, the aridity border is of no great importance ecologically or botanically. In the temperate climatic zone it coincides with the forest steppe zone, where zonobiome VI of the deciduous forests and zonobiome VII of the grassy steppe interlock in mosaic fashion; in the tropics this border lies in the region of dry forest and savanna and is not very obvious; it becomes more marked only in the semi-arid zone, through the appearance of saline soils with halophilic plant species. It is also unimportant in agriculture, since in semi-arid areas the quantity of water stored in the soil suffices for farming; the soils are not leached but are the especially fertile black earths, while the long hours of sunlight create good conditions for high crop yields. It is only when the dryness is more marked and the variation in rainfall from year to year greater, that farming even with the so-called "dry farming" method becomes unprofitable as a result of frequent bad harvests. The result is that those areas which border on the desert can be used only as grazing lands for cattle or, when still dryer, for sheep and goats.

It is similarly impossible to use the evapotranspiration maps of Thornthwaite for either ecological or agricultural purposes. One may gather from them the location of land useful for humid grassland or flooded rice fields (Talsma and Lelij 1972), requiring no irrigation at any time of the year. They even indicate, for less humid areas, those months of the year during which irrigation would be required; but for agriculture, fruit farming and forestry, knowledge about available underground water reserves is more important. This depends, however, not on potential evapotranspiration, but on the actual evaporation and the nature of the soils.

Ecological climatic diagrams likewise give no indication of whether an area is humid or arid, for they show only the *relatively humid and arid seasons for a particular climatic zone*. The comparison of the temperature curve with that for rainfall is based on the fact that annual temperature fluctuations are practically identical with those of potential evaporation, except that evaporation starts to rise sooner in spring, and declines rather later in autumn. The relationship between temperature in °C and potential evaporation in mm is not, however, constant, but increases with increasing aridity. We tested the precise relationship in Argentina. In northwest Argentina it was 1:5, in the pampas 1:7, in the other parts 1:10, and in arid Patagonia with continuous strong wind 1:15 (Walter 1967a,b). The fact that in climatic diagrams drought periods are indicated only relatively by the relation between the temperature and rainfall curves is advantageous from an ecological standpoint, since the drought resistance of plants differs between climatic zones: xerophilic species from central Europe are less

drought-resistant than those from the steppe, and the latter less than those from the desert.

A purely physically based measure of aridity is provided by the ratio Q/LN (Budyko 1980), in which Q is the net radiation on the surface of the earth, N the annual rainfall and L the heat of vaporization of water (at 25°C 583 cal g^{-1} = 2448 J g^{-1}). The border between arid and humid zones will be where the Budyko ratio had the value 1.0, that is, where AQ = LN, so that the radiant energy is just sufficient to cause all precipitated moisture to evaporate. In the humid forests of the temperate zone, the value lies between 1 and ⅓, in the tundra it is less than ⅓. In arid areas (steppe, savanna, tropical dry forests) one finds values of 1–2, in semi-deserts 2–3 and in deserts more than 3. Agreement with ecological observations seems to be generally good. On detailed examination, deviations may, however, be found in many areas; for, additionally, energy may be carried to and fro by wind; in western Europe, for example, fronts of warm and cold air masses play a major role in determining the final energy balance.

3.8 Saline Soils and the Degree of Aridity

As has already been mentioned, saline soils form only in arid areas. They are particularly evident in low-lying parts of the relief, where the water table is high; they are, in consequence, mostly wet soils. The composition of the groundwater changes with transition from a humid to an arid climate, and in the latter it is affected by the degree of aridity. Nowhere is this more clear than in the east European lowlands, where, along an axis running from NNW to SSE, the climate becomes warmer while at the same time the annual rainfall decreases, so that the terrain becomes steadily more arid (Walter 1974).

In humid forest areas with mainly podzolic soils the groundwater is poor in mineral nutrients, acidic and often brown in colour as a result of the humus colloids. In arid areas, however, the groundwater contains soluble salts, the quantity and quality of which change with increasing aridity. In wet soils, during the hot season, the groundwater rises by capillarity to the upper surface where the water evaporates and the salts accumulate, a process known as salinization. Three different forms of this can be distinguished.

1. *Calcium carbonate salinization.* In semi-arid areas of the forest steppe large quantities of $Ca(HCO_3)_2$ are dissolved in the groundwater. In alkalitrophic lowland fens with large hummocks of *Carex omskiana* (rel. *C. elata*), it reaches the surface and, as the water evaporates, accumulates there as $CaCO_3$. Peat here often contains up to 40% of chalk and may have a pH of 7.0–8.1.

2. *Sodium salinization* (solonization or solonetz formation). In the arid steppe zone small quantities of sodium salts are always present in the groundwater. The rising Na^+ ions in the humus horizon form sodium-humus complexes, which, during the wet season, when the humus horizon is saturated with water and CO_2, undergo hydrolysis. Thereafter sodium bicarbonate ($NaHCO_3$) or even sodium carbonate (Na_2CO_3) are formed.

Na-humus + H_2CO_3 → H-humus + $NaHCO_3$

or, Na-silicates react with $Ca(HCO_3)_2$

$Na_2SiO_3 + Ca(HCO_3)_2$ → $CaSiO_3 + 2NaHCO_3$

In either case, the sodium bicarbonate is dissolved in the groundwater during the wet season but accumulates in summer in the upper soil layers (solonization) which are strongly alkaline (Gedroiz 1929; Kovda 1939). Indicators of light solonization are: *Trifolium fragiferum*, *Triglochin maritimum*, *Juncus gerardi*, *Geranium collinum*, *Taraxacum bessarabicum*, *Aster tripolium*, and, because it is insensitive to brackish soil, very frequently *Potentilla anserina*. The following occur where there is heavy solonization: *Scirpus maritimus*, *Puccinellia palustris*, *Peucedanum latifolium*, *Senecio racemosus*, amongst others (Walter 1960a).

3. *Chloride and sulphate salinization* (solonchak formation). In the arid southern steppe or semi-desert, as the degree of aridity increases, the groundwater contains in-

creasing quantities of sulphates and eventually also sodium chloride, until finally these salts exceed the carbonates in quantity. In these conditions sodium salts accumulate in the soil surface, forming salt crusts, the so-called *black alkali* (soda-solonchak soils) which contain humus sol and the *white alkali* (true solonchak soils) along with Na_2SO_4 but without humus, because the high salt concentration causes the humic substances to precipitate.

Apart from undrained depressions, a high water table is found in arid areas only in river beds, so that solonchak soils follow the course of rivers. The salt crust is frequently blistered. Plant indicators are: *Salicornia europaea*, *Suaeda* and *Limonium* spp. and many other euhalophytes (see Sect. 5.2.5). Such saline soils have very characteristic profiles. Figure 8(6) shows a solonchak soil profile with a high level of brackish groundwater. In chloride-sulphate salinization, the salt concentration in the humus layers is always high, so that the humic substances are precipitated. The humus horizon, A_1, is dark black and is followed by a lighter A_2. Beneath this begins the C horizon; here the very soluble salts ($NaCl$, Na_2SO_4) rise to the surface to form a whitish crust which disappears during the wet season. Beneath the gypsum horizon begin the water-logged gley horizons, G_1 and G_2.

Different kinds of brackish soil transformation. In certain circumstances a solonchak soil may be modified. Where the water table is low, either on high ground or as a result of erosion gullys, rain affects a solonchak soil, washing salts out of the upper layers; the sodium-humus complexes dissociate and a process of solonetzing sets in. The humus-sol, together with the sesqui-oxides Fe_2O_3 and Al_2O_3, is carried down to lower levels, to be precipitated again, forming a B horizon. This process is very reminiscent of podzolisation, except that the readily exchanged Na ion plays the dispersing role of the H ion. Whereas in podzolic soils the humus-sol is formed in an acidic medium, here, as a result of the formation of sodium carbonate, the medium is very alkaline (pH = 9 or more).

This results in the formation of a solonetz soil, which, in the less arid Hungarian Puszta is called *szik soil*. Typical indicator

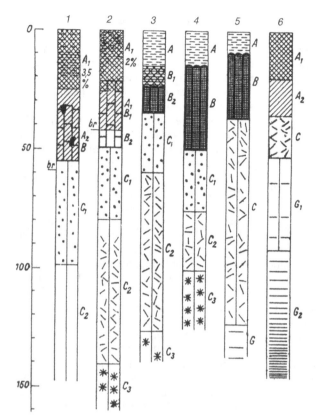

Fig. 8. Salinization of the soil with increasing aridity in eastern Europe. Typical profiles from weakly to strongly saline. 1 Southern black earth, slightly solonized, some compaction (A_2B). 2 Dark chestnut-brown soil with clearly defined B horizon; beneath the $CaCO_3$ nodules in C_1 are gypsum deposits in the form of tubules in C_2 and gypsum druses in C_3. In both 1 and 2, A_1 has a platy structure; br effervescent layer; numbers indicate percentage of humus. 3 Light chestnut-brown soil, heavily solonized; B horizon dark, columnar and very compact, A poor in humus and platy. 4 Typical columnar-solonetz soil: A ash-grey and platy, B very well-developed. 5 Solonetz soil changed by rising groundwater; gypsum nodules dissolved and gypsum deposits extend to 40 cm below the surface, beneath the gley horizon G. 6 Typical solonchak with high groundwater and dark humus horizon (A_1) and gypsum tubules, below them gley horizon; often a salt crust on the surface. (Walter 1960)

plants for this type of soil are: *Puccinellia convoluta, Camphorosma annua, C. monspeliaca, Limonium alutacea, Artemisia maritima-salina, Kochia prostrata, Linosyris villosa, Lepturus pannonicus, Gypsophila stepposa* and *Petrosimonia volvox* amongst others.

These solonetz soils are dry and the vegetation has a semi-desert character. There is thus no accumulation of humus. The soil profile (Fig. 8.4) is as follows. The A horizon is completely bleached, ash-grey, of platy structure and consists of quartz sand. Beneath this eluvial horizon, is the very dense, dark brown-black illuvial horizon (B). If, during the dry season, B shrinks markedly, it develops vertical cracks which give it a columnar structure; for this reason it is known as *columnar solonetz* soil. If the A horizon is but a few centimeters thick, it is called a *crusted solonetz* soil, and is frequently covered with *Artemisia pauciflora*. Solonetz soil is thus a structured soil, in contrast to the structureless solonchak soil.

Accumulation of salts takes place in the B horizon; in consequence of the dispersing action of Na, this horizon swells up in moist conditions and becomes completely impervious to water, so that the whole soil can be water-logged.

Below the B horizon lies the C horizon: it is subdivided into a zone with carbonate deposits in the form of chalk nodules (C1); this is followed by zones with gypsum deposits in the form of tubules (C2) and finally as druse-like deposits (C3).

It should not be thought, however, that the solonetz soils of eastern Europe are invariably formed from solonchak soils. Solonized soils are mainly steppe soils, covering extensive areas north of the Putrid Sea (Sivash). These were never affected by brackish groundwater, but were formed, according to Machow (oral communication) by the deposition of salt dust on the surface. In the dry summer months, the water level of the Sivash falls, broad expanses dry out and are covered by a white salt crust. The salt dust is blown north and deposited on the southern steppe soil. This results, in time, in an accumulation of Na ions in the humus horizon. Sodium carbonate is formed and this leads to the development of humus-sols, and the leaching of the upper horizons and development of an illuvial B horizon sets in.

The onset of early stages of such solonization can be seen in the southern part of the black earth area, in that the typical platy formation in the A1 horizon is evident, and there is a compaction of the A2 horizon (Fig. 8.1). The dark chestnut soils which extend over the southern part of this area are clearly already solonized. Here the humus horizon contains only 2% humus, it has a marked platy formation and the lowest layers (A2, B1) are already compacted and pass into the B2 horizon. Chalk nodules appear at a depth of only 50 cm (C1); below this is a horizon with gypsum tubules (C1) followed by gypsum in a druse-like formation (C3) (see Fig. 8.2). These features are even more clearly developed in light chestnut soils, which in this series are followed directly by true solonetz soil. The A horizon is strongly leached while the lower part of the B horizon shows the columnar structure (B2) (Fig. 8.3). Chalk nodules start at 30 cm depth, an indication that the rainfall here does not penetrate deeply into the soil. It is thus clear that this marine salt effect is greater the drier the climate and the greater the proximity to the area from which the salt-dust originated. This type of solonetz formation has nothing to do with leaching of solonchak soil.

The reverse process can be witnessed on the very flat north coast of the Black Sea, where solonetz soil is being transformed to solonchak (Fig. 8.5). This coastal strip is sinking slowly and steadily into the sea. Seawater often reaches to the foot of old burial mounds, barrows (Kurganes), which must have been built originally on the highest elevations. The solonetz soil thus comes here into contact with saline groundwater. The first sign of change is the formation of a gypsum horizon above the chalk horizon, until finally the entire distribution of salts and the soil profile itself have been altered.

Formation of solonchak soil can be induced also by man's activities. In areas under irrigation, where the drainage is inadequate to cope with excess water, the groundwater rises by capillary action until it reaches the upper surface. Crystallization of salts on the surface begins at once and, without expensive and difficult countermeasures, the soil is no longer fit for agriculture. Large areas of arable land are lost in this way, and it is quite likely that the widespread occurrence of saline soils in old centres of civilization, like Mesopotamia, is to be explained thus.

Solonized soils are suitable for agriculture by dry farming, but the high degree of alkalinity, caused by the presence of sodium carbonate, must be reduced. This is done by adding gypsum; Na_2CO_3 reacts with $CaSO_4$ to form the less damaging Na_2SO_4 and $CaCO_3$. Indeed, the latter even leads to an improvement of the soil structure.

Solodization. Increased action of water on a solonized soil with absorbed, exchangeable Na, leads to a further degradation to *solod soils.* When the Na-humus complex is hydrolysed, hydrogen ions replace the sodium ions and an acid humus forms. In the absence of $CaCO_3$ this causes leaching, as in a podzolic soil. Only here the adverse effect is much greater. This solodization of alkaline soils can be seen in steppe areas where rainwater runs into undrained depressions, or "pods", and only slowly seeps into the ground. These hollows, which are hardly noticeable on the terrain, may extend over 15000 ha. In wet years their vegetation has a typically fen character; in dry years, by contrast, it is a typical steppe vegetation. The plant cover is thus constantly disturbed. Further north, in the forest steppe zone, forest development takes place in such depressions with heavily leached, degraded soils. Besides willow stands, aspen groves occur, and in some cases oak and silver poplar appear.

Since solodization must be regarded as a continuation of solonization, it seems reasonable to assume that all the solod soils arose originally from saline soils, which typically arise in such depressions in an arid area. If this view is correct, solodization can only be explained as a result of an increase in the rainfall; in other words, the climate of the forest steppe and steppe zones must have become more humid. This suggestion

Table 3. Comparison of the salt distribution in solonetz and solonchak soils (Keller 1926). Water content and readily-soluble salts as a percentage of absolutely dry soil; sulphate given as SO_3. (Numbers in italics = horizons with maximal salt accumulation)

Soil type and vegetation	Solonetz soils with *Artemisia pauciflora* and *Camphorosma monspeliaca*				Intermediate soil *Atriplex verrucifera*				Solonchak soil with *Salicornia europaea*			
Depth of sampling (cm)	Water content	Soluble salts	Cl	SO_3	Water content	Soluble salts	Cl	SO_3	Water content	Soluble salts	Cl	SO_3
Surface	3.3	0.111	0.005	0.025	2.5	0.054	0.005	0.015	9.1	*18.854*	*3.005*	*7.982*
10	13.9				15.5	0.437	0.052	0.102	23.1	4.580	0.593	2.096
15		0.184	0.040	0.007								
20		0.317	0.128	0.006	18.7	2.612	0.164	1.254	32.1	3.862	0.361	1.609
30	10.3	0.466	0.187	0.030								
40					18.5	1.683	0.273	0.671	35.8	3.059	0.334	1.154
45		*1.094*	*0.260*	*0.388*								
60		0.816	0.214	0.166	21.1	2.009	0.393	0.716	35.3	2.423	0.482	1.018
90		0.496	0.202	0.051								
100	9.0				23.0	1.693	0.403	0.505	31.8	2.398	0.472	1.013

is further supported by the degradation of black earth to grey forest soil.

In semi-desert areas solonetz and solonchak soils usually occur in mosaic-like complexes, which are determined by the micro-relief; solonetz soil is found on the high ground, solonchak soil in the depressions.

The distribution of salts in solonetz and solonchak soils is shown clearly in Table 3 (after Keller, from Walter 1943, p 112).

4 Terrestrial Ecosystems and Their Special Features

4.1 The Two Cycles Within an Ecosystem

Research on ecosystems began in the field of limnology. Initially, therefore, only aquatic ecosystems were examined. The cycling of material in these aquatic systems, which can be represented as

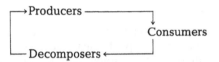

is thought of as being generally applicable. It is, however, not possible simply to extend this to terrestrial systems, which are far more complex in their structure. In terrestrial systems there are two parallel cycles, a short cycle and a long; in the short cycle, which is quantitatively the more important, consumers play a relatively minor role.

The terrestrial ecosystem hitherto most thoroughly investigated is that of the deciduous forest of the temperate zone (Duvigneaud 1971, 1974; Sukatschew and Dylis 1964; Ellenberg 1971; Medvecka-Kornas 1967; Goryschina 1974). The major produers in this ecosystem are the deciduous trees. Primary production is the annual increase in phytomass: this will include the leaf mass which is produced anew each year, the annual shoots, the annual growth in wood and the annual growth in the roots; to these must be added the generative organs, the flowers, fruits and seeds. What is measured, however, is usually only the growth in wood, and the quantity of litter which falls to the ground in a year: that is, the dead leaves, twigs, dead wood, and the flowers, fruits and seeds, as well as the dead root mass, as far as it is possible to assess this. Dead wood

accumulates in large quantities only in virgin forest. In managed, deciduous forests, therefore, the calculation is based on the average annual growth in wood. It must be remembered that the leaf mass is greater than the total mass of fallen leaves, because before leaf-fall, a part of the cell contents are removed to the axil organs and stored there.

The decomposition of litter to form mineral matter takes place in the soil as a result of the activities of decomposers (fungi, bacteria and actinomycetes). Some of the soil fauna, including larger invertebrates such as earthworms and smaller forms like mites, contribute to this process. Although their role is really limited to breaking down dead organic matter to particles of smaller size, thus facilitating the further action of microorganisms, these saprophagous animals should be regarded as decomposers, particularly since their faeces are very readily mineralized. This is thus a *short cycle:*

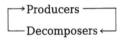

It should not be forgotten, however, that the soil fauna includes also phytophagous and carnivorous forms which play a role in maintaining a balance in the microflora.

Parallel to this is a *long cycle:* this includes the consumers with their many food chains, starting with some phytophagous form or other and ending with predators. Part of the organic content of the food is broken down and released as CO_2 in respiration, part is excreted as urine or evacuated as faeces, and a part is incorporated into the tissues of animals, a process known as secondary production. The excreta of animals are broken down by coprophages, the dead bodies by necrophages, at the same time,

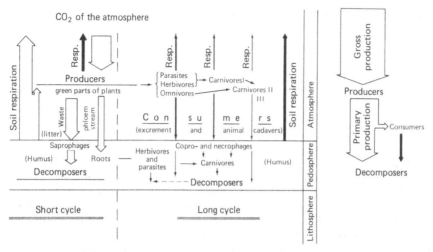

Fig. 9. Schematic representation of the short and long cycles in a deciduous forest biogeocene. *Left* material cycle; *right* energy flow (*thickness of arrows* indicates approximate relative quantitative contribution)

however, also by microorganisms, until everything has once more been reduced to mineral substances and the cycle is closed. Figure 9 shows these relationships, including the flow of energy, only a limited part of which runs via the long cycle.

The quantitative assessment of the long cycle is very difficult. The portion of the total turnover it represents is probably large in steppe and savanna ecosystems, where the number of game animals is high, but extremely small in moor lands; in either case it will, however, be quantitatively insignificant. Qualitatively, on the other hand, it is of great importance as a stabilizer and regulator of an ecosystem, as Remmert (1978), in particular, points out. The action of animals and its consequences for the ecosystem must be assessed quite differently when viewed from an ecological, rather than from an economic, man-oriented viewpoint.

The relationship between producers and the phytophagous consumers, which are usually highly specialized, feeding on one or a few plant species only, is extremely complicated, as will be clear from the following examples from the extensive vegetation region of eastern Europe (Sokolov 1980).

The two most important groups of phytophagous animals in woods are insects, which may occur in very large numbers, and mammals, in particular ungulates and small rodents. Their effect is usually assessed by calculating the phytomass consumed, attributing special significance to the consumption of the tips of shoots and productive leaf mass. It would really be more correct to take account of all observable events which contribute to the maintenance of a balance within the ecosystem.

The following three examples should make this clear. The first concerns the multistorey mixed oak woods in the forested steppe zone of eastern Europe (Jerusalimov 1980). The oaks are dominant in these forests. If they are defoliated as a result of infestation by the moth *Tortrix viridana*, the light intensity beneath the crown relative to that above it increases from 24% to 52%. The favourable light conditions thus created for the lower tree storey last for more than a month and make possible a greater production of wood by this layer. Table 4 shows the annual growth in wood of the tree stand as a whole, assessed by measuring annual rings, for the 7 years before infestation with insects (I), and for the 4 years during which infestation occurred (II).

It is clear that the reduced growth in wood of the oaks is more than compensated by the increased growth of the other woody species. Even if only the upper tree storey is considered, namely, *Quercus*, *Fraxinus* and *Tilia*, the loss to the oaks is fully compen-

Table 4. Mean annual increase in cross-sectional area (cm^2) of undamaged and *Tortrix* infected oak stands

Tree species	Mean annual increase in cross-sectional area (cm^2)		Difference (%)
	Undamaged	Infected	
Quercus robur	1348 ± 43	1101 ± 23	− 82
Fraxinus excelsior	421 ± 12	658 ± 18	+ 56
Tilia cordata	259 ± 26	325 ± 32	+ 29
Acer platanoides	143 ± 25	444 ± 22	+ 60
Acer campestre	143 ± 15	224 ± 25	+ 54
Ulmus scabra	53 ± 6	82 ± 8	+ 52
Corylus avellana	280 ± 35	287 ± 35	+ 3
Total stand	2780 ± 57	3066 ± 31	+ 10

sated; indeed, there is an increase in production of 3%. This is, however, true only when the oaks make up about 70% of the forest trees. When their proportion is less than this, the light conditions hardly change at all, while, if the oaks are more numerous, there is no full compensation by the other species. There is thus a clear difference between natural, mixed forest and managed pure stands. The herbacious layer is especially promoted in infested stands. The otherwise sterile *Aegopodium podagraria* and *Carex pilosa* may even flower; the dry weight of their phytomass increases over 2 years by more than three fold and the amount of cover from 14% to 70%. Furthermore, the frass from caterpillars on falling to the ground soon decomposes and presumably fertilizes the soil.

Even in pure stands of pine, infection with *Dendrolimus pini* can be compensated in time. Heavy loss of needles results in a reduction of wood growth in the second year of infestation, to 76% of normal, and in the third year to 56%, but in the fourth it rises to 150% and in the fifth to 194%; thus overall there is a small but discernable increase in production. The explanation lies in the fact that, owing to reduced competition from the larger, more heavily infected trees, the smaller trees show improved growth. Furthermore, in infested stands the lower, heavily shaded and non-productive whorls of branches are cast off sooner. Yet another form of compensation has been observed in stands of beech: here only part of the leaf was lost, but while the overall leaf surface was reduced, *photosynthetic activity was so greatly enhanced in the remaining leaf area* that production remained almost unaltered.

The assessment of damage to forest stands by larger grazing animals—in eastern Europe mainly elks—remains a complex problem. In natural conditions the population density of these game animals is so small that the consumption of phytomass is insignificant. A normal population density is taken to be 3–5 animals per 1000 ha. Such figures are, however, rather meaningless, for the animals are never evenly distributed, but concentrate in certain areas where suitable food is available; that is, in clearings. Damage usually occurs only when the balance has been disturbed by economic exploitation of the forest. This results in an increase in felled areas and areas with young trees, where willows and asps can flourish. These are eaten eagerly by the elk which, as a result of the plentiful supply of food, rapidly increase in number. This in turn makes it necessary to protect rejuvenating areas by fencing (Kuznetsov 1980).

In these forests, however, the more important mammals are the smaller rodents. These have their burrows under the ground, but in summer they live on herbaceous plants, in winter on nuts and seeds which they store. Their role is dealt with in greater detail in the discussion of the ecosystem of the forest of the Vorskla (Sect. 4.2.4).

Among the insects, the special group of xylophagous species should be mentioned;

these feed on the bast or woody parts of living tree trunks. These insects cannot, however, attack completely healthy trees but only those which have already been weakened. Healthy trees are only attacked by insects which eat the easily regenerating parts of the tree, mainly the leaves, or which cause the formation of galls. If insects could infect the trunks of healthy trees, the damage would be irreparable and the species would soon die out. This occurs only following introduction of an exotic pest to which the tree species is not resistant. An example is the fungus (Chrythonectria) *Endothia parasitica,* which was introduced to North America from Japan, and there destroyed the *Castanea* sp. Another example is the dying of the elms in Europe, brought about by the pyrenomycete *Ceratostomella ulmi.*

Indigenous xylophagous insects, by contrast, attack trees which have been weakened, perhaps by drought or by being too heavily shaded, those which have been damaged by forest fire, or old trees. A series of xylophagous species succeed one another, ending with saprophagous species which finally decompose the wood debris (Lindeman 1980).

In open grasslands (savanna, steppe) or in semi-desert herbivores, again mainly ungulates and rodents, play a far larger role (Abaturov 1980). In the African savanna up to 60% of the primary production is removed by herbivores. This high percentage is possible only because the vegetation consists of plants capable of compensating for the loss of their aerial parts by rapid new growth. This is particularly marked in grasses. The loss in primary products includes not only what is consumed, but also the parts that are torn off but not actually eaten by the animals while feeding and also damage caused by the hooves of large game animals or grazing cattle as they move about. In the case of the rodent *Citellus pygmaeus* the uneaten remains make up 50–60% of the plant material bitten off. In the Caspian semi-desert it has been found that their population is 70 ha^{-1}, and the loss in plant production they cause is estimated to be 20%.

The most interesting finding of these investigations, however, is that the primary production of the ecosystem did not suffer a loss as a result of the activity of these herbivores but, on the contrary, increased. Evolutionary biologists go so far as to conclude that co-evolution occurred in the development of grasses and grazing animals, forming a sort of positive feed-back system. To make an exact quantitative assessment, grazing was simulated by repeated cutting off of plants in an appropriate manner. It was found that repeated removal of 70% of the plant mass so stimulated regeneration that the total primary production increases by 40–74%. The vegetation is thus adapted to grazing, but the ecosystem remains in balance only when grazing occurs in natural conditions. Areas artificially protected from grazing degenerate as a result of the accumulation of dead plant material. On the other hand, an ecosystem also collapses when an area used for cattle farming is overstocked, which unfortunately occurs all too often. Only rational rotation of grazing can avoid this. Protected steppe reserves which are not grazed should be mown every 3 years and the hay removed.

These examples show how complex the relationships between the producers and the phytophagous consumers are. Other relationships which should be mentioned are the role of animals in pollination of flowers and in the distribution of fruits and seeds of epizoochore, endozoochore or synzoochore plant species; here birds also play a particularly important role. The balance between producers and consumers is maintained by many regulatory cycles which are geared to one another; to these must also be added the cycles involving relationships between herbivorous animals and their predators.

A special position is occupied by termites and by the leaf-cutting ants *(Atta).* While they collect both dead and living parts of plants, they do not feed on these, but on the fungi which they cultivate in their nests, and which belong to the decomposers.

We could mention further zoological investigations, but these have usually been made as isolated studies, and not as part of ecosystem research.

A rich soil fauna, likewise with many food chains, promotes breakdown of litter and the formation of mull-humus layers, while in acid, peaty or marshy soils, which have a poor fauna, accumulation of organic matter often occurs because decomposition is inhibited.

All the deciduous forest ecosystems in central Europe so far investigated are stands which have been under forestry management for centuries. Wood has constantly been removed from them, so that there is no dead wood lying on the ground, there slowly to decay and provide an important nursery for certain species of tree. Managed forests are homogeneous and differ from virgin forest, as the composition of the tree stratum is determined by the forester. The only exception to this is on the Vorskla, a left tributary of the middle Dnepr; here, on the edge of the deciduous forest zone of the steppe, far beyond the beech wood region, is almost virgin forest.

4.2 A Deciduous Forest Ecosystem (as an Example)

We will now examine in detail results from studies of the virgin-like deciduous forest already mentioned on the right bank of the Vorskla. This stand, which belongs to the Leningrad Forestry Research Station, lies 80 m above the river in 200 m NN (coordinates are 50° 38′ N and 35° 58′ E) and extends over 1000 ha. It has been very thoroughly investigated by a team from the University of Leningrad under the leadership of Goryschina (1969, 1972a, b, 1974) between 1969 and 1971; investigations of the soil were begun in 1958. Attention was paid, in particular, to water regulation in the ecosystem as a whole; to the phenology and productivity of the herbaceous layer, which is usually given relatively little consideration; to the decomposition of litter and to the role of rodents. Since these findings have been published in Russian, they are not accessible to most western ecologists.

The Ecosystem of the Forest on the Vorskla

The research area within this large forest complex was 160 ha of protected, 300-year-old oak wood, where some of the trees are 400 years old. The only interference to this forest was a long time back, when the ash trees, regarded as weeds, were cut out. Otherwise the forest has a primaeval character, with dead tree trunks on the forest floor

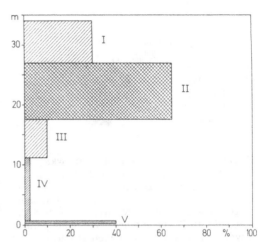

Fig. 10. Diagrammatic representation of the forest layers of the Tilieto-Quercetum aegopodiosum (300 year-old stand in the forest on the Vorskla): *I* upper tree storey of oldest oaks *(Quercus robur)*; *II* middle tree storey of *Tilia cordata* (up to 200 years), *Acer platanoides* and *Ulmus scabra (montana)*; *III* lower tree storey chiefly with *Acer* and some *Ulmus* and *Quercus*; *IV* shrub layer with tree saplings weakly developed; *V* herb layer. There is no moss layer on the ground. *Abscissa* horizontal projection of the various layers in percentage of total area; *ordinate* height in meters. (Walter 1976)

and almost no annual increase in phytomass, a sign that a state of ecological balance has been reached.

The actual measurements were made on an area of 9.6 ha. The whole research programme was conducted with relatively little expenditure on equipment. The structure of the stand is shown in Fig. 10. The following data on the abiotic factors are presented partly in abbreviated form.

a) Macroclimate

Mean climatic conditions can be seen in the climatic diagramm (Fig. 11). The dryness in late autumn is recognizable; a period of drought may occur in some years, when the summer rainfall is abnormally low (Fig. 12); this was the case in 1970. Real drought occurs in this forest steppe area on average every 6 years (1901, 1903, 1906, 1920, 1921, 1924, 1931, 1939, 1946, 1952, 1962).

b) Soil Conditions

The soil below the forest belongs to the zonal type of dark grey, lightly podsolic to

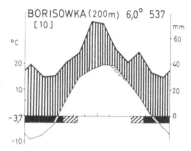

Fig. 11. Climatic diagram of Borisowka, the nearest meteorological station to the forest on the Vorskla. Absolute maximum = 40°C, mean temperature at 13.00 h in July = 25°C. Below 0°C line: *black* periods with daily average of < 5°C; *hatched* < 10°C. The absence of a drought period is typical of forest steppe, but there is a dry period in August–September. (Walter 1976)

grey (middle) podsolic forest soils. The humus content of the upper 0 to 10 cm varies between 3.8% and 8.1%, while the total humus content of the upper 50 cm is 88 to 155 t ha^{-1}. The ratio of humic acids to fulvic acids is approximately 1. The subsoil consists of deep-lying, loess-type of sandy loam. The main root mass is found in the upper 20 cm; below this it decreases rapidly. The root profile is affected to some extent by the burrowing activities of the numerous rodents.

Only after a winter with particularly heavy snow does the soil become moist to a depth of 3 m, but in the third metre the soil water content remains well below field capacity, which means that seepage of

water to greater depths does not occur. Even in exceptionally wet years, the water at 2–3 m depth is only 10% of the total water turnover of the ecosystem.

The assessment of water relationship in the soil was based on the following units of measurement:

a) minimum water capacity = field capacity (FC)

b) maximum hygroscopicity = MH

In these soils, where there was no waterlogging or seepage to groundwater, the field capacity (FC) was taken to be the upper limit of the available water (AW), the lower limit the permanent wilting point (PWP), which is taken to be equal to 1.5 MH. Thus FC–PWP = 100% AW. A distinction was made between:

1. *Readily available water*, when the water content is 100% to 50% AW. In the loess-like soil without any clear macrostructure, this is equivalent to the funicular state in which capillary water is readily moved.

2. *Poorly available water*, when the water content was 50% to 15% AW, equivalent to the pendular state in which the water-filled soil capillaries are no longer connected with one another, so that at 15% AW there is practically no capillary flow of water to the roots. This level of 15% AW is frequently encountered in the soil in late summer and results in wilting of the herbaceous plants and a partial leaf-fall by the trees. This is a special wilting point of mesophytic herbaceous plants and is equivalent to 2.3 MH

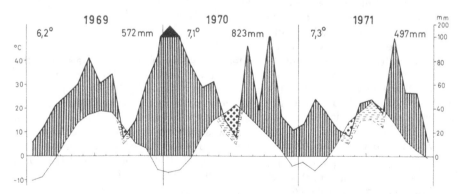

Fig. 12. Climatic diagram of Borisowka for the years 1969–1971, showing that in the steppe-forest ecotone short drought periods occur at different seasons from year to year and they therefore do not appear in Fig. 11. The winter of 1968–1969 was cold with little snow, whereas the following winter was warmer with a large quantity of snow (Walter 1976)

Table 5. Water content of the soil for differing degrees of availability (AW) at different depths. (Goryschina et al. 1974)

Depth (cm)	100% AW Field capacity	50% AW Less readily available	15% AW Almost unavailable	0% AW Completely unavailable
As a percentage of soil dry weight				
0– 5	44.7	27.4	15.1	9.9
5– 10	35.7	22.2	12.7	8.6
10– 20	25.4	16.4	10.2	7.5
20– 30	21.9	15.2	10.5	8.5
30– 40	21.2	15.6	11.6	9.9
40– 50	22.1	16.5	12.6	10.9
50– 60	22.8	17.4	13.6	12.0
60– 70	23.0	17.5	13.6	12.0
70– 80	23.2	17.6	13.7	12.0
80– 90	23.7	18.1	14.2	12.5
90–100	23.4	18.0	14.1	12.5
100–150	23.3	17.6	13.7	12.0
150–200	22.5	17.2	13.6	12.0
Recalculated in mm				
0– 20	74	46	27	18
20– 50	91	66	49	41
50–100	177	135	106	93
100–150	171	131	102	90
150–200	169	130	102	90
0–200	682	508	386	332

in the upper humus 0–5 cm of soil, 2.0 MH below this and 1.7 MH below 30 cm.

3. *Almost unavailable water*, when the water content of the soil is below 15% AW. A fall in the water content to the permanent wilting point (1.5 MH = 0% AW) was observed only in the extremely dry period 1961/62, showing that in a long period of drought the roots of woody plants are able to absorb all available water. In normal years the soil does not dry out to this extent, and this low value is reached only in very localized areas at a depth of 10–30 cm.

Table 5 shows the water content of the soil at 100%, 50%, 15% and 0% AW, expressed both as a percentage of the dry weight of the soil and in millimeters.

The water supply of the oak wood on the forest steppe, that is, on the border between the semi-humid and semi-arid climatic zones, depends entirely on the amount of water stored in the soil when the snows melt in the spring. Drought is most evident in August, when soil water content is lowest. The autumn and winter precipitation and the thaw in spring are important, and thus annual precipitation is calculated from 1st July to 30th June (hydrological year).

The isopleths in Fig. 13 show the seasonal variation in soil water content for a normal hydrological year (1958/59), for a year with evenly spread high precipitation (1966/67) and a third with high rainfall in late summer but little snow in winter (1968/69). As can be seen, in the normal year (1958/59) the soil contained in the autumn of 1958 very little water. From October, the water content slowly rose. In winter, the upper layers received some water from the lowest layers of snow which melted because they were warmed

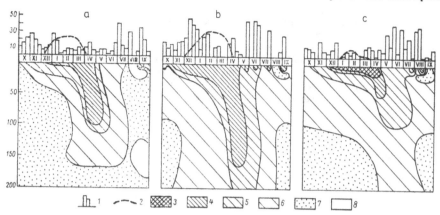

Fig. 13. Chronoisopleths of the water content of the soil of the forest floor: **a** hydrological year 1958–59, with precipitation at almost normal values; **b** 1966–67, a wet year with even distribution of precipitation; **c** 1968–69, a wet year, but with a snow-free winter and a summer precipitation maximum. *1* Precipitation in millimetres in 10-day periods; *2* depth of snow cover (in cm); *3–8* levels of water content: *3* above field capacity; *4* field capacity; percentage of available water; *5* 100–50%; *6* 50–15%; *7* 15–0% (wilting point); *8* no available water. (Goryschina 1974)

by the soil. The soil was only thoroughly moistened, however, by the spring thaw, and this continued until the end of April; slow drying out of the upper layers nevertheless soon commenced and increased during the summer.

In the exceptionally rainy year 1966/67, the course of events was similar, but the soil was moistened to a depth of over 2 m and the water content was high, even in autumn.

Conditions were very unfavourable in 1968/69. There was little snow during the winter, and in February the soil of the forest was frozen to a depth of 50 cm. In April the upper 20 cm did thaw out, but the soil below remained frozen to a depth of 70–90 cm. In consequence, the upper soil layers were wetted above field capacity, while the lower layers remained relatively dry. In summer there were heavy rains which meant that the soil did not dry out markedly in autumn—an exceptional circumstance.

Figure 14 shows the seasonal fluctuations in the water content of different soil layers. The supply of water in the upper 2 m is greatest at the end of April, when it is estimated at 660 mm, and lowest usually in August, with less than 400 mm; of this amount, 332 mm is unavailable, 56 mm is taken up only with difficulty. In September, the water content of the upper layers increases once more, but in the lower soil layers it continues

to decrease up to the end of the growth period.

The most marked fluctuation in water content is seen, of course, in the upper 20 cm of soil; below this, variation is less, but there is a minimum in summer and a maximum in very early spring, at a depth of 20–50 cm, or in deeper layers in late spring. Readily available water continued to be present in the upper 20 cm from 1967 until the end of June; in 1968 and again in 1970 until the middle of June; and in 1969 it was available right up to the end of July; but in the late summer or autumn of each year there were at least 10 to 20 days when there was only pendular water in this layer. The residual water content, with only just absorbable water in the soil, was on average 38 mm in the years 1959–1964 when precipitation was normal; but in the wet years, 1969–1971, it was as much as 76 mm. Since there is no run-off from the surface of a flat area of forest steppe, and almost no percolation to the very deep-lying groundwater, mean precipitation corresponds to the average evapotranspiration during the vegetational growth period, since it is very low in the winter months. Fluctuations in the amount of precipitation are largely buffered, because reserve water accumulates in wet years, and this excess is used up again in dry years. Water utilization is greatest in early summer, reaching 6 mm a

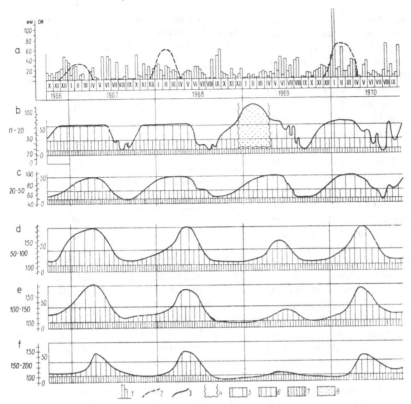

Fig. 14. Water reserves in forest soil 1966–1970: **a** rainfall (in mm) and snow cover (in cm); **b–f** reserve water (in mm) in individual soil layers: **b** 0–20 cm; **c** 20–50 cm; **d** 50–100 cm; **e** 100–150 cm; **f** 150–200 cm. *Left side* of the scale shows the total amount of reserve water; *right side* the quantity of available water: *1* rainfall; *2* snow cover; *3* reserve water; *4* soil frozen; *5* > 50% AW; *6* 15–50% AW; *7* 0–15% AW; *8* water as ice. *AW* available water. (Goryschina 1974)

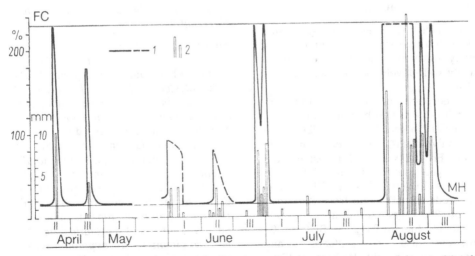

Fig. 15. Fluctuations in water content of the litter layer: *1* water as a percentage of dry wt.; *2* rainfall on individual days; *FC* field capacity of the litter layer; *MH* maximal hygroscopicity. (Goryschina 1974)

day, whereas in late summer it is only 2 mm or even 1 mm, because the plants reduce transpiration in response to the dryness.

The maximal water content of the litter layer is equivalent to 1 mm (precipitation), but in these woods the roots do not grow in the litter layer, which often dries out, so that it contains no liquid water, only water vapour; this occurs irrespective of the water content of the upper soil layer (Fig. 15). As a loose covering, with no capillary connection with the soil water, the litter prevents evaporation from the soil surface. Since the water content of the litter does not fall below MH, the air in the litter remains saturated with water vapour, so that poikilohydric microorganisms remain continuously active.

These investigations show that this mixed oak wood on the northern border of the forest steppe uses the total precipitation of the euclimatope (vide Sect. 7.1) and is thus at the limit of its existence. On southerly slopes with superficial drainage and high evaporation, the forest gives way to steppe vegetation. Further south, with decreasing precipitation and increasing evaporation, forest is found only on northerly slopes or as "gallery forests" in places where there is an additional supply of water or groundwater; in other words, forest is found only extrazonally, until it finally disappears from the landscape altogether.

c) Structure and Composition of the Forest Community

This is a *Tilieto-Quercetum aegopodiosum* forest, with a tree storey with three layers, a poorly developed shrub storey and a relatively well-developed herbaceous layer (Fig. 10).

The tree storey has the following composition (BHD = breast height diameter):

Tree layer I: height 31–33 m, BHD = 100–105 cm, only oaks *(Quercus robur)* over 230 years old. Five oaks on the 1 ha test area are about 300 years old (BHD = 140–160 cm).

Tree layer II: height 25–27 m, BHD = 50–60 cm, predominantly lime trees *(Tilia cordata)* 160–200 years old, but also oaks 90–120 years old (height 24–25 m, BHD = 30–40 cm); and also *Acer platanoides* and *Ulmus montana* (height 21–23 m, BHD = 30–40 cm).

Table 6. Average area covered by single trees and the total coverage by each species on the 1 ha test area (Goryschina et al. 1974)

Tree species	Average per tree (m²)	Total coverage (m²)	Percentage contribution of each species
Quercus robur	103.2	2937	32
Tilia cordata	32.2	1986	22
Ulmus montana	22.1	1412	31
Acer platanoides	19.6	2826	31
Total		9160	100

Tree layer III: height 15–17 m, BHD = 11–16 cm, predominantly *Acer* and *Ulmus*, with an insignificant addition of oaks (height 12–15 m, BHD = 15 cm) and lime (see also Tables 6 and 7).

The shrub storey (height 0.2–3 m) is little developed; on one hectare 177 specimens of *Euonymus europaea*, 90 of *E. verrucosa* and 14 of *Crataegus* were counted.

The herbaceous layer, 5–45 cm high, consists of many synusiae (abundance expressed on a 5-point scale and + indicating recorded present).

1. Spring geophytes (ephemeroids) (April):
   2 *Scilla sibirica*, 1 *Corydalis halleri*, 1 *Anemone ranunculoides*, 1 *Gagea lutea*, 1 *Ficaria verna*; less frequently *Adoxa moschatellina*, *Gagea minima* and *Dentaria bulbifera*.

2. Summer species (July):
   3 *Aegopodium podagraria*, 1 *Galium (Asperula) odorata*, 1 *Carex pilosa*, 1 *Stellaria holostea*, 1 *Glechoma hirsuta*, 1 *Pulmonaria obscura*, 1 *Polygonatum multiflorum*, 1 *Geum urbanum*, 1 *Viola suavis*, 1 *Asarum europaeum*; in addition as +: *Viola mirabilis*, *Galium spurium*, *Stachys sylvatica*, *Fragaria vesca*, *Poa nemoralis*, *Lamium maculatum*, *Campanula trachelium*, *Orobus vernus*, *Scrophularia nodosa*; more seldom are: *Mercurialis perennis*, *Geranium robertianum*, *Campanula rapunculoides*, *Alliaria officinalis*, *Festuca gigantea*, *Brachypodium sylvaticum* and here and there the "ruderal" species: *Urtica dioica*, *Lapsana commu-*

Table 7. Number of young trees per hectare. (Goryschina et al. 1974)

Tree species	Height up to				Total	Percentage
	1 m	1–3 m	4–5 m	6–9 m		
Acer platanoides	1104	2175	888	685	4852	68
Ulmus montana	254	418	91	77	840	12
Tilia cordata	64	240	31	14	349	5
Quercus robur	968	91	14	2	1075	15
Malus sylvestris	2	5	13	10	30	<1
Acer campestre	4	7	3	3	17	<1
Total	2396	2936	1040	791	7163	100

Table 8. Illumination at different levels in an oak wood at 13.00 h on a clear day. (Goryschina 1974)

Layer in the stand	Spring		Summer	
	lx × 10³	Rel. intensity (%)	lx × 10³	Rel. intensity (%)
Above the crowns	60–70	100	Approx. 85	100
Upper crown layer	60–70	100	10 –40	12 –45
Lower crown layer	50–60	75–90	1.5– 4.5	2 – 5
Shrub layer	30–50	45–80	0.6– 2.0	0.7– 2.5
Herbaceous layer	30–50	45–80	0.4– 1.2	0.5– 1.5

nis, *Arctium lappa* and *Moehringia trinervia*.

There is no soil plant layer.

At the base of tree trunks are found: *Brachythecium curtum, B. salebrosum, B. velutinum, Catharinea undulata, Mnium cuspidatum.*

The growth of lichens on the trunks of different species is variable. On the lowest part of the trunks *Peltigera canina, Cladonia fimbriata, Ramalina fraxinea, Parmelia sulcata,* and *Evernia prunastri* are frequently found; higher up the trunk, from a height of 50–70 cm, are added: *Parmelia scortea, P. acetabulum,* and above 1 m *Physcia stellaris, P. pulverulenta, Lecanora pallida, L. carpinea, Bacidia luteola,* and *Hypogymnia tabulosa.*

The following parasitic forms occur. *Viscum album, Fomes ignearius* and *Polyporus sulphureus.*

d) Climatic Conditions in the Forest Stand

The various factors are described here only briefly.

Radiation. Day-length increases during the vegetational period from 14 h in spring to 17 h in summer, and decreases again in autumn to 12 h. Light conditions in the stand both before the leaves were out and later were measured from a 30 m high tower. The data are presented in Table 8.

Total radiation of the herbaceous layer during the day was on average 0.08 cal cm^{-2} min^{-1} (= 0.336 J cm^{-2}min^{-1}), which is about 12% of the radiation in the open. Figure 16 shows the seasonal variation in mean illumination of the herbaceous layer for 1969, assessed by measuring light intensity at noon at 50 points, 1 m apart, on a cross. Measurements were made every 3 to 7 days.

Four stages in the period of vegetational growth can be distinguished. (1) A light

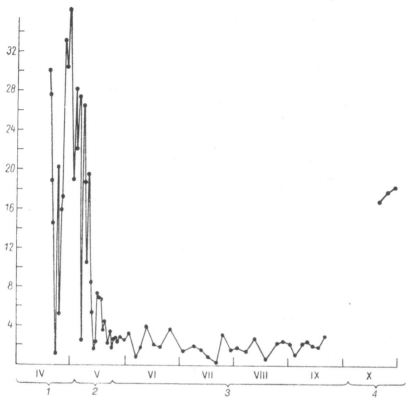

Fig. 16. Changes in light intensity (in 1000 lx) directly above the herbaceous layer of the forest around midday in April to October *(IV–X)* 1969; *1* light spring phase; *2* transitional phase (from when the buds start to open to the point when all the leaves are out); *3* summer shade period (see text for method of assessment); *4* autumn phase after leaf fall (section of curve shown above)

spring phase; (2) an intermediate phase; (3) a dark summer phase; (4) a light autumn phase after leaf fall. Light flecks in summer have an intensity of 0.6–1.0 cal cm^{-2} min^{-1} (= 2.52 J cm^{-2} min^{-1}). Those wavelengths which are important for photosynthesis are more reduced in forest shade (green shade) than others.

Temperature. After the snow has melted, the temperature, although initially still very low, rises rapidly. This is, however, usually followed by a fall in temperature in May and frequently again in June. The temperature becomes steady and warm only in July, but gradually falls again in autumn.

The upper surface of the forest litter shows very marked temperature fluctuations in spring (up to 30°C). During the day it often reaches 35°C, while in the night frosts occur, with the formation of hoar-frost on the herbaceous layer. After the trees have come into leaf, the canopy becomes the "active surface". Here, on 31.vii.1971, at a height of 22 m, a temperature of 30.5°C was recorded, while at 15 cm above the soil the temperature never rises above 25.8°C. In July, in direct sunlight, oak leaves show a rise in temperature of 2°C above air temperature.

Air Humidity. The proportion of the annual precipitation falling on this oak forest that reaches the ground is 88%. The rest is held back by the leaves, very little flowing off onto the trunks.

In forests the air humidity is higher than it is in the open; yet here it can fall to 40% or even 30% during the dry period in May characteristic of forest steppe. The humidity also falls markedly in late summer. During this season there is only occasional, light rainfall and the soil becomes increasingly dry. Air humidity is highest just above the soil and lowest in the leaf-free area just beneath the crown. Even on a rainy day, this gradient is evident at midday.

In summer, when the temperature is no longer rising, the humidity in the forest is generally 3% to 6% higher than over open steppe, and on days when the extremely dry wind, the "suchowej" blows, as much as 20% higher.

Wind. Before the leaves come out in spring, the velocity of the wind in the forest is reduced to 30–50% of that in the open. When the trees are in full leaf in summer, there is very little air movement in the forest. Since the outer canopy of the old stand has an irregular surface, the wind forces its way into the crowns of the trees which project above this surface, increasing turbulence and exchange of heat in the region of the trunk.

Seasonal Change. This is far more marked in a continental than in an oceanic climate. In spring, only the upper layers of soil warm up, so that the first plants to shoot are those with shallow roots. These plants are subject to marked fluctuations in temperature; although the sun is warm, frost is still possible at night. The water saturation deficit of the air can be high and air movement may be considerable.

Between spring and summer there is a rapid rise in temperature, even in the soil layers in which shrubs and trees root. These come into leaf, and within 2 to 3 weeks the illumination of the forest floor falls to the low summer level. Light conditions during this intermediate period are still good for the early summer species, but a sudden fall in temperature in May can inhibit growth.

In early summer, from June to the beginning of July, the weather is very unsettled. The herbaceous layer receives little light, while conditions are optimal for photosynthesis in the tree layer. The water content of the soil gradually decreases, and the first dry period occurs. Rain falls in heavy downpours and is therefore not very effective.

Dry periods of 1 to 2 weeks, or even a month or more, are characteristic of the late summer, that is, from the end of July through August and, together with high temperatures, are very unfavourable to growth.

In autumn, yellowing of the leaves of the trees starts in mid-September, when temperatures are still high. Following leaf-fall, light conditions on the forest floor improve, and so does the water supply as a result of the autumn rains. In consequence there is a period of autumnal growth in perennial herbaceous plants; this comes to a halt as night frosts become ever more frequent until, at the end of October or the beginning of November, winter finally sets in.

Thus the period during which conditions are favourable to growth is relatively short on the forest steppe. It is limited both by the late onset of spring and the dryness in late summer. To this must be added the poor illumination of the herbaceous layer throughout the summer.

e) Phenology

During the 1969–1971 investigation, attention was paid not only to the normal developmental phases, but also to shoot development and to the succession of morphologically and physiologically different leaves on one and the same plant.

In spring, observations of developmental stages were made on a daily basis; later on, on every third day; 6 tree species and 46 herbaceous plants were studied in the original work, and quantitative diagrams showing the phenological patterns of 20 species were prepared.

Spring Phase of the Geophytes. Spring geophytes are also usefully termed ephemeroids, because their growth period is, like that of annual ephemerals, limited to the brief, light, early spring period.

The first to appear above the snow is *Scilla sibirica*, which flowers abundantly at air temperatures of 0.5°–2°C. This is followed, 5 to 7 days later, at 3°–3.5°C, by the mass flowering of the purple-red *Corydalis halleriana* (= *C. solida*) and then, at 5°–6°C, *Anemone ranunculoides* opens its yellow flowers. Although these ephemeroids flower at such low temperatures, they are nevertheless pollinated by insects and produce masses of seeds. This is facilitated by the long life of each flower: 5 to 7 days in the case of *Corydalis* and 8 to 10 days in the *Anemone*. Fruit setting reaches 80%–90% in *Scilla* and *Corydalis*, but only 70% in

Gagea lutea and *Anemone*. The fruits often only ripen after the leaves have died.

Simultaneously with the ephemeroids, the winter leaves of *Asarum europaeum* and *Carex pilosa* appear. After the snow has melted, the leaves of the sedge start to grow; the petioles of the *Asarum* leaves increase in length to 10–15 cm, thereby improving their illumination. The young leaves of these two species develop between the end of April and the beginning of May, but are fully grown only by the second half of June. With the appearance of young leaves, the remarkable flowering of *Asarum* begins. The life of one flower is 10–15 days and the period of flowering lasts for more than a month. The flowers may be self-pollinated or cross-pollinated by colembolens. All the flowers set fruit.

At the end of April, young leaves of *Glechoma hirsuta* appear, as do those of *Stellaria holostea* and *Geum urbanum* and also the spring shoots of *Pulmonaria obscura, Asperula odorata, Viola suavis, V. hirta, V. mirabilis, Dactylis glomerata* and *Aegopodium podagraria*. Only 5 days later, *Pulmonaria* starts to flower; that is, at the same time as *Anemone ranunculoides,* although its period of flowering lasts far longer. Since this is a heterostylous species, only cross-pollination is possible, and this is effected by bees and bumblebees. The other species mentioned flower much later. *Aegopodium* forms only small, relatively xeromorphic leaves in spring.

Intermediate Phase. At the beginning of May, the buds of the shrubs and undergrowth open. *Ulmus* is in full flower. *Acer* starts to flower and is followed a little later by *Quercus*. The leaf buds of the tree layer open in the first few days of May. Depending on the weather, this occurs 2–4 weeks later in the oak variety *tardiflora* than in the variety *praecox*. The leaf canopy usually closes at the end of May.

In the herbaceous layer, besides *Pulmonaria*, violets flower in May; *Stellaria* and *Glechoma* somewhat later and then *Asperula* and *Geum*, but the layer's aspect depends only on *Stellaria* and *Asperula*. All these species are entomophilous and the percentage of fruit set is less.

Around this time, the ears of *Carex pilosa* form; the pollination of this anemophilous species occurs in the first half of May. The sedge frequently does not fruit, but it reproduces vegetatively. This unattractive species does not, however, contribute to the aspect of the vegetation. Almost simultaneously with the appearance of leaves on the trees, the shoots of *Polygonatum multiflorum* and *Dentaria bulbifera* appear, to be followed soon by those of *Scrophularia nodosa, Stachys sylvatica, Campanula trachelium, C. rapunculoides* and *Festuca gigantea*. *Dentaria* flowers at the beginning of May, *Polygonatum* from the beginning to the middle of May; their flowers, which are pollinated by bees, bumblebees, flies and beetles, last for only 2–4 days.

Towards the end of this intermediate phase, the shoots of the ephemeroids die off, as do the spring shoots of *Pulmonaria* and *Aegopodium*; the spring leaves of *Stellaria, Asperula,* etc. start to turn yellow; growth in the herbaceous and woody plants almost ceases.

By the end of May the entire species composition of the biogeocene can be recognized, but the climax of flowering is already over. In the flowering curve of 1969 a second maximum is apparent, coinciding with the late rain of that year.

Shade Period in Summer. By the middle of June, the herbaceous layer is fully formed, but its development continues. In better illuminated places several specimens of *Aegopodium* flower; in the second half of June the long-lasting flowering of *Campanula, Scrophularia, Stachys* and *Festuca* starts. *Glechoma, Geum* and *Lamium maculatum* likewise flower for 3 months. Certain features are general: spring leaves are replaced by the larger shade leaves of summer; fruit setting is variable, from 0% in deep shade, to 25%, 60% or even 100% with increasing light intensity; growth of rhizomes in the ground takes place actively from June to August.

The June rains frequently stimulate a secondary blooming of the tree species, while at this time the lime trees (*Tilia* spp.) are in flower.

In general, however, this is a time when growth is at a standstill, due partly to the frequent periods of dryness. The herbaceous species often wilt, becoming turgid again only after rain; the leaves even die if there is a long drought in August.

Transitional Phase to Autumn. At the end of August or beginning of September, after the start of the autumn rains, almost half the herbaceous species have a period of autumnal growth, provided it is still warm and sunny. *The autumn leaves differ markedly from summer leaves in their much smaller size and small phytomass;* they usually remain on the plants during the winter and die after the snow has all melted or when the spring leaves appear. At the same time, parts of the ephemeroids belowground start to grow; the roots of the trees also grow. Generally, however, the plants prepare for the period of winter dormancy; the summer leaves of *Aegopodium, Pulmonaria,* etc. gradually die. Seed dispersal of these species takes place at the beginning of September, but somewhat later in the late summer species, *Campanula, Scrophularia, Stachys* and *Polygonatum.* By the end of September, when leaf-fall starts, the herbaceous layer has every appearance of being 70% dead. By the beginning of October, only the leaves of *Asarum* and *Carex pilosa* are still green.

Both woody species and the ephemeroids pass into a completely dormant state. The other species vary. There is no winter dormancy at all in *Aegopodium, Asperula, Stellaria* or *Viola;* they can recommence growth any time the temperature rises, although, in contrast to western Europe, this hardly ever happens in eastern Europe, because low temperatures persist.

Winter Phase. This starts with the establishment of a permanent snow cover. *From mid-January to the beginning of February, the true winter dormancy of the species ends and is replaced by a quiescent period due to the extremely low temperatures.* Below the snow some development takes place in the ephemeroids, which thus appear to some extent before the disappearance of the snow cover.

On the basis of the developmental periods described above, the herbaceous plants of the oak forest can be separated into several synusiae or ecological groups:

1. ephemeroids or spring geophytes;
2. hemi-ephemeroids, like *Dentaria* or *Adoxa,* which flower shortly before the leaves come out on the trees but cease growth only in mid-summer;
3. early summer plants with renewed growth in autumn—a very characteristic group for this biocenose, including *Aegopodium, Pulmonaria, Stellaria, Asperula,* etc.; these may flower again in autumn and are not dormant in winter;
4. late summer species, such as *Polygonatum, Campanula, Scrophularia, Stachys,* the aboveground parts of which die in September, so that there is no autumnal growth;
5. the two evergreen species, *Asarum* and *Carex pilosa* with leaves which remain on the plant for 14 months.

Not all species can be readily fitted into only one of these synusiae. *Campanula trachelium,* for example, forms leaf rosettes very early; these soon die and are replaced by petiolate leaves, while in autumn leaf rosettes are occasionally formed again. The grasses *(Festuca, Dactylis)* are also difficult to place, for they flower in summer and resume growth again in autumn.

4.2.1 The Synusiae of the Herbaceous Layer—Photosynthesis and Assimilation

The five herbaceous synusiae of the *Tilieto-Quercetum aegopodiosum* are only partial ecological systems. There is, however, competition between their roots and those of the tree layer for soil water and mineral nutrients; the tree layer, with its root system, is here the dominant of the two (vide Sect. 7.3). Synusiae which develop simultaneously are also strongly competitive. Nonetheless, it is necessary to examine the ecological characteristics of the individual synusiae separately.

The short period of vegetational growth of the herbaceous layer during favourable conditions must be fully exploited to achieve maximal productivity.

Photosynthetic activity was estimated by an improved form of the colorimetric method of Čatsky and Slavik. Measurements were made at the time of greatest activity, that is, from 10.00h to 14.00–15.00h each day throughout the growth

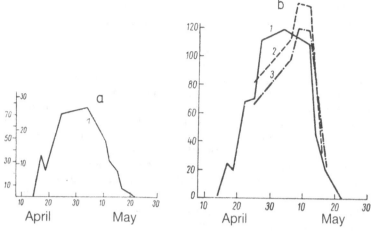

Fig. 17. a Photosynthesis of *Scilla sibirica* measured in its habitat in 1964; **b** potential photosynthesis of *1 Scilla; 2 Corydalis; 3 Anemone. Left ordinate scale* assimilation in mg CO_2 g^{-1} (dry wt.) h^{-1}; *right ordinate scale* in **a** is in mg CO_2 dm^{-2} h^{-1}

period. In addition to these measurements in the field, in particular cases, the potential assimilation capacity in constant conditions was assessed with the ^{14}C method. Finally, changes in the starch content of storage organs was established quantitatively and by microscopic examination of sections.

4.2.1.1 Ephemeroids or Spring Geophytes
(Goryschina 1969, 1972a, b)

In these plants photosynthesis starts immediately after the unfurling of the already green leaves when they emerge from the snow in mid-April. It reaches a maximum at the beginning of May and ceases at the end of May when the leaves turn yellow (Fig. 17a). Considering the low air temperatures, the maximal values of 70–80 mg CO_2 $g^{-1}h^{-1}$ in *Scilla* can be regarded as high. If net assimilation dm^{-2} is calculated rather than g^{-1} dry weight, hourly rates of over 25 mg CO_2 are attained.

The entire growth period of the ephemeroids lasts 4–6 weeks; the period of intensive photosynthetic activity is even more brief. Higher values were of course obtained for photosynthetic potential at favourable temperatures, measured with the ^{14}C method (Fig. 17b). The net assimilation of *Ficaria verna* (Fig. 18) is relatively low.

At the start of development, starch reserves from storage organs are used for growth and for the high rate of respiration,

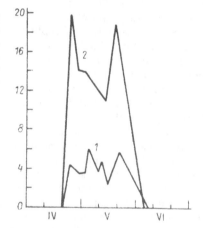

Fig. 18. Photosynthesis of *Ficaria verna* in mg CO_2 dm^{-2} h^{-1} from the end of April *(IV)* to beginning of June *(VI)* 1969. *1* In forest; *2* in a forest clearing

because the net assimilation is inadequate to meet these demands. This was found to apply also to the American ephemeroids *Claytonia* and *Erythronium* (Risser and Cottam 1968). It is only at the end of the period of flowering and at the onset of fruit formation when net assimilation reaches its maximum and when growth ceases and the respiration rate falls that there is surplus assimilation. Reserves in the form of starch are again built up for the following year (Fig. 19) and approximately half the total assimilation of the vegetational period is

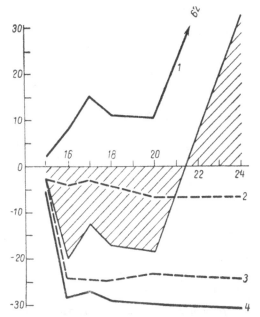

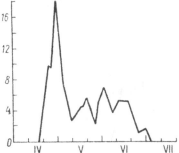

Fig. 20. Photosynthesis curve for *Dentaria bulbifera* from April to July 1970. *Ordinate* mg CO_2 $dm^{-2} h^{-1}$

Fig. 19. Assimilation economy of *Scilla sibirica* between 14th and 24th April. *Ordinate* kg dry wt. per ha. *1* Gain through photosynthesis; *2* loss through respiration; *3* utilization for shoot formation; *4* total loss. *Cross-hatching* indicates net loss or gain

utilized for this purpose. The fall-off in photosynthesis in May is not due to increased shade as a result of the trees coming into leaf, but to rapid ageing of the leaves, that is, to internal factors. Indeed, our observations in the University of Hohenheim showed that the leaves of plants kept in constantly favourable light conditions turned yellow earlier than would normally be expected.

4.2.1.2 Hemiephemeroids

The net assimilation curve during the vegetational growth period for the hemiephemeroid *Dentaria bulbifera* is shown in Fig. 20. At the beginning of May, values are reached which are just as great as the photosynthetic maxima attained by ephemeroids; thereafter there is a decline as a result of decreasing light intensity once the trees have come into leaf; the curve is, in consequence, asymmetrical. In contrast to the ephemeroids, the reserves are immediately replenished: between the 20th April and the 7th

May, the starch content of the rhizomes increased from less than 0.1% to 18.8%, but by the 16th July had increased further only to 19.5%.

4.2.1.3 Early Summer Species

This, the largest group amongst the herbaceous plants, is represented by the dominant species *Aegopodium podagraria*. Its growth, which lasts 5 months, starts before the trees come into leaf and continues until the first snow falls. The net assimilation curve starts with a steep rise to a clearly marked maximum at the beginning of May, followed by a gradual decline until autumn is reached (Fig. 21). The curve for potential photosynthesis is very similar (Fig. 22). When the assimilation curves for different years are compared, occasional, individual, smaller maxima can be distinguished on the declining part of the curve; these are due to short, warm, moist periods of weather during the summer months. They do not, of course, occur on the curve for potential photosynthesis. During the late summer drought period when the leaves wilt, CO_2-assimilation stops altogether. In autumn, under more favourable light conditions, photosynthetic activity starts again, but it is very low, being detectable only on clear days. This is because at this time the large summer leaves are replaced by the smaller autumn leaves and the leaf surface of these is so small that the assimilation is of little significance.

The somewhat xeromorphic spring leaves have a higher temperature quotient ($Q_{10} = 1.6$) for photosynthesis than do the summer

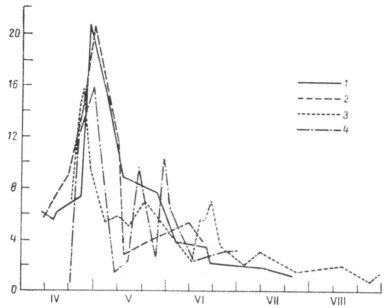

Fig. 21. Photosynthesis curves for *Aegopodium podagraria* from April to August in 4 years. *1* 1966;
2 1967; *3* 1969; *4* 1970. *Ordinate* mg CO_2 dm^{-2} h^{-1}

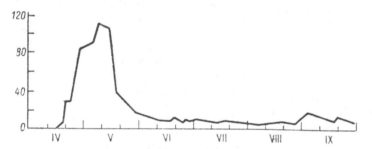

Fig. 22. Potential photosynthesis (estimation with ^{14}C) of *Aegopodium podagraria* in 1964. *Ordinate* mg
CO_2 g^{-1} dry wt. h^{-1}

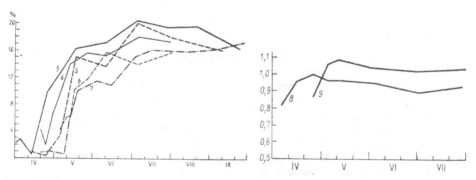

Fig. 23. *Left* starch content as a percentage dry wt. of the rhizome of *Aegopodium podagraria*; *5* 1966;
6 1967; *7* 1969. *Solid line* old rhizomes; *dashed* or *dot-dash* (in *7*), young rhizomes. *Right* specific weight
of the rhizomes in 1966: *8* old; *9* young

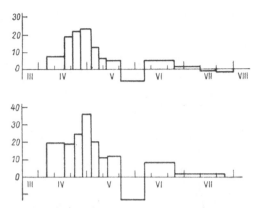

Fig. 24. *Aegopodium podagraria. Above* net photosynthesis in mg CO_2 dm^{-2} of leaf surface per day. *Below* dry wt. increase in a plant in mg per day. There is close correlation between the two curves

leaves (Q_{10} = 1.1). *The spring leaves behave like sun leaves, the summer leaves like shade leaves.* The photosynthetic apparatus is thus adapted to conditions prevailing at different stages of vegetational growth.

The productivity of the spring leaves is of particular importance for, as Fig. 23 shows, the starch content of the rhizome, very low at the beginning of April, rises markedly in April and May, and then remains more or less constant until the autumn. The starch reserves are then used up slowly during the winter and are just adequate to support the development of the first leaves in spring. The calculations based on measurements of net assimilation were corroborated by dry weight measurements carried out at short

intervals during the period of vegetational growth (Fig. 24).

Aegopodium thus behaves quite differently from *Scilla.* Growth of the organs of assimilation does not take place at the expense of reserves built up in the previous year; instead, assimilates formed during the period of maximal photosynthetic activity are used for this purpose. The overall gain is so great that starch is stored first in the old rhizome and then in the newly formed one (Fig. 23). The high productivity in spring is not a reflection of very intensive photosynthesis, but of a rapid increase in leaf surface. The leaf surface index is 1.2–1.3 in a rapidly growing *Aegopodium* stand, whereas it is only 0.7–0.8 in ephemeroids.

The assimilation pattern in *Pulmonaria* is like that of *Aegopodium* (Fig. 25). The seasonal dimorphism of the leaves is still more marked and similarly linked with physiological differences.

Asperula and *Stellaria* also show the same general pattern. Here the replacement of one leaf form by another takes place when, on one and the same shoot, the early spring leaves die, while at the same time larger summer leaves develop towards the apex.

4.2.1.4 Late Summer Species

In species belonging to this group (*Festuca gigantea, Stachys sylvatica* and *Scrophularia nodosa*) growth starts only at the end of the transitional phase, that is, at a time when illumination of the forest floor has already been considerably re-

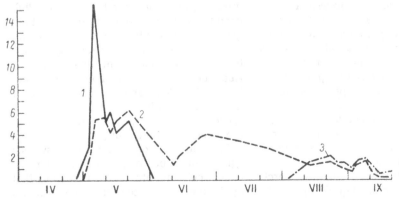

Fig. 25. *Pulmonaria obscura:* photosynthesis curve in mg CO_2 dm^{-2} h^{-1}. *1* Spring leaves; *2* summer leaves; *3* autumn leaves

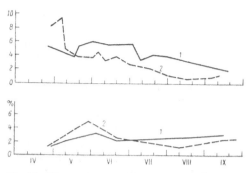

Fig. 26. *Above* photosynthesis curves in mg CO_2 dm^{-2} h^{-1} for *1 Festuca gigantea* and *2 Stachys sylvatica. Below* curve of starch content in percentage dry wt. in the tiller nodes of *F. gigantea (1)* and the rhizomes of *S. sylvatica (2)*

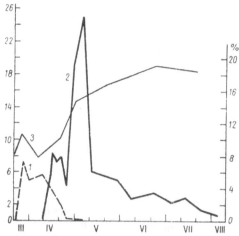

Fig. 27. *Asarum europaeum.* Photosynthesis curves of *1* overwintered leaves; *2* young leaves (in mg CO_2 dm^{-2} h^{-1}); *3* starch content of rhizomes (in percentage of dry weight), between March and August

duced. In consequence, there is no photosynthetic maximum after they have come into leaf; instead, the curve remains on an even, low level throughout the period of vegetational growth (Fig. 26). Underground organs, including even the large tuber of *Scrophularia nodosa*, contain hardly any reserve materials and are completely empty immediately after the emergence of the leaves. For shoot formation, therefore, currently produced assimilates have to be used. With the small net assimilation in forest shade, surpluses are not large, and growth in leaf surface takes place very slowly. The reproductive phase extends over 2.5–3 months.

4.2.1.5 Summer Species with Evergreen Leaves

Asarum europaeum was investigated in detail. In this species assimilation starts in the winter leaves in March, immediately after the snow has disappeared. The leaves are, however, brownish in colour, and their photosynthetic activity is not great; it declines during April and stops altogether at the end of the month when new leaves have been formed (Fig. 27). Nevertheless, the net assimilation of the old leaves is of great importance. It suffices to provide material for the growth of new leaves and even allows for an increase in the low starch reserves of the rhizome (Fig. 27, 3). Photosynthesis in the young leaves reaches the same level as in the ephemeroids at the

beginning of May, with the same sharp maximum, but even by mid-May it is no greater than the maximum reached by the weakly assimilating old leaves in April. Starch accumulation in the rhizome increases slowly until the autumn. These reserves are then largely used up during the winter. *Carex pilosa* shows a pattern closely similar to *Asarum*.

Concluding Remarks

Figure 28 shows, in a summarized form, the photosynthetic curves of the various groups in relation to the four phases of illumination of the forest floor. The question as to which group is best adapted to the particular light conditions on the floor of the oak forest biogeocene can best be answered by pointing out that on the edges of deep forest shade the most common plants are species of *Aegopodium, Asperula, Polygonatum* and *Viola suaveolens*, apart from weak individuals of *Carex pilosa*; that is, mainly early species. Late summer species, by contrast, grow mainly in better illuminated spots.

The ephemeroids from a special synusia, which cannot be compared with others, for its particular adaptation lies in the fact that, by the time-sequence of their development, its members avoid the effects of forest shade. In Hohenheim we observed that *Ficaria*

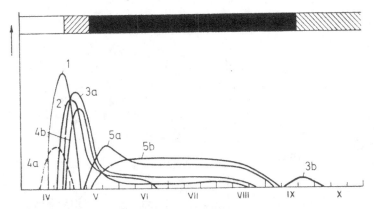

Fig. 28. Schematic comparison of the course of photosynthesis in the various synusiae of the herbaceous layer of a mixed oak forest: *1* ephemeroids; *2* hemiephemeroids; *3* early summer species (*a* spring leaves; *b* autumn leaves); *4* evergreen species (*a* leaves from previous year; *b* young leaves); *5* late-summer species (*a* and *b* slightly different types). *Ordinate* relative intensity of photosynthesis. *Above* different illumination phases on the forest floor: *white* light phase; *hatched* spring and autumn transitional phases; *black* shade phase. (Goryschina, from Walter 1976)

verna, for example, developed luxuriantly in spring in places that in summer were in deep forest shade and lacked angiosperm species.

A preliminary calculation was made of the assimilated mass of representatives of the various synusiae during the vegetational growth period; in Table 9 these very approximate estimates are compared with productivity measured directly on the basis of dry weight.

The difference between actual assimilation and primary production, 30% in *Scilla* and 60% in *Aegopodium*, is due to loss through respiration. In *Aegopodium* losses due to leaves dying or to damage by consumers are also significant. The respiratory losses in *Aegopodium* during the long, unproductive summer months in deep shade must have a greater effect than they do in *Scilla*, which has a very short growth period.

The phytomass of the herbaceous layer was estimated by weighing those parts above ground on each of 20 areas, 0.5×0.5 m in size, distributed according to a definite plan. The accuracy of the estimation of air-dried mass was ± 6–13% for total weight, and ± 7–17% for the weight of the individual species when fully developed.

Figure 29 shows the change in aerial mass during the period of vegetational growth. The shape of the curve is very similar from year to year; 1969 stands out because spring production was delayed by 7–10 days as a result of unfavourable weather. Maximum phytomass of the herbaceous layer as a whole is reached in the first half of May, when the new shoots of *Aegopodium* and *Carex pilosa* have almost completed their growth and the ephemeroids are just beginning to wither. At this stage, the phytomass reaches 0.67–0.69 t ha^{-1}, 30% of which is

Table 9. Comparison of assimilated mass and primary production in kg ha^{-1} in some herbaceous plants of the oak wood biogeocene. (Goryschina et al. 1974)

Plant species	Assimilated mass (I)	Primary production (II)				Ratio II:I
		Above-ground	Belowground			
			Increase	Reserve	Total	
Scilla bifolia	985	220	30	452	702	0.71
Aegopodium podagraria	2100	640	30	130	800	0.38
Carex pilosa	1020	230	–	–	–	–

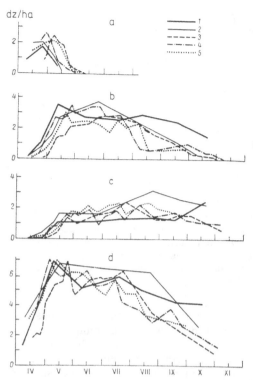

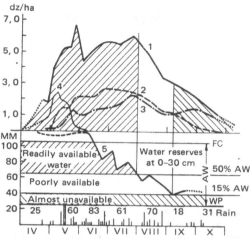

Fig.30. Dependence of the phytomass of the herbaceous layer on the water stored in the upper 0–30 cm of soil in mm (5), and the availability of the water: *FC* field capacity; *WP* wilting point; *AW* available water. Phytomass: *1* herb layer; *2 Aegopodium podagraria; 3 Carex pilosa; 4* ephemeroids. (Goryschina, from Walter 1976)

Fig.29. Changes in phytomass of the various synusiae and of the total herbaceous layer from April to September in the years 1967–71 (air-dry mass in dz ha^{-1} = 100 kg ha^{-1}: **a** ephemeroids; **b** *Aegopodium podagraria;* **c** *Carex pilosa* (young shoots); **d** total herb layer. *1* 1967; *2* 1968; *3* 1969; *4* 1970; *5* 1971). (Goryschina, from Walter 1976)

accounted for by *Aegopodium,* 25–27% by ephemeroids, 12% by *Carex pilosa* and 7–10% by other species. There may be a second, but less well defined maximum in the second half of July. At this time, 45–50% of the phytomass is accounted for by *Aegopodium,* 30–40% by young *Carex pilosa* shoots and 14–17% by other species. The phytomass then starts to decline and by the end of Semptember, it is only 0.25 t ha^{-1}, with *Carex pilosa* making up 40–65% of this, *Aegopodium* only 17–25% and the other species 20–30%.

The phytomass of the herbaceous layer shows a clear relation to the amount of available water (AW) in the soil (Fig. 30). *Aegopodium* is particularly dependant on soil moisture; as soon as this falls, the old leaves die.

Only living phytomass was taken into account (maximally 0.27–0.35 t ha^{-1} for

Aegopodium). If the dead leaves are added, the values increase by 15–20%. On plots with almost pure *Aegopodium* they reach 0.85 t ha^{-1}.

The underground structures of the herbaceous layer lie no deeper than 30 cm in the soil; the main mass is at a depth of 10–12 cm and is equivalent to about 1 t ha^{-1} dry matter; of this, 76% is formed from the rhizomes of *Aegopodium,* 2% those of *Carex pilosa* and 22% of the organs of the ephemeroids.

Thus the total phytomass of the herbaceous layer is 1.7 t ha^{-1} (0.7 aboveground and 1.0 belowground). Although this is not large, it nevertheless plays an important role in the cycling of matter because turnover is very rapid, particularly in the case of the ephemeroids, the aerial parts of which die and decay after a short period of growth.

4.2.2 Phytomass and Primary Production in the Mixed Oak Wood of the Forest Steppe

The contribution of the herbaceous layer to the phytomass and to productivity of the forest biogeocene as a whole is relatively small; in this forest on the Vorskla this applies also to the shrub layer and the young tree layer, for they both grow under the

Table 10. Photosynthesis by shrubs and young trees on cloudless days. (Goryschina et al. 1974)

Species of woody plant	Intensity of photosynthesis in mg CO_2 dm^{-2} h^{-1}				
	10. v	14. v	28. vi	11. vii	2. viii
Euonymus verrucosa	3.4	9.5	1.5	0.6	0
Tilia cordata	7.0	7.1	0.8	1.0	0
Acer platanoides	16.0	9.3	1.7	0.9	0
Ulmus montana	11.2	–	0.8	1.2	0

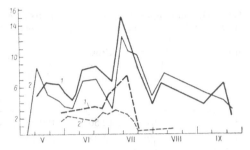

Fig. 31. Photosynthesis by the leaves of a 300-year-old oak (mg CO_2 dm^{-2} h^{-1}) between May (V) and September (IX). Solid lines 1970; dashed lines 1971; 1 from upper part of the crown; 2 from lower part of the crown

same unfavourable light conditions as the summer plants of the herbaceous layer. Although the shrubs and young trees come into leaf 1–2 weeks before the tree layer, the young, still pale leaves are not able to exploit these favourable weeks, because there is no net assimilation of CO_2: instead, they give off CO_2. A few additional measurements made in 1971, admittedly at high temperatures of 25°–27°C, only partially confirmed these data. Moreover, in individual cases a definite but very small spring maximum of photosynthesis was noted, to which, however, no great significance can be attached (Table 10).

Summer values for photosynthesis of the woody undergrowth were very low, and in the dry period (August) it was not possible to demonstrate CO_2 assimilation at all. This confirms the depressed state of this layer. Only in better illuminated spots did growth proceed.

The curve for net production by trees was quite different from that for the forest floor species; this is because the leaves in the upper part of the crown are always exposed to the full light of day. CO_2 assimilation began in May when the leaves came out and continued with high values right into September. Even after the leaves had started to turn yellow, positive values were still recorded. Maximal values were obtained in high summer (Fig. 31). If they are calculated per dm^2 of leaf surface, they are somewhat lower than the values obtained for most of the herbaceous plants; it must, however, be

borne in mind that about 30% of the surface of an oak leaf is formed of veins, which are comprised of non-photosynthetic tissues, while in herbaceous plants the veins make up only 10–15% of the leaf surface.

If the oak forest is looked at as an ecosystem, investigations of photosynthesis show that this is coordinated in terms of time and space in the different synusiae, and this results in maximal productivity being achieved.

The following were measured: (1) phytomass and its fluctuations during the year as well as from year to year, (2) increase in phytomass (that is, net productivity), (3) the litter, that is, the portion of the primary production which dies; this, together with growth in wood, represents net primary productivity. Losses through respiration were not accounted for, so that gross productivity could not be calculated.

Estimation of phytomass and of the growth of trees was made by comparing measurements with standard trees (that is, trees of average dimensions). Measurement of the phytomass of the herbaceous layer is difficult since there is no time of the year when the entire phytomass of this layer can be determined, since, in summer, the ephemeroids are no longer there. Added to this is the great heterogeneity of the herbaceous layer of deciduous forests. For this reason, each group of plants was evaluated separately, at the time of its maximal development. The major contribution is made by the ephemeroids, Aegopodium and Carex pilosa (Fig. 29).

It is even more difficult to establish a value for phytomass belowground; the lack

Table 11. Standing phytomass of different tree storeys including young trees (in t ha^{-1}) and the primary production (in t ha^{-1} per year). (Goryschina et al. 1974)

Storey	Number of trees (ha^{-1})	Cross-sectional area of trunk	Standing phytomass				Total	Primary production
			Trunk	Twigs and branches	Shoots (1 year old)	Leaves		
I	15	12.50	100.85	46.65	0.18	1.16	148.84	2.49
II	90	17.97	105.10	18.35	0.72	1.58	125.75	2.94
III	495	4.67	22.06	2.93	0.10	0.86	25.95	2.81
Saplings	7130	0.97	2.83	3.21	0.02	0.10	6.16	0.70
Aboveground phytomass			230.84	71.14	1.02	3.70	306.70	8.94
Belowground phytomass							124.93	Not estimated
Total phytomass (young trees of less than 0.01 t omitted)							431.63	

of precision of such measurements is well known. In standard trees of different ages, relationships were established between the woody mass of the trunk, the bark, the branches and the roots. Besides the old stand, a second stand in the immediate neighbourhood which had been managed for 80 years was used as a reference stand. The results for the old forest are summarized in Table 11.

The relative proportion of woody mass represented by trunks is small in layer I because the mass of the larger branches is very considerable in these older oak trees; indeed, a single branch may often weigh as much as the trunk of a younger oak, but is not, of course, included in the estimate of trunk mass.

The limes in layer II show very marked branching: one result is that the total mass of 1-year shoots is greater than the increase in trunk and older branches of the trees of this species (= 0.64 t ha^{-1}).

The total phytomass above ground of the tree layer of the old oak-lime stand is very great, namely, 307 t ha^{-1}; so, also, is the annual new production of phytomass in the form of leaves, twigs and wood which, without taking account of new young plants, amounts to 8.24 t ha^{-1}. This, however, only slightly exceeds the amount of litter which falls each year. The net increase in wood mass is thus not large in the old stand, because here there is a considerable loss of

dead wood from trunks and branches. On the basis of 10 years of observation, this can be put at a mean value of 2.2 t ha^{-1}, which is approximately equivalent to the increase in trunk and branch wood in layers I and II. This means that the old forest stand approaches the conditions of virgin forest, in which the standing phytomass remains constant over a long period of time. By contrast, in the 80-year-old, managed stand there was a respectable growth in wood of 4.16 t ha^{-1}, without significant loss of dead wood.

The phytomass belowground, calculated on the basis of measurement of the root systems of standard trees, isolated from large earth monoliths, was found to be 125 t ha^{-1} dry weight. A more detailed analysis was made of the different components of the underground phytomass to a depth of 170 cm; these were large roots (diameter more than 10 mm), thin roots (diameter 1–10 mm) and very fine roots (diameter about 0.5 mm). Of the latter, two thirds occur in the upper 20 cm of the soil, three quarters in the upper 50 cm. The trees account for 80–90% of the total mass of fine roots, and of these, 76% is concentrated in the upper 20 cm of soil.

The above- and belowground phytomasses of the tree layer in the old mixed oak stand together amount to 432 t ha^{-1} dry weight; of this, 30% is accounted for by the phytomass belowground.

The young trees have only a very small phytomass, namely, 1.6 t ha^{-1}, the largest

Table 12. Composition of the litter in different years. (Goryschina et al. 1974)

Test area	Year	Litter from trees (kg ha^{-1})			Litter from herbaceous layer (kg ha^{-1})					Total weight of litter	Proportion comprised of herbaceous layer in %
		Leaves	Branches	Total	Ephemeroids	Aegopodium	Carex pilosa	Other herbs	Total		
300-year-old oak forest	1966	3760	230	3990	250	270	70	120	710	4700	15.1
	1967	3660	230	3890	160	320	120	180	790	4670	16.7
	1968	3550	220	3770	190	340	290	90	910	4680	19.4
	1969	3057	225	3282	220	270	220	110	729	4102	20.0
	1970	3523	268	3791	250	310	200	80	840	4631	18.2
	1971	3568	373	3941	200	240	220	100	769	4701	16.2
80-year-old oak forest	1966	3560	149	3700	80	120	400	80	680	4380	15.2
	1967	3470	139	3600	70	120	570	80	840	4440	18.9
	1968	3530	149	3670	70	70	650	40	830	4500	18.4
	1969	3016	68	3084	90	70	550	70	780	3864	20.1
	1970	2957	188	3145	70	80	660	50	860	4005	21.4
	1971	3380	170	3550	100	120	530	140	890	4440	20.0

part of this being accounted for by young maple trees. The annual production of 0.7 t ha^{-1} is only about equal to that of the herbaceous layer. The shrub layer is completely without significance.

Measurements were also made every 10 days of the diameter of tree trunks. In May the increase in diameter was very marked as a result of the activity of the cambium. Fluctuations during the growing period show a close correlation, not with rainfall, but with the amount of available soil water. In late summer, in consequence of the dryness, there is a clearly marked shrinkage of the trunks.

For the sake of comparison, the following data are given for the 80-year-old stand: the total phytomass of the tree layer, in which oak is completely dominant, amounted to 314 t ha^{-1}, of which 64 t ha^{-1}, or 20%, was accounted for by the underground mass. After thinning in 1955, the crown cover was reduced to 80–90%. The annual increase of trunk and branch wood was 4.16 t ha^{-1}, and with the leaves and new shoots 7.76 t ha^{-1}. Young trees developed well in the relatively well-lit forest. Their phytomass was 7.9 t ha^{-1}, and this consisted mainly of lime and maple trees. The shrub layer, with a mere

0.1 t ha^{-1}, was, however, very poorly developed. The phytomass belowground is far less than in the old stand, for the large old lateral roots were absent. The herbaceous layer was better developed with a phytomass above ground of 0.8–0.9 t ha^{-1}. Carex pilosa accounts for 40–45% of this, Aegopodium for 8–10%, the ephemeroids 6% and other species 10–12%. In open, well-lit stands Carex pilosa is always dominant.

4.2.3 Short Cycle: the Litter Layer (Composition and Decay)

In a mixed oak wood the annual amount of litter which falls to the ground varies, according to weather conditions and age of the stand, from 3.5 to 5 t ha^{-1}; 15–20% of this is litter from the herbaceous layer (see Table 12) and is produced throughout the growing period. Since it decays rapidly, it never actually forms part of the litter layer on the ground.

As can be seen from the table, there was very little quantitative difference in the litter produced annually by the two stands; only in 1969 was the amount less: this was true also for the 80-year-old stand in 1970. The

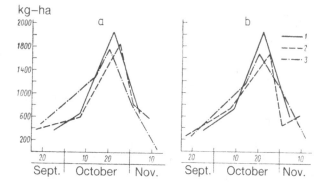

Fig. 32. Total fall of litter in three consecutive years in oak tree stands: **a** 300-year-old stand; **b** 80-year-old stand: *1* 1969; *2* 1970; *3* 1971

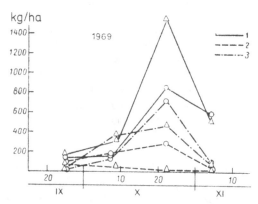

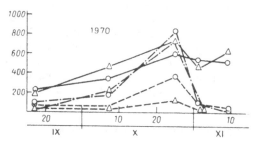

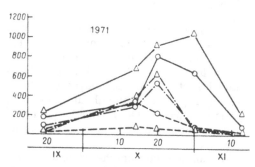

Fig. 33. Quantities of litter for three consecutive years from different tree species: ○ in 300-year-old stand; △ in 80-year-old stand; *1* oak; *2* lime; *3* maple

proportion of litter arising from the herbaceous layer varied between 15.1% and 21.4%. Small twigs made up 2–9% and fruits 1–5%; the latter figure was always higher for the younger stand.

In the old stand the amount of total litter accounted for by the oaks was 45–50%; this is less than in the 80-year-old stand, where it is 64–68%. The lime trees accounted for 16% of the litter in the old stand, maples 12%. The corresponding figures for the younger reference stand were: lime 3–5%, maple 20–26% and elm 4–6%.

Figure 32 shows the start and the maximum of litter fall, while Fig. 33 provides data on leaf-fall for different tree species. Oaks usually lose their leaves somewhat later than do the other species, often only in November.

The chemical composition of the litter, its content of nitrogen and of various mineral salts, can be seen in Table 13. The order of abundance of the minerals was:

$$CaO > SiO_2 > N > K_2O > P_2O_5 > MgO =$$
$$= Al_2O_3 > Fe_2O_3 = MnO$$

In the old stand freshly formed litter was richer in N, P_2O_5, K and Ca, but poorer in SiO_2 and Al_2O_3. A greater N content and less SiO_2 facilitates decay. For this reason the litter was never as thick in the old stand as it was in the 80-year-old stand. The litter from the herbaceous layer was very rich in potassium and was indeed the source of 58% of the potassium released from the litter to the soil.

The maximum dry mass of litter was reached in April and is 4.7–10.0t ha⁻¹. The quantity of litter gradually decreased during

Table 13. Annual quantities of chemical compounds and elements released from the litter in kg ha⁻¹. (Goryschina et al. 1974)

Test area	Type of litter	Dry mass	N	SiO_2	Fe_2O_3	Al_2O_3	P_2O_5	CaO	MgO	K_2O	MnO	Total ash
300-year-old oak forest												
	Tree layer:											
	Leaves	3520.0	51.0	61.6	5.6	7.7	18.0	109.5	10.2	28.9	4.9	246.4
	Branches	258.0	1.9	1.3	0.4	0.9	0.5	9.2	0.1	1.9	0.2	14.6
	Total	3778.0	52.9	62.9	6.0	8.6	18.5	118.7	10.3	30.8	5.1	261.0
	Herb layer:											
	Ephemeroids	212.0	10.8	2.9	0.2	0.7	2.0	3.2	1.0	11.8	0.2	22.0
	Aegopodium	292.0	6.4	4.7	0.2	1.7	3.5	4.6	1.2	19.8	0.1	35.8
	Carex	187.0	4.9	7.6	0.2	0.9	1.3	1.6	0.7	4.7	0.1	17.1
	Others	114.0	3.2	2.2	0.1	0.6	1.0	2.2	0.7	6.3	0.0	13.1
	Total	807.0	25.3	17.4	0.7	3.9	7.8	11.6	3.6	42.6	0.4	88.0
	Total litter	4585.0	78.2	80.3	6.7	12.5	26.3	30.3	13.9	73.4	5.5	349.0
	Percentage of herbaceous litter	17.6	32.3	21.7	10.4	31.1	29.6	8.9	25.9	58.0	7.3	25.2
80-year-old oak forest												
	Tree layer:											
	Leaves	3320.0	44.5	61.1	5.0	9.6	15.6	97.9	10.6	28.9	5.3	234.0
	Branches	140.0	1.0	0.7	0.2	0.5	0.2	0.2	0.1	1.0	0.1	7.8
	Total	3460.0	46.5	61.8	5.2	10.1	15.8	102.9	10.7	29.9	5.4	241.8
	Herb layer:											
	Ephemeroids	80.0	4.1	1.1	0.1	0.3	0.8	1.2	0.4	4.4	0.1	8.4
	Aegopodium	96.0	2.1	1.6	0.1	0.6	1.2	1.5	0.4	6.5	0.1	12.0
	Carex	560.0	14.8	22.7	0.5	2.7	3.8	4.8	2.0	14.2	0.2	50.9
	Others	76.0	2.1	1.5	0.1	0.4	0.7	1.5	0.4	4.2	0.1	8.9
	Total	812.0	23.1	26.9	0.8	4.0	6.5	9.0	3.2	29.3	0.5	80.2
	Total litter	4272.0	69.6	88.7	6.0	14.1	22.3	111.9	13.9	59.2	5.9	322.0
	Percentage of herbaceous litter	19.0	33.2	30.3	13.3	28.4	29.1	7.8	23.0	49.5	8.5	24.9

Table 14. Quantities of elements and organic compounds reaching the soil from the litter layer (kg ha^{-1}). (Goryschina et al. 1974)

Time of year	Ca	Mg	K	P	Organic substances
May–September 1968	56	8	42	9	309
September–April 1969	31	12	12	4	90
Total in 1 year	87	20	54	13	392

the growing period until, by October, only half or a third remained (2–4.5 t ha^{-1}). A certain amount of litter was decomposed under the snow cover during the winter (up to 2.5 t ha^{-1}). Only when the soil was frozen did the entire litter mass remain unchanged until the spring.

In order to measure the rate of decay of litter from different tree species, litter-bags of loosely woven synthetic fibre were laid out in the litter layer of the forest and their contents analyzed after 4, 7 and 12 months. The losses increased in the following order: oak < maple < lime < ash.

The litter of the herbaceous layer is far more rapidly decomposed and shows a greater loss of organic matter than of mineral salts. Potassium is particularly rapidly washed out of the litter, while nitrogen and, to an even greater degree, calcium are retained for a longer time.

Furthermore, it was possible to determine the chemical composition of soil water draining from the upper horizons by using samples collected with a lysimeter. In summer there was usually no seepage of water from the humus horizon into deeper layers, and even within the humus horizon the concentration of mineral salts decreases rapidly with increasing depth. Almost nothing is leached out of the B horizon. The lysimeter water was slightly acidic and the concentration of dissolved substances very low (150–300 mg l^{-1}); that is, two to three times lower than the concentration found in river and groundwater in the study area. Potassium content was relatively high, while the ratio Ca/K was the same as that in the litter. Hydrogen carbonate was the most common anion. The solution was usually pale yellow because of the presence of humus colloids, generally at a concentration of 150 mg l^{-1}, but in summer the liquid might even be

brown, when the colloid content is particularly high.

Non-volatile acids, characteristic of podzolic soils, were found in the soil water only in autumn at low temperatures. Table 14 shows the quantities of mineral nutrients and organic acids reaching the soil from the litter between May 1968 and April 1969.

These quantities of ash elements are approximately the same as the amounts contained in the litter as it falls; only about 10% of the organic matter of the litter is, however, washed out by the gravitational water; the other 90% undergoes a slow process of mineralization, with humus as an intermediate stage.

The mineralization of nitrogen-containing compounds leads, as is to be expected, to the formation mainly of ammonium salts. The enrichment of the soil with mineral nitrogen was not determined. Ellenberg (1977) in particular has emphasized the importance of this edaphic factor.

4.2.4 The Long Cycle: the Herbivores of the Mixed Oak Wood

In this study of the mixed oak wood of the forest steppe special attention was paid to the rodents, as the unquestionably dominant group of herbivores. Their numbers fluctuated from year to year, from 70–90 up to 400–500 ha^{-1}.

Particular attention was paid to the effect of rodents on the herbaceous layer and, as a matter of secondary importance, their influence on the shrub and tree layers. Some recourse had to be made to laboratory work as well. Investigations in the forest were conducted on a 9-ha area, 300 × 300 m, immediately adjacent to the study area.

Between 1967 and 1969, a total of 5178 rodents were marked, of which 4283 were of the species *Clethrionomys glareolus* (bank voles), 2 were *Microtus (Pitymys) subterraneus* (common pine vole), 864 were *Apodemus (Sylvaemus) flavicollis* (yellow-necked mouse) and 30 were *Apodemus agrarius* (striped field mouse). The number of rodents counted per hectare and month between April and October 1968 was only 22–110, while *Microtus* and *Apodemus agrarius* were absent altogether.

A study of the species of plants eaten and of the quantities consumed each day was made in cages, sited in natural conditions as far as was possible. Since feeding behaviour is, however, very dependant on sex, age, and physiological state of the animals, and also on the freshness of the food and the availability of water, these results are only of limited applicability to conditions in the forest.

Clethrionomys was offered 35 herbaceous forest and field plants. It fed eagerly on green fodder, but would not touch *Galium verum;* the flowers of all the other plants were preferred to their leaves, while their stalks were not readily eaten. The favourite species were: *Polygonatum, Anthriscus sylvestris, Glechoma hederacea, Stachys sylvatica, Taraxacum officinale, Cichorium intybus* and *Lapsana communis.*

Apodemus flavicollis, when offered the same species of plants, left 11 of them untouched.

Clethrionomys ate the leaves of all the types of tree and shrub offered to it, but showed a preference for *Tilia, Crataegus, Euonymus* and *Viscum.* Only the bark of lime and ash was gnawed; this was inadequate as the sole source of food, and the animals died after a few days. It may be more nutritious in winter. *Apodemus flavicollis* refused the leaves and bark of trees and shrubs almost completely.

It is known that rodents destroy the seeds of woody plants. Seeds offered in these experiments were eaten eagerly; only those of *Crataegus* were refused by *Apodemus flavicollis.* Fresh seeds were preferred to dry ones, while berries, fruits, bulbs, tubers and mushrooms (the cap only) were all highly favoured foods. The moss *Leucodon* and galls on oak leaves were hardly eaten at all; the latter only by *Clethrionomys.*

Quantitative estimation of the food consumed is especially difficult. From this investigation it was calculated that the rodents eat only 1% of the phytomass of the herbaceous layer, a very small proportion. The amount may be much larger in the wild, however, where the animals require a lot of energy for movement. On the other hand, in the forest the animals feed preferentially on fruits and mushrooms, so that consumption of green phytomass may possibly be less. Damage to the herbaceous layer is, however, not limited to that which is eaten; the feeding activities of animals may cause damage which leads to later wilting and death of the plants. Furthermore, animals build up food stores which are only partially eaten or may be left untouched, and a cache contains 20–45 g (on average 30 g) dry weight. In 1968, there were estimated to be 250 such stores on the 9-ha study area.

The burrowing activity of the rodents is also important. From June to August 1967, the following average counts per hectare were recorded: 3350 openings to underground burrows; 2525 heaps of excavated soil with a total volume of 2.84 m^3 and covering 1.18% of the area. The 2900 m of tracks on the surface and 7900 m of tunnels below the ground altogether covered 3.4% of the area and had an aboveground volume of 2.33 m^3 and below the ground of 6.35 m^3. This is equivalent to 0.87% of the volume of a soil layer 10 cm thick.

The many passage ways may cause drying out of the soil and of the plant roots; on the other hand, they facilitate seepage of rainwater into the soil. Burrowing activities also alter soil structure and chemical composition. The excavated soil is richer both in mineral nutrients and humus; its pH is neutral and the decomposition of organic matter accelerated by the better soil aeration, so that there is mull formation. This, too, influences the forest vegetation.

Both the destruction of seeds and the damage to young trees caused by rodents have an adverse effect on the rejuvenation capacity of the oak forest. Observations in the study area showed that in August any unripe acorns which fall from the trees are scarcely eaten. As they ripen, an ever-greater proportion of acorns is consumed, particularly in mid-September. Later on, the quantity of ripe acorns falling is so large that the loss has hardly any effect. In late autumn, from the end of October–November until just before the first snow falls, acorns are collected for storage with ever-greater intensity.

Laboratory experiments showed that an animal consumes 9g of acorns per day. In the field, acorns serve as the main food for 210 days (September until March). In 1968 there were 25 hibernating rodents per hectare. This means that 47 kg ha^{-1} of acorns was consumed. Unfortunately, no estimate was made of the total quantity of acorns which fell from the trees, but it must be borne in mind that only part of the acorns collected and stored are actually eaten. A rodent is capable of carrying off 10–13 kg of different seeds into its storage chambers.

As far as damage to the usually too numerous saplings is concerned, it was established that up to 46.5% of the maple saplings died as a result of damage, for oaks the figure was 30%, for lime 27% and for elms 8.4%. These observations were made in the forest steppe in the area of the Tellermann forest. A large increase in the numbers of rodents may even lead to 15-year-old trees being damaged.

Damage to older trees arises from the activities of some rodents which nest in holes in the base of the trunk, in hollow tree trunks, or in hollows in the forks of branches; frass and decaying remains of food and litter all cause the wood to decay, and this finally leads to the death of branches or of whole trees.

In an area of 0.5 ha of an old oak stand, 16 nests were found 8–14 m above the ground. A tree which had been blown over in a gale contained a nest 23–24 m from its base in a hole 65–75 cm deep; this was one-third filled with excrement of *Apodemus flavicollis* and the remains of acorns.

Similar conditions are encountered in other types of forest. In evergreen pine forests the different phases of illumination of the forest floor do not occur. The number of synusiae in the herbaceous layer is smaller. Summer green or evergreen dwarf shrubs with mycorrhizae, adapted to nutrient-poor soil, usually dominate. The soil is not covered with a layer of leaf litter; as a result

a thick layer of moss forms, the productivity of which must also be taken into account.

In a tropical jungle it is very difficult to make an estimate of wood production, because the trees have no annual rings and it is therefore necessary to measure the increase in thickness of the trunks. In the canopy thick clumps of epiphytes form partial ecosystems with their own recycling process. Furthermore, the increase in phytomass of the individual epiphytes must not be neglected. The main difficulty, however, is the great heterogeneity of these primaeval forests. They are usually a macromosaic of separate developmental phases (see Sect. 8.1). The optimal phase (p 182) is the most suitable for investigation.

4.3 Ecosystems with Herbaceous Vegetation

Early investigations were mainly concerned with the ecosystems of deciduous forest, but more recently ecosystems with herbaceous vegetation have also been studied. The most important of these in terms of area are the grass lands which, as zonal vegetation, form prairie or steppe. Meadows and marshes with their graminids, on the other hand, are pedobiomes.

In contrast to forests, the entire phytomass of herbaceous ecosystems of the temperate zone dies every year and grows again in the following spring. In a deciduous forest only the leaves fall off, while the woody parts, consisting increasingly, as the trees age, of dead wood, are retained for a long time.

In most deciduous trees the leaf surface is formed during a short period in the spring and then remains constant, apart from the summer shoots, and these are not significant. Yellowing of the leaves and leaf fall begin very slowly in early autumn, and leaf fall occurs very suddenly in mid-October. In some species, such as *Salix* spp., the annual shoots continue to grow throughout the summer, forming new leaves in the process, until the first frost kills the shoot tips.

In ecosystems with herbaceous plants the shoots also develop rapidly in spring, but continue to grow throughout the summer; growth usually stops when the fruits start to ripen, and this occurs at different times in different species. During the first vegetative phase, new leaves are constantly formed, but the first spring leaves die progressively and are replaced by summer leaves, which are often xeromorphic.

The maximal standing aerial phytomass (B_{max}) is usually reached in summer, shortly before or at the time of flowering, but a considerable proportion of the leaves is already dead, so that primary production is greater than B_{max}. Since B_{max} is reached at different times by different species, the B_{max} for the entire stand is much less than the sum of the values for all the individual species. Iwaki (1959) gives the following example of a grassland in Japan.

B_{max} of the dominant grass species was reached in August, B_{max} of *Hemerocallis*, *Convallaria* and *Disporum*, on the other hand, as early as June, while for yet other species it was reached in July. B_{max} for the grasses was $180\,g\,m^{-2}$, for the early species $34\,g\,m^{-2}$ and for the summer species $118\,g$ m^{-2}. Together this makes $332\,g\,m^{-2}$; that is, 23% more than B_{max} for the entire stand in August with only $270\,g\,m^{-2}$. Other authors have reported a difference of as much as 43%.

Even the separate estimation of B_{max} for the separate species is, however, inadequate; strictly speaking, the leaves and buds which have died before B_{max} is reached must be included. If this is done, the total aerial production is often found to be 2–5 times the B_{max} of the total stand. In addition, losses in leaf surface occur during the summer as a result of the activities of phytophagous consumers. Losses brought about by arthropods are normally only a few percent, but those due to rodents of steppe and desert are usually much greater, often more than 10%. We have seen, however, that such losses affect productivity in very different ways (p 44 ff).

It is still more difficult to form an estimate of the productivity belowground of ecosystems with herbaceous plants. Even a rough estimate of its magnitude has been made successfully in only a few cases. It may be ¼ to ½ of the B_{max} of the whole stand or ¼ to ⅕ of the living phytomass below ground (vide examples of Iwaki 1959). For this reason, values are usually only given for aboveground production in herbaceous stands.

Bazilevich and Titljanova (1978) give a brief summary of the ecology of grass ecosystems. This is based on very extensive investigations made between 1964 and 1974 at the experimental station "Karachi" in the Baraba basin in western Siberia, but the authors have also taken into account the literature available on grass ecosystems of the whole of the temperate zone of the northern hemisphere.

The relief of the Baraba basin is subdivided in a complex manner, with many different soil types (Kovalev 1976). On the higher ground there is zonal meadow steppe on black earth, while in the lake area there is a pedobiome of boggy grassland, consisting of both fresh and lightly halophilic meadows, for this is a slightly semi-arid area.

The mean temperature over the year as a whole is $-0.5°C$, for July it is $18°–19°C$, in January $-20°$ to $-21°C$. The growth period when the temperature is above $5°C$ is 160 days. Annual rainfall is $440–450\,mm$, 60–70% of which falls in summer, that is, May to September, with a maximum in July. Potential evaporation is slightly higher than the rainfall, being lower only in rainy years. Alkaline saline soils occur here.

In this summarizing review the following types of grassland, in order of increasing aridity, are listed:

a) bogs, which periodically dry out
b) steppe-like meadows
c) halophilic meadows
d) steppe.

In the following discussion the symbols a–d will be used to indicate the different types of grassland. The numbers represent dry weight in $g\,m^{-2}$. The phytomass which survives the winter alive is indicated by R (roots and rhizomes), although it is very difficult to distinguish living organic matter from that which is dead (D). The total underground matter is thus R + D. It contains the reserves needed for the formation of new shoots in spring; it shows less variation from one grass type to another than does B_{max}. The general observations made by these authors are:

1. The drier the grassland ecosystem, the greater is R (g m^{-2} in the upper 50 cm of

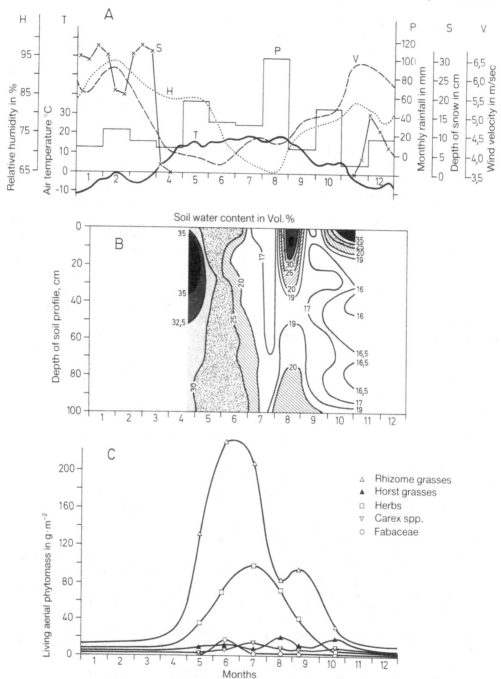

Fig. 34. Simultaneous measurements, made in 1957, of abiotic (**A, B**) and biotic factors (**C–H**) in a grass-steppe ecosystem in the Central Black Earth Reserve. **A** Meteorological factors; **B** water content of the soil; **C** aboveground phytomass; **D** phenology of the plants; **E** dead aerial plant parts; **F** number of invertebrates; **G** humus mass; **H** number of rodents (the most common vertebrates). (Fedorov and Gilmanov 1980)

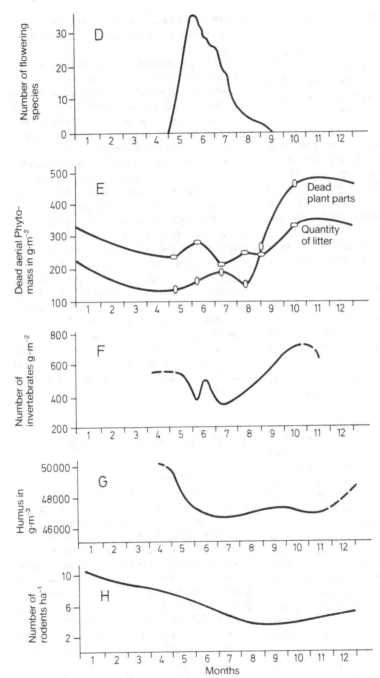

Fig. 34 D–H

soil): a = 330, b = 410–515, c was not assessed, d = 670.

2. The ratio $R : B_{max}$ increases in the same order, with a = 2.9, b = 3.7, c = 4.3, d = 6.0; in contrast, the total above- and belowground phytomass changes very little from one type of grassland to another, from a = 600 to d = 900.

3. The ratio of dead underground material, D, to living, R, that is, D/R, depends on the rate with which D is decomposed. This is promoted by salinization (in (c) $D/R = 0.5$), inhibited by dryness (in (b) $D/R = 1.0$, in (d) $D/R = 1.2$) and even more strongly inhibited by wetness (in (a) $D/R = 2.8$) where peat may form. For this reason, the ratio between total standing underground phytomass (R + D) and B_{max} is 11.2 : 1 for (a), 5.9 : 1 in (b), 6.4 : 1 in (c) and 8.6 : 1 in (d).

The distribution of living organic matter in a grassland is thus very different from that in an ecosystem with woody plants, where organic matter accumulates aboveground as wood; this is particularly true of forests.

4. As far as productivity is concerned, the aerial primary production of forest communities is only a fraction of the phytomass aboveground because the photosynthetically important leaf surface comprises only a small part of the total phytomass. The aerial primary production (P_0) of grassland, on the other hand, always exceeds B_{max} by as much as 1.2–2.2 fold; this excess is greatest in grassy moors and least in steppes and halophilic stands. The average values of P_0 expressed as g m^{-2} per year (and highest and lowest values in brackets) are as follows: (a) 370 (214–600), (b) 259 (126–508), (c) 146 (36–224) and (d) 150 (44–360). The corresponding values for primary production belowground are: (a) 1880 (850–2920), (b) 650 (126–1240), (c) 320 (120–490) and (d) 356 (200–560). It can be seen that the range within a group is very large. The lowest recorded value was for halophytic grassland (= 466) and the highest for grassy moorland (= 2250 g m^{-2} yr^{-1}).

5. Renewal time for the phytomass (turnover of carbon) is only about 0.4 years for (a), 0.9 for (b), 1.2 for (c) and 1.3 for (d).

In the east European deciduous forest zone every gram of CO_2-assimilating plant tissue gives rise to 4–6 g new biomass per year, whereas in the steppe this value is 10 g.

This rapid turnover of the carbon content of the phytomass is, in our opinion, the reason why in the grass steppe zone, where grassland and forests grow in the same climatic conditions but on different soils, primary production of the steppe is more or less the same as that of the forests.

Production is more efficient in ecosystems with herbaceous vegetation because they do not form unproductive woody tissues which have only a mechanical supporting function. Thus the highest annual values for primary production (albeit over a small area) of approximately 40 t ha^{-1} yr^{-1} were obtained for the giant herbs along river banks on Sakhalin; and such values are possible only because, as a result of lateral light, a leaf surface index of 21 can be attained (Morosov and Belaya, quoted in Walter 1981, which provides literature references).

The abiotic and biotic conditions of the grass steppe of the reserve in the central Black Earth zone for 1957 are shown in Fig. 34.

4.4 Ecosystems of the Desert

Particular difficulties are encountered in attempting to estimate the primary production of desert areas with woody plants. In extremely arid climates precipitation is not only low, but varies greatly from year to year. A year with almost no rainfall may be followed by one in which it is usually high; occasionally there may even be several rainy years in succession. The perennial plants must be able to survive the drought periods. Where these are not succulents but woody species without reserves of stored water, this is achieved by a reduction of the transpiring phytomass aboveground. Most of the branches die. Only a few remain alive and it is their regenerative buds which then produce shoots in rainy years and lead to a renewed growth of aerial phytomass. Thus on every shrub there are many branches which have died in periods of drought, but which remain attached to the bush for many years (standing litter). A positive primary production is thus achieved only in good, rainy years, while in drought years it is frequently negative; that is, there is a decrease in living phytomass. The mean value over many years of the primary production of perennial plants is thus only just above zero; small dwarf shrubs may be very old, even well over 100 years.

In less extreme deserts, with alternating drought and rainy years, Russian ecologists use the following method to determine the average primary production of a drought year and of a rainy year. At the end of the vegetative growth period, the new annual growth (Russian: prirost), and the quantity of both fallen litter (Russian: opad) and litter which remains attached to the plants (Russian: otpad) are measured over an area of 1 ha. Half of the sum of these three values approximates to the average primary production, if indeed averages can be considered to have any meaning in desert conditions. Apart from the perennial plants, there is also the ephemeral vegetation—annuals (ephemerals) and geophytes (ephemeroids)—which is absent altogether in drought years, but in rainy years can be very well developed. The primary production aboveground of this ephemeral vegetation is easier to determine. It is approximately equal to the standing dry weight at the end of the very brief period of vegetative growth.

Since deserts differ markedly in climatic and soil conditions, each has certain aspects which are peculiar to it alone. In Volume 2, therefore, deserts are not treated in a general manner, but each one separately.

4.5 Ecosystems with Mineralization as a Result of Fire

When, as a result of great dryness, the activity of decomposers is inhibited, litter will accumulate in an ecosystem; a fire will bring about very rapid mineralization. Mineral nutrients stored in the litter are in this way made available to plants once more, and the cycling is resumed.

Fire is in this case an important factor for the ecosystem, and indeed a natural one, even if today in certain areas fires are started by man either deliberately, as on savannas, or through negligence.

Some old statistics on the causes of 1535 fires in pine forest of eastern Europe are as follows:

58% — carelessness in handling fire;
30% — encroachment of fire from neighbouring terrain;
10% — arson;
 2% — lightning.

Woods in areas with extreme summer drought, such as the mediterranean zone, are particularly endangered by unextinguished cigarettes thrown away by tourists, but natural fires caused by lightning in less populated areas are and were, even before human habitation, far more frequent than has been assumed. Indeed, the fact that there are in certain climatic zones many pyrophytes, that is, plants adapted to the effects of fire, shows that *fire is a natural climatic factor*.

A particularly impressive example from Australia shows that the cycling of matter in certain ecosystems comes to a stop without the periodic fires. This heath ecosystem will now be discussed in detail and evidence will be presented from all over the world to show how frequent are fires caused by lightning. One prerequisite for such natural fires is an accumulation of combustible material which is readily flammable during the dry period.

4.5.1 The Fire-Dependent Ecosystem of a Heath in Southern Australia

This is the so-called 90-Mile Plain, a heath area rich in Proteaceae with such poor, sandy soil that it remained unsettled. Specht (1957, 1958) investigated this ecosystem very thoroughly. We summarize his results.

Ecological Conditions

The sandy soils of the heath are acidic, poor in nitrogen and phosphorous, and often also in potassium, copper, zinc and molybdenum. In rainy areas there are podzolic soils; with increasing aridity the podzolization is reduced, until finally there is a solonization with a well-defined horizon where accumulation of Na^+ ions occurs. The poverty of the soil is related to the poverty of the mother rock, for in places where this consists not of quartz sand but of silicates, it gives rise to clays when weathered, and grassy savanna develops instead of heath.

In the area investigated by Specht, annual precipitation is 450 mm; 70–75% of this falls in the winter months, that is, from May to November; approximately every fifth year is a drought year with 7 drought months in which there is no rainfall worth mentioning.

Soil temperatures fluctuate in the course of the year:

at 7.5 cm depth from 1.9° to 45.0°C
at 15 cm depth from 4.1° to 36.0°C
at 30 cm depth from 5.8° to 29.0°C.

Ground frost may occur in any month, but is rare between November and February. Dew also falls frequently.

Heath develops where the depth of the sand is at least 120 cm. The upper 15–25 cm (A_1 horizon) of the sand is coloured grey by humus. There is no intact litter layer on the surface. A_2 is yellow and contains no humus, A_3 shows a few rust flecks and lies immediately above the solonized clay of the B horizon with its columnar structure. The pH values of the profile vary between 5.8 and 6.4. Typical sclerophils include *Eucalyptus baxteri*, 9 Proteaceae, 2 *Casuarina*, *Xanthorrhoea* and several Leguminoseae. Their main period of growth is in summer, the sandy soil remaining moist until January; even in February and March the B horizon still contains water.

The relief is altered slightly by rolling dunes, which results in some grouping of the plant cover; there are, however, no clearly demarcated plant communities.

The root systems of 91 species were examined. In 25% of these species the roots grow both downwards to penetrate the B horizon, as well as laterally in the upper 25 cm; from the latter, root shoots develop in some species. In small perennials, which make up 42% of the plants examined, the root system is shallow and lies within the upper 30 or 60 cm. These species have their main growth period in spring. The roots of all plants with rhizomes and tubers lie in the upper 30 cm only, as is also true of *Stipa semibarbata*. *Xanthorrhoea australis* has a very finely divaricate root system, which penetrates the soil over a radius of 1 m and to a depth of over 2.5 m. The roots of *Drosera* and the orchids reach a depth of only 5–7 cm; these are ephemerals or ephemeroids. An estimation of the dry weight of the roots in the soil shows that the main mass lies in the upper 25 cm, with a maximum at a depth of approximately 15 cm.

The heath vegetation is often subject to fire, which destroys the aerial parts of the plants; the root systems remain intact, however, and 70% of all species can develop again from basal buds or as root shoots. Nutrients are absorbed by the roots

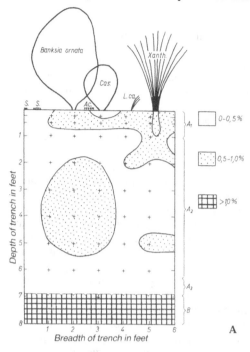

Fig. 35 A–D. Results of observations by Specht (1958) on the water content (percentage dry wt.) of sandy soil below heath vegetation on Dark Island Heath (Ninety-Mile Plain), South Australia. The effect of the vegetation on the water content of the soil beneath is shown. *Crosses* indicate points at which soil samples were taken. Species illustrated are: *Ac Acrotriche affinis; Ast Astroloma conostephioides; B Baeckea ericaea; Banksia ornata; Bor Borronia caerulescens; Cas Casuarina pusilla; Hib Hibbertia stricta; Hib. ser Hibbertia sericea; Hyp Hypolaena fastigiata; L. ca Lepidosperma carphoides; L. co. Lepidosperma laterale; Phyl Phyllota pleurandroides; S Schoenus tepperi; Xanth Xanthorroea australis.* **A** February 3: at the end of the drought period and before the start of the rainy season. **B** March 20: after 24 mm of rain had fallen on almost completely dry soil. **C** July 8: during a very moist winter. **D** December 13: at the start of the summer drought

mainly from the B horizon, and accumulate in the litter on the A_0 and A_1 horizon, in this way becoming available also to the shallow-rooting plants.

The distribution of water in the soil is very interesting. Sandy soil is regarded as a homogeneous soil, but as far as water content is concerned, this is by no means the case. Because of the vegetation, rain does not wet the upper soil surface evenly; any rain which falls on the aerial

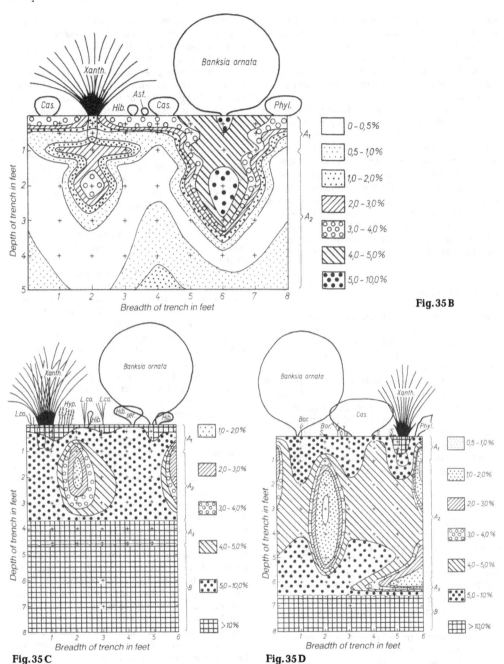

Fig. 35 B

Fig. 35 C

Fig. 35 D

parts of plants flows either to the stem or to the periphery of the plant, so that more water reaches the soil at these points. The result is a very complex pattern of soil water content. When the rain stops, some of the water sinks deeper into the soil, some of it is taken up by the root tips. Since the latter are not evenly distributed, the pattern of water distribution is made still more complex. At monthly intervals over a 2-year period, Specht made closely spaced soil profiles along a section through the vegetation; the water content was expressed as a percentage of soil dry weight. The wilting point of the soil (water potential = −15 bar) was 1% in the A_1 horizon, 0.7% in the A_2 and 7–17% in the clay layers. The results are shown in Fig. 35. The crosses mark the points at which samples were taken. It can be seen how difficult it is to make good average samples for determining water content even in sandy soils, and how unrepresentative single samples are.

The large species such as *Banksia ornata* and *Xanthorrhoea australis* carry the rainwater to the stem; soil water content is thus usually higher below these plants, and this is favourable to their growth. With increasing age they represent an ever-greater threat to the small-leafed species, which have to make do with less water. Curves of transpiration measurements for the larger species made during the rainy season show a single maximum around midday; during the summer drought there are two peaks, but these become ever less pronounced as drying out of the soil progresses. *Xanthorrhoea* is characterized by particularly low rates of transpiration. The main growth period is nevertheless the summer, at temperatures over 18°C.

The Effect of Fire

The composition of the heath vegetation is determined by the frequency of fires. After a fire a very characteristic succession takes place. *Xanthorrhoea* makes the most rapid reappearance and forms 90% of the aerial vegetation during the first year after a fire. This proportion is rapidly reduced over the next few years as a result of the development of the other species and then remains constant at 5–10%. The actual number of plants also remains constant, for *Xanthorrhoea* only flowers after a fire and most of its seeds are destroyed by insects, so that even in areas where there has been fire no seedlings are to be found. By contrast, *Banksia ornata* is rejuvenated only by seedlings. Its proportion in the vegetation therefore increases very slowly, reaching 50% in the 15th year and then remaining constant at this level. The main mass of dry substance in 25-year-old specimens, which reach a height of 2.5 m, is comprised of woody fruits. These fruits only open after a fire. Nevertheless, in areas which had been free of fire for 50 years and where dead bushes were to be found, opening of the fruits and scattering of the seeds had occurred. Yet only a few seedlings could be found.

Thus *Banksia ornata* belongs to the ecological type known as a *pyrophyte*; plants of this habit, especially Proteaceae, are widely distributed in Australia. These are species in which the fruits do not open on the living plant. If, for example, a branch of *Hakea platysperma* bearing old, woody, closed fruits is held for about 10 min over an open fire, then a short time later the two lids of the fruit burst open and beneath each of these is a seed. The seeds do not fall out immediately, however, for each has a hook with which it remains attached to the lid. Only hours later are the seeds

set free by the wind; they thus land on ash which has in the meantime cooled down, and in this way find a suitable seedbed. The same behaviour is shown by *Xylomelum pyriforme* (Proteaceae) with its large, woody, pear-shaped fruits, by the Australian conifer *Actinostrobus* and by many other species. Rejuvenation of the population of such pyrophytes is possible only after fire. Thus fire is an important natural factor.

Other heath plants, such as *Casuarina pusilla* and *Banksia marginata* regenerate from underground organs and reproduce mainly vegetatively. In *Leptospermum myrsinoides* both vegetative and sexual reproduction are equally important, but in *Phyllota* only the latter.

The following can be distinguished among smaller species of the undergrowth: (1) herbs, which disappear again after only 3–4 years; (2) dwarf shrubs, the proportion of which is greatest after 10 years; (3) plants which reach their maximal development after 15 years, and (4) those species which continue to increase in abundance relative to the rest even after 25 years. Only two annuals were observed in the first year. Figure 36 shows clearly the relative abundance of different species year by year after a fire. Figure 37 shows: (a) the increase in dry substance of the aerial parts of the most important species, (b) the litter production and (c) dry substance of the underground parts, the greatest part of which remains alive after a fire. Even 25 years after a fire, the dry substance belowground is still three times greater than the aerial dry substance. On part of a 50-year-old stand, 15 000 kg of aerial dry substance was weighed; of this, *Banksia ornata* contributed 12 920 kg, *Casuarina* only 10 kg. In six test areas, each 5 × 10 yards in size (1 yard = 0.94 m), the number of species after a fire was 36; this number fell to 20 after 25 years and to about 10 after 50 years. Conditions are most favourable for growth during the first year after a fire when, owing to the small transpiring surface area, water supply is good and more nutrient minerals are available from the ash. Competition soon commences, however, with gradual dominance of only a few large species, while the smaller forms are repressed. The dead shoots of the latter are found under the larger bushes; in the intervening open spaces they still survive. The material production per hectare over 25 years is as follows:

Aboveground	27 500 kg ha^{-1}
Belowground	13 500 kg ha^{-1}
Together	31 000 kg ha^{-1}
As litter	7 500 kg ha^{-1}
Total	38 500 kg ha^{-1}

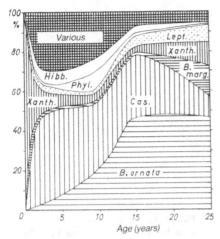

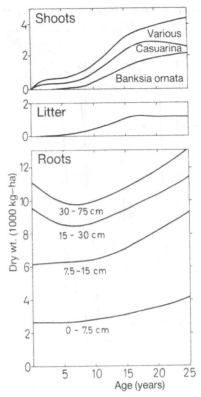

Fig.36. Percentage contribution of individual species to the total phytomass (dry wt.) on a sandy
soil during the 25 years following a fire (Specht
1958). *Hibb. Hibbertia* spp.; *Phyl. Phylota remota;
Cas. Casuarina pusilla; Lept. Leptospermum
myrsinoides; Xanth. Xanthorrhoea australis; B.
marg. Banksia marginata; B. ornata Banksia
ornata*

Fig.37. Change in dry weight of aboveground
plant parts, the litter layer and the roots at different depths during the 25 years following a fire.
(Specht 1958)

Beside dry substance production, the cycling of
the most important elements, such as C, N, P, K,
Na, Ca, Mg, Zn, Cu and Mn was investigated.
Like the dry substance, the content of these
elements in aerial organs generally increases until
the stand is 15 years old, and then remains constant. Only in the case of manganese is there any
marked decrease between the 15th and 25th
years. In *Banksia ornata* this is related to decrease
in leaf mass. In the course of time, nutrient substances are found mainly in the fruits of *Banksia*
and in the dead leaves of *Xanthorrhoea* (about ¼
to ⅓ of the total amount). The quantity of nutrient
elements in the litter also increases continuously,
while particularly large amounts are contained in
underground organs. The only additional external
source of nutrients is that of nitrogen, which is
bound by the root tubers of *Phyllota* and the
mycorrhizae of *Casuarina*. In older stands, however, both these species are displaced.by others.
Small amounts of the other elements can be taken
up from the B horizon. Generally, however, the
supply of nutrient minerals in the upper soil layers
is rapidly reduced in the first years after a fire as a
result of regeneration of plant cover, and it then
remains fairly constant. *Increasing incorporation
of nutrient elements into particular plant organs
above and below the ground and the absence of
decomposition of litter must lead, over a period of
50 years, to a degeneration of the vegetation.*
Before this age is reached, however, a fire usually

brings about mineralization of the organic matter
so that the succession can begin all over again.

The conditions described here can also be considered to apply to macchia-like vegetation of the
other winter rainfall areas with poor soil (e.g. the
Cape in South Africa, with its many Proteaceae).
It is for this reason that they have been described
in such detail.

4.5.2 Natural Fires in all Parts of the Earth

In the particular case of the Australian
heath, the mineralizing action of decomposers is replaced by natural fire. The question
arises, however, as to whether in the past
fire in general played a greater role as a
natural factor, that is in earlier times,
without the participation of man.

A very interesting observation was made
in the unpopulated Muddus virgin forest in
Swedish Lappland; this is the most northerly

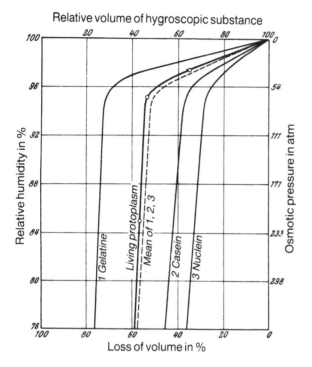

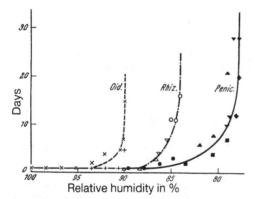

Fig. 40. Curve showing the relationship between the relative humidity (hydrature) and the state of hydration of the protoplasm of the vacuole-free carpospores of *Lemanea*. Comparable curves are shown for the hydration of *1* gelatine; *2* casein; *3* nuclein; *broken line* shows the mean for these three substances. (Based on data from Walter 1923)

Fig. 41. Dependence of the germination period (in days) of the spores of mould on relative humidity (hydrature) at 20°C. *Oid: Oidium (Oospora) lactis; Rhiz: Rhizopus nigricans; Penic: Penicillium glaucum.* (Heintzeler, from Walter and Kreeb 1970)

rated with water vapour; that is, with a hydrature of 100% hy. When the humidity falls, the greater part of the absorbed water is lost between 100% and 96% hy and then there is a sharp inflexion in the curve; the rest of the absorbed water is lost far more slowly with decreasing hydrature (Walter 1923; Walter and Kreeb 1970; Walter and Wiebe 1966).

By means of volume measurements of non-vacuolate protoplasts of algal cells (carpospores of *Lemanea*), it has been possible to show that living protoplasm behaves exactly like non-living hygroscopic substances (Walter 1923). The swelling curve for these cells corresponded to that for a mixture of gelatine, casein and nuclein (Fig. 40).

By growing poikilohydric species in air of different humidities (hydrature), it is possible to determine the limits of hydrature within which growth and cell division are possible; these data are of course valid, irrespective of the water content of the nutrient surface in which the organisms are cultivated (Walter 1924; Heintzeller 1939; Burcik 1950). Thus with a hydrature of 75%, a humidity at which there is no development of mould, the water content of egg powder is 11.6%, of noodles 13.0%, oats 13.6%, pea or bean meal 15.3%, dried potatoes 15.9%, rice or rusks 16.6%, dried stone mushrooms 19.5%, tobacco 23.4% and of dried vegetables 26.7% (Walter and Kreeb 1970).

The limiting values are specific for the individual species of growing microorganisms; the values (in bar) corresponding to the hydrature values are given on p 88.

pyrophilic *Pinus sylvestris* of Eurasia corresponds to *Pinus contorta* of North America. Fires caused by lightning are natural environmental factors also in the steppe of Euro-Siberia and the prairies of North America, and prevent accumulation of too much litter in the form of "steppe felt".

In extreme drought years, there may even be moor fires which go on smouldering for a long time.

There is a further consideration which speaks for the great importance of fires caused by lightning. When James Cook, during his first voyage around the world in 1768–1771, saw Australia from the sea, he noticed the clouds of smoke from forest fires. It was assumed that the natives of Australia had already taken to burning the forest. The flora of Australia is, however, characterized by a large number of pyrophytes, which are dependent on fire for their reproduction. Since these species arose in the Tertiary, long before man could have caused fire, they must have evolved this habit in response to natural fires. Indeed, the *Eucalyptus* forests of Australia are pyrophytic (Mount 1969). The seeds of *Eucalyptus* are shed abundantly following a fire, and encounter a very favourable seedbed, free of litter. The shedding of seeds, normally very slight because the fruits remain closed, becomes so pronounced that their usual total destruction by phytophagous organisms does not take place and sufficient seeds remain to germinate and produce seedlings. In areas where there are frequent naturally occurring forest fires, the *Eucalyptus* seedlings develop in the first few years an underground woody swelling, formed of the lowest internodes and the upper root neck, a lignotuber, in which reserve substances are stored. After fires which destroy the shoot, new shoots arise from the lignotuber (mallee-vegetation) and these grow rapidly in the absence of competitors which have been destroyed by the fire. This is thus a clear case of adaptation to frequent, natural forest fires.

Lightning must certainly have been, in earlier times, the most frequent cause of natural fires. This is supported by observations made in Africa, in Liberia, Tanzania, Zimbabwe, the Transvaal and Natal. Here it has also been found that fires may be started by sparks struck by falling rocks (Killick 1963). Phillips (1965) observed in 1938 in the Transvaal how in this way a falling diabase block kindled a fire on a steep slope. In other cases quartzite rocks were involved. In another instance it was suggested, rightly or wrongly, that a fire had been caused by self-ignition of dead and rotting plants.

The most precise data on the frequency of lightning-caused fires are those given by Komarek (1966, 1967, 1971) for North America; these are based on statistics for the National Forests of the USA. During the 20 years from 1939 to 1958, 132000 fires caused by lightning were reported in the western states, including Alaska. In the Rocky Mountains 70% of all fires were caused by lightning. During 10 days in July 1940, 1488 lightning fires occurred in Montana and North Idaho. In the Californian National Forests there are on average 775 fires caused by lightning per year. We will take as an example the reports for 1965.

During the summer in North America, sharp fronts are often formed between the cold polar air flowing from the north and the maritime, tropical, warm air advancing from the south; as a result, heavy thunderstorms develop. Between the 1st and 5th of May, 1965, such a front stretched from South Dakota to Philadelphia, then shifted further south to reach Florida on May 15th. In these few days, one or more lightning fires were reported from nine forestry districts in South Dakota and Nebraska, from one in New Mexico, from 13 in the Appalachian mountains, from one in the lake district and from 13 in Florida, making a total of 37 districts affected by fires. These are all indisputable cases in which the lightning and the resulting smoke cloud were directly observed.

The forested sand dune area of Nebraska, being surrounded by a natural long-grass prairie, is particularly endangered. Prairie fires are so frequent that during any storm the glassed-in and well-isolated fire watch-tower is manned. The first author of this book observed a storm himself from this tower in 1930. Hardly had the storm passed than a glow of light spread rapidly in the east. A prairie fire! The forester rushed to the telephone to alarm the fire brigade. But it was the rising full moon.

On May 6, 1965, a prairie fire which had been started by lightning spread to a forest stand, advanced rapidly 20 km westwards and then, after a change in the wind, spread northwards over a 20-km-long front, until it finally extinguished at the edge of a heavily grazed area. According to reports by the forestry office, there are on average seven prairie fires caused by lightning per year, but in one year there was one lightning fire for every 5000 ha. There can thus be no doubt that in the original prairie, grass fires were a widespread phenomenon, even without the helping hand of man. Today, on the farm-land which the prairie has become, they are, however, very rare.

Forest fires resulting from lightning are also frequent in the forest area of the Appalachian mountains. In the Mononghela National Forest (West Virginia), 2 of 30 fires per year are brought about in this way. The extensive spread of pine forests in the eastern USA is thus comprehensible.

The weather conditions described above for the summer 1965 can be regarded as typical for North America, and they may be repeated several times a year.

It is of importance that similar storm fronts may also form in the arid western states of North America. This occurs when warm air from the Gulf of Mexico streams into the mountainous area and there encounters the cold air on the crests of the mountains, so that localized storm fronts build up. The summer rains of Arizona are brought about in this way. Between the 7th and 16th July, 1965, 536 lightning fires were counted in Arizona and New Mexico.

The occurrence of such storms during the otherwise very dry summer period has an important bearing on the chaparral formation stretching from California as far as the Canadian border. On 24th to 25th July, 1965, many forestry districts reported more than 5 fires, some as many as 28, which had been caused by lightning.

On 14th August, 1965, there were 30 lightning fires in the Elko district of Nevada, and these joined to form 5 very large-scale fires, causing one million dollars worth of damage.

In the Teton National Park, Wyoming, there were no forest fires in 1965, following the very heavy snows of the preceding winter, although the average over many years was 10 lightning fires per year.

Examination of forestry records shows that for the mixed pine forests, the grasslands and the sclerophyllous vegetational areas, as well as for the semi-deserts of

North America, fire can certainly be regarded as a natural climatic factor. By counting the annual rings of very old trees bearing fire scars, it was possible to show that even in the 16th century the forests of the Sierra Nevada in California were subject to fire on average every 8 years (Wagner 1961). It is possible to speak of a "fire-climate" in California.

Komarek (1973) points out that in all old coal seams, even in those of the 400 million year old Carboniferous, "fossil charcoal" is found: this is known to geologists as "fusain", and is assumed to have been formed by natural forest fires resulting either from lightning or spontaneous combustion. Mägdefrau (1968) mentions that this has been observed in coal seams of other geological periods as well. Although the carboniferous forests grew in swamps, the occurrence of forest fires was not precluded, for even the extremely wet *Taxodium* and *Nyssa* bog forest in the southern part of North America may, during periods of drought, dry out to such an extent that they catch fire from lightning and a large quantity of charcoal is formed. During wet periods this charcoal is covered with water and is preserved between the dead wood, just as it was in the Tertiary when brown coal was formed with fossilized *Taxodium* remains. Fusain occurs in distinct layers of brown coal, suggesting periodic growth of forest and a high frequency of forest fires.

Komarek (1971) has suggested that tropical Africa, with its summer rain period, is a climatic area predestined to be affected by lightning fires. Here much combustible plant material is formed, and this is readily inflammable during the dry period. At the onset of the rainy season, there are frequent storms with much lightning but hardly any rain, so that fires may be caused by the lightning. Such fires are, indeed, observed time and again in tropical Africa, but statistical data are available only for the forests of South Africa, where in the years 1957–1970 there were on average 29 lightning fires a year which had to be extinguished. Other fires which went out on their own were not included in the record. This figure represented 11% of all larger forest fires.

For the years 1955–1959, the Forestry Department in Pretoria gives more precise data. These relate to the area controlled by Forest Stations, which comprises 5% of the total surface area of South Africa. According to Killick (1963), the number of fires caused by lightning and the percentage (in brackets) which this represents of all fires was as follows:

1955	1956	1957	1958	1959
10	27	35	22	36
(6.3%)	(11.7%)	(16.9%)	(13.2%)	(13.5%)

We must therefore regard fire as an important ecological factor in Africa, even at a time when man did not exist. Natural fires have, since primaeval times, determined the distribution of grassland and woody plants in Africa.

Although plant geographers have tended to regard most savannas in Africa as anthropogenous in origin, this viewpoint will have to be revised. The annual burning of savannas may be an adaptation on the part of the native population to the fire-climate of Africa; that is, apart from any intention of facilitating hunting, improving grazing, and so on, it may have arisen as a means of protecting houses and food stores from destruction by the surprise occurrence of natural fires. Radical prevention of fire in protected nature reserves, such as the Kruger National Park, has led to development of an undergrowth which is unfavourable to the game animals. Correct ecological use of burning can be a cheap way of keeping African ecosystems in balance while at the same time having a favourable effect on farming. There is, however, no golden rule in this respect; the most favourable time for burning and the correct interval between fires must be established separately for each area.

The ecological importance of natural fires in Africa is shown also by the large number of fire-resistant plant life-forms in African savanna areas, forms which have not arisen subsequent to the appearance of man, but certainly are of Tertiary origin.

We may thus assume that even without any action on the part of man, fire has played a part in determining the character of the vegetation, while artificial prevention of fire in nature reserves should be regarded as an unnatural interference.

The forests in the Grand Teton National Park (USA) were continuously plagued by bark beetles, which had to be controlled at great expense and effort, until it was recognized that this calamity was the result of the prevention of forest fires: in the protected area too many old and sick trees remained standing and these greatly enhanced multiplication of the bark beetles. In natural conditions the old wood is destroyed by fire; for this reason it was decided not to fight natural fires, and in this way the great increase in numbers of the bark beetle has been prevented (oral communication).

In the reserve established in California for the protection of the giant sequoia, *Sequoiadendrum giganteum* and of *Sequoia sempervirens* fire prevention has had an unfavourable influence on the propagation of these species, since their seedlings grow well only on areas which have been scorched by fire (Hartesveldt and Harvey 1967).

Komarek further points out that fire is of very great importance for the indigenous game animals. The zonal vegetation is often to be regarded as a "biological desert". The game finds sufficient food only where the plant cover comprises the more open succession stages which develop after a fire.

In areas affected by frequent fires, tree ferns and tree-like monocotyledons (palms, *Yucca brevifolia*, *Xanthorrhoea*) spread. These have no cambium and are thus fire-resistant (Vogel 1967).

Rudel (1981) has compiled data from observations made all over the world on the temperatures reached during fires, both above and below the soil surface, and of the effect of fire on plants, soil and microorganisms.

Summing up, it can be said that fire is a natural ecological factor which has had a very strong formative influence on many vegetation types, such as the sclerophyllous vegetation of the winter rainfall areas, the grasslands of the temperate and subtropical zones and, to some extent, also the tropical zone and, further, has had a marked effect in determining their distribution. While the far more frequent fires caused by man have made a further spread of this vegetation possible at the cost of the fire-sensitive forests, it would be incorrect to regard these vegetational forms as only anthropogenic in origin. Even natural fires alone, without the helping hand of man, would have resulted in the replacement of the zonal vegetation by "pyrophils" (plants in which growth and development are enhanced by fire) to a certain degree. The various types of *Pinus* forests and of sclerophyllous shrubland bear witness to this. In considering the origin of grasslands or sclerophyllous shrubs, the role of natural fires must always be borne in mind.

4.6 The Shortened Material Cycle and the Role of Mycorrhizae

We understand by this a material cycle in which there is no full mineralization of the litter. It may be important in tropical rain forests, but this has not yet been demonstrated.

In tropical forests there is an apparent contradiction, for the vegetation is almost unimaginably luxuriant on the poorest soils, soils which are practically unusable for agriculture. The first author became aware of this in 1934 when he was working in East Africa, at the research station Amani, in the tropical rainforest area of the East Usambara mountains. During the period of German colonial rule, the first coffee plantations were established in precisely this area, because it was assumed that the abundant vegetation was an indication of particularly rich soil. The forest was cleared and the wood burned. The new crops gave good yields during the first few years, but soon failed completely.

G. Milne, a soil scientist, was working at Amani at this time. His investigations showed that even on steep slopes, the soil was a very thick red-brown clay, lying above gneiss, granulite or pegmatite rock. The high temperatures, humidity and CO_2-production favour the breakdown of silicates and the leaching of bases and silicic acid. What is left are mainly the sesquioxides (Al_2O_3 and Fe_2O_3). The soil profile shows very little differentiation and it is acidic (pH = 5.3–4.6). There is no litter layer and the soil surface is covered only by rapidly rotting leaves and twigs. No humus horizon is to be seen, although the quantity of organic matter on wet burning may amount to 2.5–4%. The nitrogen content decreases rapidly from 0.364% N in the upper 8 cm to 0.074% at a depth of 1 m. N is only present in organic form; NH_4^+ and NO_3^- are found only in traces. Phosphorous, which is soluble in weak H_2SO_4, is found at a rate of 18 ppm on the surface.

The water content of the soil is, at 27%, always high. The groundwater which flows off into the streams is coloured light brown by the humus sol, but is so poor in mineral salts that its conductivity is equivalent to that of distilled water. Further leaching of the soil does not occur, despite the high rate of precipitation of 1948 mm per year.

The conclusion which, as an ecologist, one had to draw when confronted with these facts was that all the resources of mineral nutrients of this complex ecosystem must be contained within the phytomass itself. The dead plant parts which are constantly falling off are immediately mineralized and the nutrients liberated are taken up again by the roots without any losses being incurred (Walter 1936, 1936a). Only in this way is it possible for the tropical forest to continue indefinitely.

Upon clearance of the forest and burning of the wood, the nutrients contained in the aerial phytomass are almost completely washed out by the rain and only a soil very poor in nutrients remains. This explanation has since found general acceptance. Apart from areas with volcanic soils of recent origin, the tropics are at a serious disadvantage from an agricultural point of view (Weischet 1977).

This very rapid nutrient cycle, in which there is no loss of material, is difficult to understand. The observations of Went and Stark (1968) in the forests of the Amazona basin are therefore of particular importance. They found that the hyphae of mycorrhizal fungi form a direct link between the roots of trees and the dead plant material comprising the litter. Thus it is possible that nutrient elements are taken up directly in organic form by the mycorrhizae, and a complete mineralization of the litter need not take place. There may thus be a "direct mineral elements cycle". The many holosaprophytes among the Burmanniaceae, Orchidaceae and Monotropaceae and also *Voyria* among the Gentianaceae, which absorb all the organic compounds required for their development from the soil with the help of their mycorrhizae, all show that this sort of nutrition is possible, both in the case of endotrophic as also of ectotrophic mycorrhizae. Since the other representatives of these families, which contain chlorophyll, also have mycorrhizae, it is possible, even if not proven, that they are to some extent hemisaprophytes. Björkman (1956) states that *Monotropa hypopitys* makes contact through the hyphae of the mycorrhizal fungus with the roots of the spruce *Picea abies* on which it is clearly parasitic.

Direct utilization of the litter by tree mycorrhizae would make it easier to explain the fact that, despite the high rainfall in tropical rain forests and the weak sorption of the ferrallitic soils, leaching of nutrient elements does not occur, even with very rapid mineralization. The growth of *Pinus sylvestris* and *Calluna vulgaris* on very poor sandy soils would also be more readily understood if the same assumption can be made for their mycorrhizae. The very marked competition of *Picea abies* on raw humus soils demonstrated by Karpov (vide Sect. 7.3), could be explained in the same way, for this tree shows such a strong absorptive capacity for the nitrogen of the litter that only undemanding plants, *Vaccinium* and mosses, can grow in the undergrowth. If root competition is eliminated without changing the light conditions, even nitrophilic species such as *Chamaenerion angustifolium* and *Rubus idaeus* occur.

According to Meyer (1974), mycorrhizal fungi, in contrast to the usual saprophytic fungi, are incapable of breaking down lignin and can digest cellulose only to a limited extent; they require simple carbohydrates. European forest fungi can be divided into five groups, on the basis of their nutrition:

1. Completely saprophytic fungi (e.g. *Mycena*).

2. Mainly saprophytic but partly mycorrhiza-forming (e.g. *Phallus impudicus*, *Collybia peronata*).

3. Unspecialized mycorrhizal fungi, which can form fruiting bodies even when living entirely saprophytically (e.g. *Scleroderma*, *Xerocomus*).

4. Fungi which form mycorrhizae with various tree species and fruit only when

in symbiosis (e.g. *Amanita muscaria*, *Russula* spp.).

5. Narrowly specialized mycorrhizal fungi (e.g. *Boletus elegans*, *Lactaria porninsis*, *Suillus tridentinus*, all of which grow with *Larix*).

Nothing is yet known about the mycorrhizal fungi of tropical species. We know only that most tropical trees are mycotrophic. It is thus possible that these fungi are capable of supplying the roots of higher plants with the necessary nutrient elements also in organic form, as is the case with mycorrhizal fungi of holosaprophytic flowering plants.

Promising investigations are being undertaken on an international basis in San Carlos on the Rio Negro (Jordan and Herera 1981).

A quite different endotrophic vesicular-arbuscular (VA) mycorrhiza plays an important role in tropical herbaceous crop plants. This was investigated very thoroughly at the Institute for Tropical and Subtropical Agriculture at Göttingen, West Germany. The symbionts in the roots and also in the carpophores of the peanut (Graw and Rehm 1977) are species of *Glomus* (Endogonaceae, Phycomycetes). These mycorrhizae enable the plants to absorb phosphate from phosphate-poor soils with poorly soluble compounds (Al and Fe phosphates of the latosols), and this is important for tropical agriculture. It applies also when fertilizers are used containing the poorly soluble hydroxylapatite or monocalcium phosphate.

Up to now, however, all experiments have been carried out in greenhouse conditions (Graw 1978; Moawad 1979; Graw et al. 1979; Sieverding 1979; SAIF 1981; in more detail in the Institute's annual report of theses for the years 1977–1981).

4.7 Dependent Ecosystems Without Producers

A unique example of such an ecosystem is the vast sand-dune area of the Namib fog-desert on the coast of Namibia. This 30–130 km wide strip extends uninterruptedly over 400 km from Walvis Bay to Luderitz Bay. The sand lies above the Namib platform. Only a few isolated mountains rise up out of the sand. The dunes, which may be 240 m high, are probably the highest on earth.

The whole dune area consists of longitudinal dunes, running mainly in a north–

south direction, with an average of 1.8 km between the dune chains. The dune base is effectively stable, while the crest can shift, and has a steeper slope on the lee side on which sand grains blown over the crest slide down. The Namib fog-desert is almost completely without rainfall. Only two to three times a century does heavy rain fall, as happened, for example, in March 1934 with 113 mm rain and in April of the same year with 27 mm, making a total of 140 mm, which led to great flooding. The next occasion was in 1976 with 118 mm between January and March. Foggy days are, however, very frequent. This fog, which forms over the off-shore Benguela current, is driven by the south-westerly wind over the dunes, moistening the upper surface of the sand. Precipitation from the fog averages about 0.2 mm per day, but varies from 0.01 to 0.7 mm. The condensation can be markedly higher, however, when the driving fog encounters a vertical obstacle.

As soon as the sun appears, the upper surface of the sand dries out again. The precipitation is inadequate for higher plants, but sufficient for those animals which are adapted to such conditions. The dunes therefore accommodate a relatively rich, mainly endemic fauna, which attracted the attention of C. Koch (Koch 1961), an expert on the Tenebrionidae.

The tenebrionids and other animals with a smaller surface lose only little water by evaporation from the body surface and at the same time form metabolic water during the respiratory breakdown of carbohydrates and fats; thus even a very small uptake of water suffices for survival.

Small animals are able to lick up the droplets of water which form on sand in fog, or they may enhance condensation of the fog by some or other means: the beetle *Lepidochora,* for example, throws up a little dyke of sand, perpendicular to the direction of the wind (Seely and Hamilton 1977); others stand on their heads so that the fog condenses on their backs and the water droplets run down to the mouth (Hamilton and Seely 1976). Predatory animals obtain the water they need from their prey, the water content of which is relatively high. The psammophile fauna remains above the ground during fog and at night, but belowground on sunny days.

The food supply of phytophagous species, normally provided by the producers, is replaced in this ecosystem by dead plant material blown from elsewhere; this consists mainly of fragments of grass. This ecosystem thus depends on producers in neighbouring ecosystems. The plant detritus collects on the crests of the dunes where it forms clumps which slide down the steep slope, to accumulate at the base; as a result, the sandy soil in some places contains up to 36% organic matter, including the protein-rich remains of insects which, either actively or passively, have reached the area and there died. The further food chains are described by Kühnelt (1975) as follows: the detritus is eaten almost exclusively by psammophilic forms among the tenebrionids, but to a small extent by termites *(Psammotermes granti).* The next link in the food chain is formed by the "small predators"—solifuges and spiders—which also feed upon flies blown in from the Kuiseb Valley. Larger predators are the diurnal lizard, *Aporosaurus anchietae,* and the nocturnal golden mole, *Eremitalpa granti namibiensis.* These animals have to bury themselves when the surface of the sand becomes hot from the sun, and often spend only a few hours a day on the sand. Breathing when buried in the sand is possible because the sand consists of on average 0.5 mm quartz grains without any dust and with an air-filled pore volume of 50%.

Very interesting detailed adaptations and modes of behaviour have been observed in the different psammophilic animal species. There is even a species of snake which can move very rapidly in the sand and holds only its head above the surface (Seely 1978).

All organic remains are ultimately mineralized by decomposers, for when there is fog the moisture content of the upper sand layers is so high that bacteria and fungi occur in an active state (see p 92–95). No microbiological examination has, however, yet been made. Seely and Louw (1980) provide quantitative data on production and the amounts of energy in this unusual ecosystem of the Namib dunes during drought years (average rainfall 14 mm) and the rainy year 1976 when 118 mm rain fell. These authors distinguish three different biotopes in the dune area:

1. The dune valleys, which make up 55% of the total area and have a substratum of gravel or outcrops of sand stone. Here, during the rainy year, there was a single flowering of ephemeral grasses, *Stipagrostis ciliata* and *S. gonatostachys*.

2. The firmer dune slopes on the windward side or at the base of the lee slope, comprise 44% of the total surface area; on these, even in rainless years, isolated clumps of *Stipagrostis sabulicola* or cushions of the succulent *Trianthema hereroensis* form a sub-biotope.

3. The steep lee-side slopes from the crests of the dunes, on which sand grains blown over the crests slide downwards. This comprises only 1% of the surface.

The starting point for the material cycle and energy flow of this ecosystem is formed, as has already been mentioned, by the dry mass of detritus blown thither from adjacent areas and consisting primarily of dead grass. To this must be added the production by the few *Stipagrostis* clumps and *Trianthema* cushions. In rainy years, the ephemeral grasses of the dune valleys also contribute.

During the good rainy year 1976, the total organic dry mass increased 2.7 fold in biotope 1, in biotope 2, with its ephemerals, 53 fold, and in biotope 3, as a result of an increase in blown detritus, 9.4 fold. Similarly, primary production of the sclerophyllous *Stipagrostis* clumps was 2.4 times the normal value, that of the *Trianthema* cushions 19 times greater.

This organic material is, however, sporadically distributed and limited to certain places. On the dune slopes the living clumps and cushions cover hardly 1% of the area of biotope 2, although they make up 59% of the organic mass. In absolute terms, the total living and dry mass of a *Stipagrostis sabulicola* clump above the ground is 354 g, below the ground 394 g, while for *Trianthema* the values are 408 g above ground and 699 g below the ground. Furthermore, 99% of the detritus is found on the lower part of the steep lee slopes.

If the total organic mass found for the whole of the dune area examined be calculated, this is found in dry years to be only $3.0 \mathrm{g} \mathrm{m}^{-2}$ phytomass and $0.01 \mathrm{g} \mathrm{m}^{-2}$ zoomass; that is, the ratio of phytomass to zoomass is $300 : 1$.

After the rainy year, there was a ninefold increase in phytomass and a sixfold increase in zoomass. The phytomass is broken down only slowly, because moisture conditions are favourable for the poikilohydric decomposers only when there is fog. Destruction of the phytomass is thus due mainly to the saprophagous tenebrionid beetles which are detritus feeders. They can digest cellulose, but the low nitrogen and phosphorus content of the phytomass is limiting. The excreta of these animals and their dead bodies, unlike the phytomass, are readily broken down.

It is not known to us whether another ecosystem like this, for which frequent fog is a prerequisite, exists elsewhere. The only possibility is the dune areas of the Chilean-Peruvian desert, where there is even more marked formation of fog.

Further examples of dependent ecosystems are the coastal areas of the Guano Islands, the nesting areas of sea birds which obtain their food from the sea. The breeding grounds of seals form a similar example. Further details of these special cases are to be found in Volume 2.

5 The Fundamental Requirements for Active Living Processes: Temperature and Hydrature, Their Absolute Limits

The physiologist studies organisms in the laboratory under strictly controlled external conditions; the ecologist, however, makes his investigations in the field under constantly changing and often very extreme conditions as, for example, in the desert or the Arctic.

The life of all organisms is bound to the living substance, the protoplasm, with its complex submicroscopic structure, various subdivisions and many organelles. Protoplasm is only in an active state if two fundamental requirements are fulfilled:

1. a certain degree of warmth, that is, a certain temperature (°C);
2. a particular level of water activity in the protoplasm; that is, a certain level of hydrature (% hy).

The term "hydrature" was suggested by Walter (1931) as a rough analogy to "temperature", at a time when the thermodynamic approach to plant physiology had not yet been introduced (Walter and Kreeb 1970).

Hydrature corresponds in thermodynamic terms to the "relative activity of the water",

$a_w = \frac{p}{p_0}$, where p = the vapour pressure, p_0 = saturation pressure at the same temperature and pressure; the hydrature gives the activity in % (pure water = 100%).

The hydrature (hy) of the living protoplasm of a plant cell is equal to the water activity of the cell sap in the vacuole; this can be determined by the osmotic potential of the cell sap which is expressed in bars or, in older works, in atmospheres (1 atm = 1.013 bar).

The potential osmotic pressure π^*, is numerically equal to the osmotic potential expressed as a negative value. It is related to water activity, $a = \frac{p}{p_0}$, which expressed as percentage is equal to hydrature, by

$$\pi^*_{20°} = -3067 \log \frac{p}{p_0} \quad \text{(atm)}$$

$$\pi^*_{20°} = -3107 \log \frac{p}{p_0} \quad \text{(bar)}$$

From these equations it is possible to calculate the values for the osmotic potential, that is, the water potential, in −(minus) bar, which correspond to the hydrature values in % hy. In the following table we list such hydrature values (% hy) with the corresponding percentage relative humidity, and osmotic potential or water potential at 20°C expressed in −bar.

% hy	−bar		% hy	−bar
100	0		85	220
99	13.6		80	301
98	27.2		75	389
97	41.1		70	481
96	55.2		65	580
95	69.3		60	688
94.	83.5		50	925
93	98.0		40	1235
92	112.0		30	1625
91	127.0		20	2160
90	143.0		10	3110

5.1 Temperature Ranges

It has long been known that there are thermophilic organisms among the lower plants. Brock (1978) was the first to be able to demonstrate, on the basis of his investigations in Yellowstone Park, USA, seen by the first author in 1969, that bacteria can reproduce in boiling water. Within a few days, they form a reddish layer on immersed glass slides. These are heterotrophic rod-shaped bacteria which may grow to form threads.

Because of the high altitude of Yellowstone Park, the boiling point of water there is 92°C, but these bacteria also grow in the slightly super-heated water of the geysers, that is, at 93.5°–95.5°C. Brock also reported the presence in New Zealand of bacterial films in hot springs at 99°–100°C. It has not yet been possible to obtain these bacteria in pure culture, although this has been achieved with the sulphur-oxidizing autotrophic *Sulfolobus* sp. This species is considered to belong to the Gram-negative, chemolithotrophic bacteria. Its optimal temperature for growth lies between 63° and 80°C, but it still grows at temperatures above 90°C. Being thermophilic organisms these bacteria have thermostable enzymes, as has been demonstrated for aldolase.

In fumaroles from which hot steam emerges, no bacteria could be isolated. Brock concludes from this that liquid water is necessary as a medium.

According to present information, the maximum temperatures at which individuals of various groups of organisms can develop fully are as follows (see also Aragno 1981):

Animals
- Fishes and other aquatic animals   38°C
- Insects   45–50°C
- Ostracoda   49–50°C

Higher plants
- Vascular plants   45°C
- Mosses   50°C

Microorganisms (eukaryotic)
- Protozoa   56°C
- Algae   55–60°C
- Fungi   60–62°C

Prokaryotic microorganisms
- Cyanophyta (autotrophs) *Synechococcus* and the N-binding *Nostocaceae Mastigocladus*   70–73°C
- Bacteria capable of photosynthesis (*Chloroflexus*)   70–73°C
- Chemolithotrophic bacteria (*Sulfolobus*)   90°C
- Heterotrophic bacteria of hot springs and geysers up to   100°C

The only places in the geobiosphere where high temperatures preclude the existence of any living forms are the small areas which are strongly heated by volcanic activity.

The lower temperature limit for active life is determined by the point at which ice formation in the cells commences. Freezing of living cells always starts somewhat below 0°C. Added to this is the fact that living organs can be supercooled to a greater or lesser extent without any formation of ice and concomitant damage.

If the cells of those lower plants which lack vacuoles are dehydrated, that is, there is a great reduction in the hydrature of the protoplasm, then the freezing point is correspondingly low. The lower temperature limit for active life is determined in this case by the lower hydrature limit. The absolute limit for active life is 70% hy (Sect. 5.2.1), which is equivalent to a freezing point of −30°C. It thus becomes comprehensible that Lange (1965) was able to demonstrate slight photosynthesis in dried out lichens even at −24°C, provided they were not covered in ice and exchange of gases was still possible.

The reddish-coloured snow alga *Chlamydomonas (Haematococcus) nivalis* is distributed over the ice and névé areas in polar zones and in the nival belt of high mountains.

In a latent state of anabiosis lower plants and even some frost-hardy higher plants were able to survive freezing in liquid nitrogen, that is, a temperature of −190°C, even −235°C (Tumanov 1979). This is possible only if there is an initial sudden cooling, which causes a "glazing" of the protoplasm, the structure of which thus remains intact, and secondly, step-wise warming up during the thawing, to avoid the formation of ice crystals which are damaging to the protoplasm.

In general there is a decrease in average annual temperatures from the equator towards either pole. In the hot tropical zones the plants are adapted to high temperatures, and the minimum temperature at which damage occurs is very high. For example, a minimum temperature of 21°C is given for *Pilea obtusata* (Urticaceae), *Guarea trichilioides* (Meliaceae) and the mangrove *Avicennia nitida* (Verbenaceae) on Puerto Rico, while the temperature maxima of these plants lie between 45° and 48°C (Altmann and Dittmer 1966).

In tropical montane and subtropical areas the temperature minima are only just above

0°C. For *Psychotria berteriana, Cordia borinquensis* and *Ceropegia peltata* the temperature minimum is said to be +4°C and for the tropical-montane ferns, *Alsophila, Gleichenia* and others, +2°C, while there also the temperature maxima are 45°–48°C. A study of the germination minima for seeds of tropical species from different altitudes is urgently required. Larcher (1980) has examined the frost resistance of palm leaves by exposing them to frost temperatures in enclosed chambers. Damage occurred to feathered palms at −3° to −10°C, to fan palms only at −15°C. It is, however, likely that the minimal temperatures for survival in the open are not the same as these values. We will return to all these questions in special sections in Volumes 2 and 3. In areas with cold winters damage can also occur as a result of the desiccating effect of frost (Walter 1929; Larcher 1982).

The situation is far more complex amongst animal organisms for, besides the cold-blooded or poikilothermal species, in which body temperature is, as with all plants, entirely dependent on that of the environment, there are also warm-blooded or homeothermal species. The latter are able, provided they have sufficiently energy-rich food, to maintain an almost constant body temperature, independently of the external temperature. They therefore remain active even at the pole of coldness in Siberia, at winter temperatures of −60°C, as is the case with reindeer, wolves and others; others pass the winter in a less active state of hibernation, as do bears and marmots. We are not able here, however, to go into the very extensive literature on this subject.

5.2 Hydrature Limits of Living Processes

The water or hydrature factor is ecologically of particular importance for plants, since plants depend on the water that is available in the place where their seeds (or generally their distribution units) fall and germinate. The more mobile animals, by contrast, can visit often distant water holes to drink and can thus maintain their body hydrature in otherwise waterless areas. Where this is not possible, water may be obtained by eating parts of plants which contain much water, while some animals may obtain metabolic water by the respiratory metabolism of carbohydrates or especially fats and in this way cover the low water losses. This plants cannot do, for they have a far greater external surface area and lose a great deal of water through transpiration.

The importance of water balance is more clearly seen from ecological observations in the field than from physiological investigations in the laboratory, because physiological experiments are often performed over brief periods and the plants are usually well supplied with water. It is therefore necessary here to consider the ecological importance of hydrature in more detail.

First, we must distinguish two groups of plants which have fundamentally different types of water economy.

1. *Poikilohydric* plants: the hydrature of these is completely dependent on that of the aerial environment and they must therefore be able to survive complete drying out (lower plants).

2. *Homeohydric* plants, which maintain a certain hydrature of the protoplasts of their cells and are largely independent of the hydrature of their environment (higher plants).

Within the second group there is a special group of land plants which are able to grow on salty soil; these are known as *halophytes*, and we shall return to them in Section 5.2.5.

5.2.1 Poikilohydric Plants

The first living organisms arose in water and the first stages of their evolution occurred in water. The hydrobiosphere is older than the geobiosphere, but we may assume that very early in the earth's history, lower organisms also settled on the surface of the land, especially wherever there was frequent rainfall or where the land surface was periodically flooded. Such organisms would have to be capable of entering a condition of suspended animation (anabiosis) so that they would not die on being dried out and could become active again when wetted once more (vide Sect. 2.2).

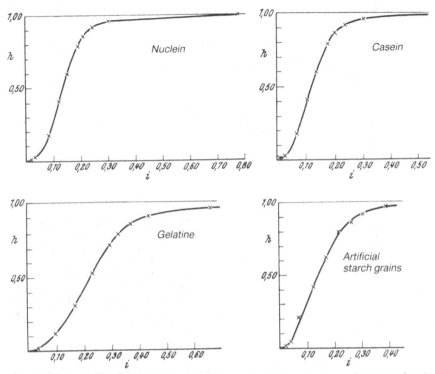

Fig. 39. Typical absorption curves for "nuclein" and other hygroscopic substances showing the relationship between their water content, i (g H_2O g^{-1} dry substance) and relative water vapour pressure, h (equal to the hydrature expressed as a percentage). (Katz 1918)

This ability we still find today in lower plants such as the cyanophytes, bacteria, simple green algae (Chlorophyta) and others with vacuole-free cells, but also in lower fungi and lichens.

In the oldest rocks impressions are found which could have been made by such organisms, but finds of actual fossils cannot be expected because the cells would have been too soft. We can today, however, study analogous extant communities in places which are only periodically moistened and where there are poikilohydric lower plants (aerial algae, bacteria, fungi and lichens), which certainly settled the earth's surface before there were any true land plants (cormophytes). Today these lower plants are dominant only in places where higher plants cannot grow (Friedmann and Galun 1974). Examples of such communities are the "ink stripes" on vertical limestone walls in the Alps, the "window algae" beneath transparent stones and the "takyres", clayey

deposits from rivers which flow periodically down slopes into deserts.

In the extremely dry Transaltai-Gobi desert, where the soil is covered with a stony pavement, humus has been found; closer examination has shown this to be formed by the photosynthetic activities of 40 algal species in the superficial soil horizons (Rachkovskaya 1977). Such communities of soil algae in the whole of the Saharo-Gobi desert area have been very thoroughly investigated by Novichkova-Ivanova (1980).

The hydrature of these poikilohydric species changes with the humidity of the air, that is, the hydrature of the air, just as is the case with hygroscopic substances such as "nuclein", gelatine, starch, casein, etc. In moist air, water is taken up, in dry air it is lost, until in either case an equilibrium is reached. Figure 39 shows the swelling curve for "nuclein" which, as in the case of all such compounds, is sigmoid. Maximum swelling (hydration) is reached in air satu-

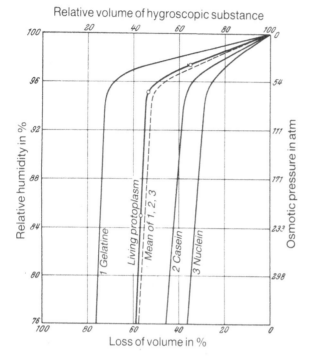

Fig. 40. Curve showing the relationship between the relative humidity (hydrature) and the state of hydration of the protoplasm of the vacuole-free carpospores of *Lemanea*. Comparable curves are shown for the hydration of *1* gelatine; *2* casein; *3* nuclein; *broken line* shows the mean for these three substances. (Based on data from Walter 1923)

By means of volume measurements of non-vacuolate protoplasts of algal cells (carpospores of *Lemanea*), it has been possible to show that living protoplasm behaves exactly like non-living hygroscopic substances (Walter 1923). The swelling curve for these cells corresponded to that for a mixture of gelatine, casein and nuclein (Fig. 40).

By growing poikilohydric species in air of different humidities (hydrature), it is possible to determine the limits of hydrature within which growth and cell division are possible; these data are of course valid, irrespective of the water content of the nutrient surface in which the organisms are cultivated (Walter 1924; Heintzeller 1939; Burcik 1950). Thus with a hydrature of 75%, a humidity at which there is no development of mould, the water content of egg powder is 11.6%, of noodles 13.0%, oats 13.6%, pea or bean meal 15.3%, dried potatoes 15.9%, rice or rusks 16.6%, dried stone mushrooms 19.5%, tobacco 23.4% and of dried vegetables 26.7% (Walter and Kreeb 1970).

The limiting values are specific for the individual species of growing microorganisms; the values (in bar) corresponding to the hydrature values are given on p 88.

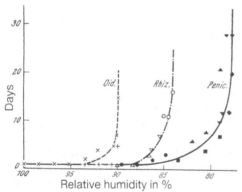

Fig. 41. Dependence of the germination period (in days) of the spores of mould on relative humidity (hydrature) at 20°C. *Oid: Oidium (Oospora) lactis; Rhiz: Rhizopus nigricans; Penic: Penicillium glaucum.* (Heintzeler, from Walter and Kreeb 1970)

rated with water vapour; that is, with a hydrature of 100% hy. When the humidity falls, the greater part of the absorbed water is lost between 100% and 96% hy and then there is a sharp inflexion in the curve; the rest of the absorbed water is lost far more slowly with decreasing hydrature (Walter 1923; Walter and Kreeb 1970; Walter and Wiebe 1966).

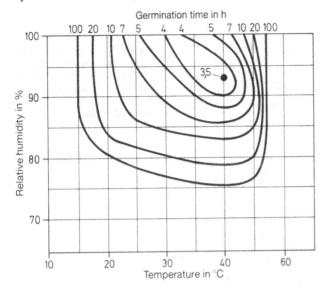

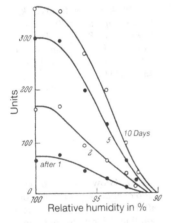

Fig. 42. Dependence of germination period (in h) of *Aspergillus niger* on temperature and hydrature (% R.H.). (Bonner 1948, from Fedorov and Gilmanov 1980)

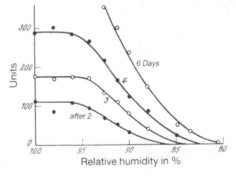

Fig. 44. Curves showing the growth of *Penicillium glaucum* on a solid substrate at different hydratures (R.H.). *Ordinate* diameter of a colony; *1 unit* = 0.038 mm

Fig. 43. Dependence of the growth rate of *Oidium (Oospora) lactis* (on a solid substrate) on hydrature (% R.H.). (Heintzeler 1939, from Walter and Kreeb 1970; also the source of Figs. 44–47)

Knowledge of the effect of hydrature on the growth and development of bacteria and fungi is particularly important, because, as decomposers, these organisms are responsible for the breakdown of organic matter in the litter layer and the soil.

Figure 41 shows the germination curves for several moulds; the minimum hydrature lies between 90% hy for the milk mould *Oidium lactis* and 75% for *Penicillium glaucum*. It depends to some extent on temperature and is always lowest at the optimum growth temperature (Fig. 42).

From the investigations so far made it should appear that *a hydrature of 70% is the absolute limit for all active life.* Snow (1949) did find that the conidia of *Aspergillus glaucus* germinated after 2 years even when kept at 66–62% hy, but this is of no practical importance. The only exception thus far is *Xeromyces bisporus* which forms an almost invisible coating on dried prunes. It is said to germinate at 60.5% hy and to form asexual spores at 66.3% hy [Fortschritte der Botanik (1972) 34:104]. In practical terms, however, this feeble growth is of no significance.

Figures 43 and 44 show the dependence of growth rate on hydrature for two different

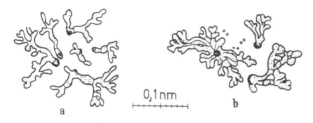

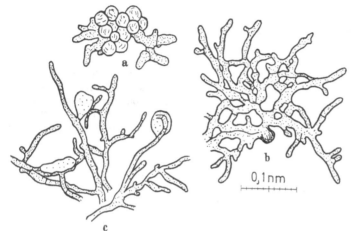

Fig. 45. Stunted growth close to minimum hydrature: **a** *Aspergillus glaucus* after 7 weeks at 72.5% R.H. and 30°C; **b** *Aspergillus niger* after 10 weeks at 79% R.H. and 25°C

Fig. 46. Abnormal forms of *Rhizopus nigricans* at 20°C: **a** after 6 weeks at 84.1% R.H.; **b** at 84.6% R.H.; **c** formation of bladder cells after 5 days at 88.3% R.H.

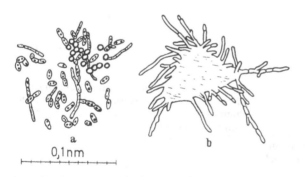

Fig. 47. Growth forms of *Ustilago avenae:* **a** budding colony after 5 days in a 10% sucrose solution; **b** formation of hyphae (recognizable on the edge of the colony) after 10 days in a 20% sucrose solution

species of fungi kept on a solid gelatine medium. This dependence is found also with fluid media of optimal composition for the growth of fungi.

It is frequently observed that growth is slightly less at a hydrature just below 100% hy. The reason for this is that the content of soluble nutrients in the medium is too low. In pure water growth can occur only when there are reserves of nutrient substances available in the cells.

Otherwise the growth rate of fungi which have low hydrature minima shows hardly any decrease as the hydrature falls from 100% to 95%, but below 95% hy growth rate falls rapidly and reaches zero asymptotically as the hydrature minimum is approached.

At low levels of hydrature, growth is not only quantitatively, but also qualitatively affected: xeromorphic stunted forms develop (Fig. 45) and in phycomycetes cross-walls are formed (Fig. 46). The smut fungus *Ustilago avenae* reproduces at a high hydrature by budding, but at low hydrature by forming hyphae (Fig. 47) on both solid and fluid substrates.

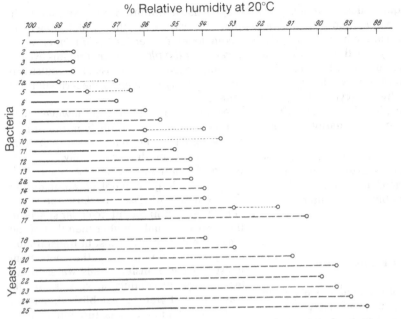

Fig. 48. Hydrature limits to the growth of bacteria and yeasts. ——— Normal or not continuously inhibited growth; − − −○, markedly inhibited growth; ○ - - - - ○, fluctuating limits. The species investigated, with the number of strains examined, in brackets, were: *1, 1a* two hydrature types of *Bac. mycoides* (5); *2, 2a* two hydrature types of *Pseudomonas pyocyanea* (4); *3 Bac. luteus* (2); *4 Bac. asterosporus* (2); *5 Bact. radicicola* (3); *6 Azotomonas insolita* (1); *7 Pseudomonas tumefaciens* (3); *8 Bac. mesentericus* (3); *9 Bact. vulgare* (3); *10 Bact. coli* (3); *11 Bact. subtilis* (2); *12 Bact. prodigiosum* (4); *13 Bact. aerogenes* (4); *14 Mycobact. siliacum* (1); *15 Pseudomonas inigua* (1); *16 yellow air sarcina* (5); *17 Micrococcus roseus* (3); *18 Torula utilis* (1); *19 Schizosaccharomyces jörgensohnii* (1); *20 Willia* sp. (1); *21 Saccharomyces cerevisiae* (3); *22 Zygosaccharomyces polymorphus* (1); *23 Oospora lactis* (2); *24 "rose yeast"* (1); *25 Endomyces vernalis*. (Burcik 1950, from Walter and Kreeb 1970)

Increase in the size of colonies of bacteria and yeasts at different hydratures have been measured (Fig. 48). Most bacteria develop only at a very high hydrature of 100–98% hy (= 0 to − 27 bar); streptomycetes will, however, still grow at 91.5% hy (= − 120 bar) (Jagnow 1957), whereas yeasts are less demanding. In a concentrated sucrose solution (85% hy = − 220 bar), however, they no longer develop.

It can be seen, therefore, that bacterial activity in litter and soil soon ceases after slight drying out, while the fungi continue to be active even when there is no water available to higher plants. This difference is of very great importance in the decomposition of organic substances by fungi, a fact not previously pointed out (vide p 50, 86).

Several practical tips may be added. The most widely distributed moulds are the green moulds *Penicillium glaucum* or *Aspergillus glaucus* because these two species require the lowest humidity (hydrature) for their development. The formation of conidia, which leads to their rapid distribution, occurs only in air of more than 80% humidity. Since cooking salt (NaCl) is deliquescent only at a humidity of 76%, there is no danger of mould forming in rooms in which pure cooking salt remains dry. In the moist tropics it is warm and the air humidity is usually over 80%. As a result, almost everything soon becomes covered with green mould. By slight heating and a rise in temperature of 1–2°C, the air becomes drier and the danger of mould forming is overcome. Herbaria in the humid tropics are for this reason often slightly heated.

It is far more difficult to establish the hydrature limits for aerial algae, which must be cultivated in light, because the warming makes the maintenance of a constant humidity of the air impossible. Two wettable aerial algae which grow in moister places are *Trentepohlia* and *Prasiola*, for which the hydrature minima are 95% hy and 90% hy, respectively; for the non-wettable pleurococcid algae on the bark of trees it is lower, being about 70% hy, while it is particularly high, 99% hy, for cyanophytes (for literature references see Walter and Kreeb 1970). We return to "secondarily" poikilohydric plants on p 98.

We know of no investigations of the hydrature limits to life activities of poikilohydric animal organisms (Protozoa, Tardigrada). It is probably just a little below 100% hy.

5.2.2 Homeohydric Plants

True colonization of the land was achieved only by the higher plants which developed a quite different type of water economy, enabling them to be relatively independent of air humidity, and to maintain the hydrature of their protoplasm even in very dry air. These are the cormophytes, homeohydric plants which die when they dry out.

Basically, the protoplasm of all plants is capable of withstanding desiccation. Obvious examples are the non-vacuolated spores of the pteridophytes and the cells of the seeds of spermatophytes. The protoplasm of non-vacuolated cells is not subject to any mechanical effects on drying out: it shrinks with all its organelles, and the fine structure is hardly damaged at all, or only temporarily.

After germination of the spores or seeds, however, the cells form large vacuoles. If there is a large vacuole in a cell and the protoplasm forms only a thin layer on the inside of the cell wall, then, on desiccation, the vacuole does not decrease in volume as it does in plasmolysis, because it remains trapped as a result of adhesion to the wall; strong tensions develop, causing coagulation and death of the protoplasm.

On the other hand, the vacuole forms an inner, watery medium for the protoplasm, with which it is in hydrature balance as a result of the semi-permeability of the tonoplast. The hydrature of the protoplasm is always determined by the concentration of the cell sap, that is, its osmotic potential, and

not by the hydrature of the external medium. The "vacuome", that is, all the vacuoles taken together, *represents an inner water reserve for the plant,* which it retains even in dry air, because water loss to the outside is very markedly reduced by the formation of a cuticle.[1]

An autotrophic plant can, however, only temporarily close itself off from the atmosphere, for it must normally take up CO_2 from the ambient air during the day, and this occurs through the stomata. A plant loses a relatively large amount of water during this exchange of gases, because the water vapour tension in the intercellular spaces in the leaves is usually higher than that of the atmosphere. The water lost in transpiration is replaced by uptake by the roots from moist soil and transfer of this water to the leaves through conducting vessels. In this way a flow of water is established, leading from the soil into the roots, through the conducting system to the leaves and finally, through the stomata into the atmosphere. This largely physical process has been given much attention by physiologists, particularly since a simple method for measuring water potential by the pressure-bomb method has been available. The water potential gradient determines the direction of the water flow.

For an ecologically oriented biologist who is mainly interested in the adaptation of plants in the course of their development to changing external conditions, the transpiration stream is of lesser importance; *above all, it is the state of hydration of the protoplasm, that is, its hydrature (= water activity). This is determined not by water potential, however, but by the osmotic potential of the cell sap.*

The turgor mechanism of a plant cell is a sort of buffer system which protects the protoplast from daily fluctuations of the hydrature and thus of its state of hydration.

---

1 To some extent, the higher plants on land have remained aquatic plants, because the protoplast is always in contact with the watery medium of the vacuole (Walter 1967c). Thus even in deserts the higher plants are able to maintain a high level of water activity of their protoplasts (usually over 97% hy; that is, at an osmotic potential greater than −40 bar): as a result, complete dormancy never occurs; only growth ceases.

While the water potential of a cell and of the water in the conducting vessels shows marked fluctuations from zero with water saturation up to low values when there is strong transpiration in the afternoon, the hydrature of the protoplasts never reaches 100%, even with complete water saturation of the cells. When transpiration is high, reduction of hydrature of meristem cells also fluctuates relatively little. In buds they are protected from transpiration, while the effect of brief fluctuations of water potential in the conducting vessels is much reduced, because the ends of these vessels are separated from the meristem cells by many layers of still undifferentiated cells. We may assume that the hydrature of the meristem cells, which we cannot measure, is roughly equivalent to the average hydrature of the protoplasts of differentiated leaf cells; this means that they will be affected to the same degree by a prolonged water shortage.

A slight reduction in the state of hydration of the protoplasts affects all organelles to the same extent, because they are all subject to the same turgor pressure and are largely in the same state of hydration; it has an inhibiting effect on metabolism and also *affects differentiation of the meristem cells, both quantitatively and qualitatively.* This results in an altered structure of newly formed organs and thus to adaptation to the changed conditions (Walter 1972).

Cybernetically, this can be seen as a controlled feedback system. The factor to ˙be regulated is water balance; the disturbing factor is dryness, which leads to increased transpiration; the steady state is a steady water balance with the highest possible hydrature of the plasma, while the living protoplasm must be seen as the sensor; the adjustable factor is represented by the meristem cells of the bud tip from which the leaf primordia develop with a particular, predetermined structure.

The sequence of the feedback process: increase in dryness → disturbance of the water balance as a result of increased transpiration → increase in cell sap concentration → decrease in hydrature (shrinkage) of the protoplasm including that of the meristem cells → formation of new organs with xeromorphic structure → fall in transpiration rate, decreased water losses → adjustment of the water balance to a new and somewhat lower level of hydrature.

In this process one link in the chain—the formation of xeromorphic structures—can-not be explained in purely physicochemical terms; it is a biological process, one of qualitatively altered growth.

These very important ecological relationships only become clear when water balance as a whole is considered and not simply its separate aspects, only when the plant is seen as a complete unit, as an organism which is constantly changing and thus to be studied over a long period of time; that is, as a type of behavioural research (Walter 1972).

The differences between the xeromorphic sun leaves and the hygromorphic shade leaves of trees such as the beech are well known. These differences, too, are determined by changes in water availability and not by light conditions; as early as 1912, Nordhausen showed that the structure of these leaves is not determined during development in spring, but during bud formation in the summer of the previous year, when a direct effect of sunlight on the leaf primordia in the buds is not possible; the meristem of the buds on branches exposed to sunlight in summer is subject to greater cohesive tension than that in the buds of the shaded branches. Schröder (1938) was able to induce experimentally the formation of sun leaves on beech branches growing 2 m above the ground in deep shade by reducing the water supply to these branches at the time of bud formation by means of deep cuts into the wood. Further examples with data on quantitative xeromorphic values and cell sap concentration are given by Walter (1960b, pp 225–231).

Isolated cells from sun- and shade leaves show the same differences in CO_2-assimilation with changes in light intensity as do whole leaves (vide Sect. 8.3.1). Thus differences in characteristics of the protoplasts must be involved (Harvey 1980). Sun leaves are more xeromorphic than shade leaves, the osmotic potential of their cells is lower, with the result that the protoplasts, including the chloroplasts, are more dehydrated, and this in turn brings about a change in metabolic reactions.

The adaptation of plants to life on land occurred very gradually. Large vacuoles and a thin layer of protoplasm containing chloroplasts lying just beneath the cell wall represented a competitive advantage even to aquatic algae in comparison with non-vacuolated algal species; this is because cells with large vacuoles will have a greater surface area which can be exposed to light and in this way a greater production is achieved without the greater utilization of nitrogen and phosphorus which would be required for the formation of the equivalent amount of protoplasm. At the same time, the turgor mechanism gives the cell greater firmness. In the more highly evolved green algae, therefore, vacuolated species predominate. The first terrestrial cormophytes can be traced back to such green algae, although the actual ancestors are not known; it might be supposed that they were among the original tropical Chaetophoraceae, for the extant *Fritschiella* puts out lateral branches into the air, and these develop a thick cuticle; this latter is also characteristic of the first land plants.

At any event, the precursors must have been Chlorophyta (probably freshwater forms; vide 2.2), because they have the same chloroplast pigment, the same assimilation product (starch) and the same cell wall substance (cellulose) as the cormophytes, and this is not the case with any other algal group.[2]

The central problem for land plants is the maintenance of *the greatest possible hydrature of the protoplast and its organelles, that is, the lowest possible concentration (or highest possible osmotic potential) of cell sap*, despite the heavy water losses, especially in dry air, which are a consequence of the large CO_2-assimilating surface area. Only then is an optimal level of metabolic activity possible and maximal productivity achieved. *Water balance must always be considered in relation to photosynthesis.*

All the structural adaptations of land plants serve this purpose. In the course of evolution from the pteridophytes to gymnosperms and angiosperms these adaptations have complied ever more closely with this requirement.

In the pteridophytes the sporophyte is homeohydric (more rarely secondarily poi-

kilohydric); the gametophyte, or prothallus is, however, still very poorly adapted to land life and is therefore limited to places which are very moist at least part of the time; water uptake is by short rhizoids. Yet even in the hardy sporophytes, the tracheidal conducting system is not very efficient. In places where there is temporary dryness only species with a reduced leaf area, such as some *Equisetum* species, or secondarily poikilohydric species (see below) can survive.

The tree species of the coal forests, the lepidodendrales and sigillares, too, had a small leaf surface and were limited to moist places with a warm climate.

5.2.3 Secondarily Poikilohydric Plants

The pteridophytes which today inhabit dry areas, or places subject to periodic dryness, including species of such genera as *Ceterach*, *Notholaena*, *Cheilanthes*, *Pellaea*, *Actiniopteris* and some *Selageinella* have all become secondarily poikilohydric (Ziegler and Vieweg 1970). They are found on rocks or as epiphytes even in wet areas; in Zambia, for example, there are more than 15 poikilohydric species (Kornas 1977).

Desiccation tolerance is achieved by reduction in the size of the cells and their vacuoles or, alternatively, by filling up the vacuole with phloroglucin tannins, which set firm on desiccation (Rouschal 1939). Mechanical damage to the protoplasm is also prevented by a certain elasticity of the cells when the leaves curl up (Walter and Bauer 1937); the prothallus of *Ceterach* even has the capacity to recover from complete desiccation (Oppenheimer and Halevy 1962). Amongst the mosses, species which live in wet places, such as *Hookeria splendens*, have very large cells, but those subject to temporary desiccation usually have small, long, narrow cells, which are easily deformed. These, too, are secondarily poikilohydric species, which, in contrast to primarily poikilohydric species, *do not regain their activity in moist air but only after being wetted with water.*

This was recently confirmed in a most impressive way by experiments conducted in Lange's laboratory in 1980. Photosynthesis at 96% air humidity of the secondarily poikilohydric moss

---

2 This applies also to the mosses (Bryophyta), but nothing is known about their evolution or their role in the colonization of the land. They have an archegonium in common with the Pteridophyta, suggesting that they arose at the same time. It is possible that they formed a more important part of the soil covering of vegetation during the pteridophyte period, but there are no fossil remains.

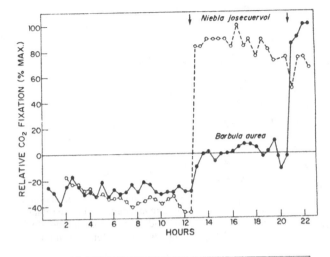

Fig. 49. CO$_2$ turnover of the moss *Barbula aurea* and the lichen *Niebla josecuervoi* at a temperature of 10°C and 100% R.H. The experiment started with air-dry thalli in the dark. *First arrow* indicates the start of illumination (175 μE m^{-2} s^{-1}); *second arrow* the point at which the thalli were wetted with mist. In air saturated with water vapour the lichen swells up; it starts to assimilate CO$_2$ as soon as it is exposed to light; the moss starts assimiliation only after it has been wetted. (Rundel and Lange 1980)

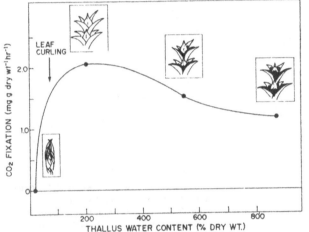

Fig. 50. Dependence of net photosynthesis (CO$_2$ assimilation) in the light at 20°C on water content of *Barbula aurea* shoots. The maximum rate is reached when the shoots are saturated with water and no water is held by capillary action by the leaves. Water is shown *black* at points *3* and *4*. (Rundel and Lange 1980)

Barbula aurea from the Sonora desert was compared with that of the typically poikilohydric lichen *Niebla josecuervoi* from the fog-desert of lower California. The rate of CO$_2$-assimilation of the latter plant was very high, whereas it only commenced in the moss after the plant had been sprayed with water (Fig. 49). Figure 50 shows the dependence of photosynthetic rate in the moss on water content. It is maximal with a water content of 200% (with maximal hydrature but without capillary water, because the latter inhibits CO$_2$ diffusion in the leaves). In fog deserts such as the Namib where there is no precipitation, only lichens are found, while in the rainy climate of Iceland, the pioneers on the volcanic island of Surtsey were mosses (vide p 192).

In the course of the Earth's history, the pteridophytes were replaced by the gymnosperms. Amongst these no secondarily poi-

kilohydric species are known. *Welwitschia*, which can endure long periods of dryness, is a homeohydric desert species (see Vol. 2).

All spermatophytes, that is, gymnosperms and angiosperms, have in their development a poikilohydric stage which makes distribution possible, namely, the embryo in the seed; this, with its non-vacuolated cells, can survive desiccation. It is only when the seed takes up water and swells (that is, at a higher hydrature) that respiration first commences; subsequently, cell division and increase in cell size with the formation of vacuoles occur. This is the process of germination, and with it the capacity to resist desiccation is lost.

The rate of respiration in a swelling seed depends on the degree of swelling, that is,

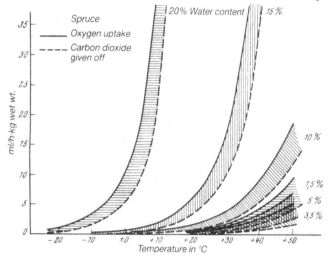

Fig.51. Dependence of oxygen uptake and carbon dioxide release in seed of spruce on temperature and water content. (Schönborn 1965)

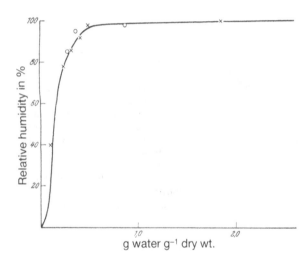

Fig.52. Relationship between water content of moss and relative humidity. X, values from experiments of Mayer and Plantefol 1924; O, values obtained by Walter 1925. The value for 100% R.H. lies outside the figure (cf. Fig.53)

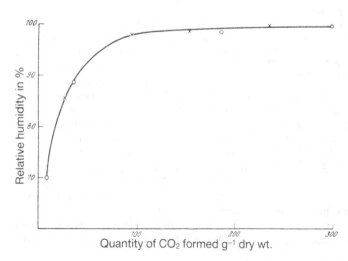

Fig.53. Relationship between respiration and relative humidity in mosses. X, values from experiments by Mayer and Plantefol 1924; O, values obtained by Walter 1925

on the hydrature, but also on the temperature. This is shown in Fig. 51 for spruce seeds. What is striking is that the O_2 uptake is greater than the CO_2 output, so that the respiratory coefficient is less than 1. This shows that the respiratory substrate is the fat stored in the endosperm. At different degrees of swelling or imbibition the level of the hydrature is the same in all the various tissues of a seed.

The rate of respiration seems to rise linearly with increased hydrature, for the curve showing the relationship between respiratory rate and degree of swelling has the same shape as that for the relationship of swelling to hydrature: it rises slowly at first, but then more rapidly (cf. Fig. 52 with Fig. 53). Since it is possible to demonstrate a small quantity of respiratory activity in spruce seeds with a minimal water content of 3.5%, and in barley seeds at only 1.5%, the anabiotic state of seeds cannot be an absolutely latent condition; this explains why seeds do not maintain their capacity to germinate indefinitely.

It should be mentioned that the non-vacuolated pollen grains of the spermatophytes also behave like poikilohydric cells; that is, they can withstand complete drying out, and only germinate after swelling at more than 96–98% hy.

Further details on the behaviour of seeds as well as references to the literature are to be found in Walter and Kreeb (1970, pp 109–116).

Unlike the gymnosperms, the angiosperms include some secondarily poikilohydric species. These grow on flat rocky surfaces where, during drought periods, there is no possibility of water uptake; nor do they store water like succulent species. These include the only European genera of Gesneriaceae, namely, *Ramonda* and *Haberlea* (Iberian peninsula, Balkans); the cyperacean *Afrotrilepis pilosa (Eriospora pilosa)* found on granite summits of the Sierra Leone (West Africa); the most thoroughly investigated species, *Myrothamnus flabellifolia*, found in South Africa, as well as a second species in Madagascar and the scrophulariacean *Chamaegigas intrepidus* (Ziegler and Vieweg 1970), a water plant occurring in small water bodies which dry out periodically—only the small basal

leaves of this plant can tolerate desiccation, not its floating leaves (vide Vol. 2). In addition there are several rock-colonizing Velloziaceae and also, in South Africa, the scrophulariacean *Craterostigma* (two species), the gramineaen *Oropetium capense* and the cyperaceaen *Coleochloa setifera*. According to Gaff (1980) there are 83 species of seed plants which can tolerate desiccation; in some, however, this does not apply to all organs. Monocotyledons become pale when dried out, but turn green again when wetted. The literature on desiccation tolerance has been reviewed by Bewley and Krochko (1982), but they make no distinction between primarily and secondarily poikilohydric plants.

5.2.4 Special Adaptations of Homeohydric Angiosperms to Hydrature Conditions

Today the pteridophytes do not play an important part in the formation of the plant cover. Tree ferns are found only in the fog belt of tropical mountains and in very wet climates in the southern hemisphere.

As early as the Triassic, the pteridophytes were replaced by gymnosperms. In the Jurassic, the ginkgoales were important in the northern hemisphere, but today they are extinct except for one species in China, which also grows well in botanical gardens. The gymnosperms were superior to the pteridophytes in having no prothallus stage, which is badly adapted to life on land. Their vascular system, however, still consists of long tracheids; related to this is the fact that the leaves of conifers usually consist of needles or scales and retain this very xeromorphic appearance, even in moist biotopes.

The plants best adapted to terrestrial life are the angiosperms. In these the prothallus is extremely reduced and enclosed in the embryo sac, while even fertilization takes place within the closed ovary. The vascular system often has vessels with a wide lumen and in moist tropical forests the leaves may be extremely large; in general, leaf surface is adapted to water supply. The angiosperms replaced the gymnosperms except in the cold boreal climate, which is unsuitable for most deciduous trees and in the higher mountain zones of the northern hemisphere where conifers are still dominant in the tree layer of forests. In the most extreme continental climate of the "light taiga" in eastern Siberia, *Larix* grows, but it loses its needles in winter, as it does also along the tree line in the central European Alps.

Conifers are even less important in the southern hemisphere. The Araucariaceae and Podocarpaceae have survived only in small areas where conditions are barely suitable for evergreen angiosperm trees. *Araucaria* is found in the transitional area to the pampas in southern Brazil and occurs together with *Austrocedrus* in high mountain areas of Chile, in Australia and on Norfolk island, in the Pacific east of Brisbane. *Podocarpus* is found in the tropics, usually near the alpine tree line, and various other conifers are found on tropical peat and heath soils.

The angiosperms conquered practically the whole of the geobiosphere. Their ancestral form seems to have been an evergreen tree, a characteristic also of the conifers. A wide variety of life-forms with highly specialized ecological adaptations has evolved, however, and this has occurred independently in different families and genera. The most important aspect is adaptation to particular conditions of hydrature. A cold season has an effect similar to that of drought: during the dry season in the tropics, deciduous trees are leafless, while in the temperate zone this is the case during the cold season. Among the Raunkiaer life-forms, the therophytes, geophytes or chamaephytes may similarly be adaptations both to drought in arid areas and to the cold season in northerly climatic zones.

In the ecological types known as lianes and epiphytes, which have become adapted to the particular light conditions of a tropical rainforest, problems relating to the water supply have also played an important role. In the lianes this finds expression in special features of the vascular system, and in the epiphytes in the structure of aerial roots, water-storage organs, and so on.

Only among secondarily aquatic angiosperms, that is, the water plants, is a high hydrature of the protoplasts (high osmotic potential of the cell sap) always guaranteed. This aquatic way of life may be found in whole families, for example, the Potamo-

getonaceae and the Nymphaeaceae, but also in single genera, such as *Myriophyllum* (Haloragaceae) or *Jussieua* and *Trapa* (Onagraceae). In the genus *Ranunculus* aquatic plants are found only in the section *Batrachium*. There are 45 species of angiosperm, members of the families Zosteraceae and Hydrocharitaceae, which have even returned to life in the sea. They are able to tolerate the full salt concentration of seawater. The number of species which are found in brackish water is even greater. These fresh- and seawater hydrophytes belong, however, to ecosystems of the hydrobiosphere with which we are not concerned.

Very specialized structural adaptations had to be evolved by plants in arid areas, to maintain a high hydrature in spite of an inadequate water supply; as has already been mentioned, adaptation to a secondarily poikilohydric life-form is rare amongst angiosperms. These species *do not, however, have any particular physiological resistance to dryness, as is generally assumed.*

These adaptations evolved independently in different families. The resulting multiplicity of forms is very great, although there are cases of convergence. We will get to know these in detail in Volume 2. Here only the main forms will be discussed, and these are often linked by transitional forms.

A distinction has been made between drought-evading and drought-resistant species, but in arid areas all plants are exposed to drought in that their uptake of water from the soil is made difficult. The critical question is the state in which they exist during the drought period. This occurs, for example:

1. as seeds in the *ephemerals*,
2. as underground storage organs in the *geophytes* (= *ephemeroids*),
3. as whole shoots in an anabiotic or latent state in the *poikilohydric species*,
4. as living shoots in a certain state of dormancy in *xerophytes* which do not store water,
5. in an active state as a result of stored water reserves in the case of *eusucculents*.

Of these five types we need say little about the ephemerals and ephemeroids. Their shoots and flowers develop in arid areas only when the ground has been soaked by good rains, and as a result they show no particular morphological or physiological adaptations to drought. They may even include plants which are annual weeds of humid areas, such as *Erodium cicutarium* which is found in Tucson, Arizona (USA). Other winter annuals here are reminiscent of the spring plants of temperate zones, the summer annuals of the moist tropics. Their development is, however, very dependent on the amount of rain which falls in any particular year. In the most extreme deserts with very infrequent, episodic rain, the only species found are ephemerals, with seeds which can remain dormant for many years. But they cannot evade the drought; that is possible only for the mobile animals which can seek out moist habitats.

The poikilohydric species have already been discussed (Sect. 5.2.3).

The xerophytes, with which we do not include the water-storing succulents, can be divided into three types, according to increasing drought resistance:

a) malacophyllous xerophytes,
b) sclerophyllous xerophytes,
c) stenohydric xerophytes.

a) The *malacophyllous xerophytes,* as the name suggests, have soft and often very hairy leaves. At the start of the dry period, the leaves are relatively large; they have a high rate of transpiration with the result that, as the drought continues, the water balance is disturbed and cell sap concentration rises; in other words, the hydrature falls and newly formed leaves are smaller and still more hairy (more xeromorphic). The larger leaves die and, in a long-lasting drought, so do the smaller leaves, so that only the shoot tips with small leaf primordia remain and water losses are then small. After rain, the plants take up water rapidly, cell sap concentration falls to a normal level, and in the now favourable conditions of hydrature the surviving leaf primordia develop into large, more hygromorphic leaves.

This group includes many herbaceous steppe species (Lamiaceae, Asteraceae), the semi-shrub of the semi-desert (*Artemisia* spp.), *Cistus* spp. of the mediterranean area, and *Encelia farinosa* of the Sonoran desert (vide Vol.2).

Figure 54 shows the seasonal fluctuations in cell sap concentration (potential osmotic pressure) for three malacophyllous species of the mediterranean area. These species are "hydrolabile", that is, the hydrature shows marked fluctuations during a year. The plants produce leaves only in spring, after the winter period, and again in autumn, at the beginning of the rainy season.

b) The *sclerophyllous xerophytes* include both woody plants with leaves hardened by the presence of woody tissue which gives mechanical support, and secondly, broom species, which have green photosynthetic axial organs, sprouts or twigs and cast off their leaves at the onset of drought; these plants remain active during the drought, but by closing their stomata, increasingly reduce transpiration as water uptake from the soil becomes more difficult. They are found in places where the soil always contains available water in the deepest layers to which the roots penetrate. This form is found particularly in zonobiome IV, which has heavy winter rain and long summer droughts, but also to some extent in deserts. These species regulate their water balance by means of the stomata. They are "hydrostable", that is, cell-sap concentration or the hydrature of the protoplasm fluctuates little, as can be seen in Fig.54.

c) The *stenohydric xerophytes* are characterized by an almost constant cell-sap concentration and thus hardly any fluctuations in hydrature. Although they are typical of extreme deserts with infrequent rain, they have not yet been investigated in detail. At the onset of drought, they close their stomata; leaves, if present, turn yellow and photosynthesis ceases. Cuticular transpiration is, as a result of the xeromorphic anatomy of the axial organs, extremely low, so that these plants need take up only minimal amounts of water from the soil and with the available reserves, although often small, these plants, although in a state of starvation, may survive for years. As a result of the reduction in the sugar content of the vacuoles, the cell-sap concentration may even fall a little during the drought. They do not survive complete drying-out of the soil, however.

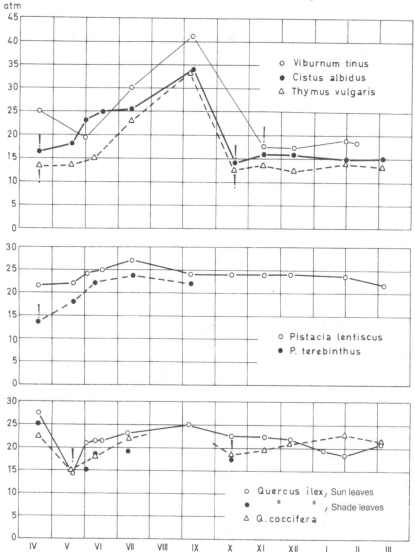

Fig. 54. Curves showing annual fluctuations in potential osmotic pressure ($-$ osm. potential in atm) in sclerophyllous (*centre* and *below*) and malacophyllous species (*above*) (Braun-Blanquet and Walter 1931). *Abscissa:* months. Time of new growth indicated by !; drought from June to September

The *eusucculents* form a special group; they differ from the other three xerophytic groups in their ability to survive for a long time on completely dry soil. When conditions are favourable, they store large amounts of water and then use it very sparingly during the drought period. The root system is very shallow, lying in the upper soil layers which dry out during droughts. The small absorbing roots die, and only robust, non-absorbing roots remain alive. In this way the plant insulates itself from the soil. As soon as sufficient rain falls to wet the upper soil layers, new absorbing roots are formed, often within 24 h, in order to refill the water-storage organs. As a result of the stored water, cell-sap concentration in succulents is very low. Photosynthesis is maintained even during times of drought by means of diurnal acid metabolism (vide Sect. 6.1).

In the possession of these special adaptations, in particular their water-storage capacity, the succulents differ so markedly from the other xerophytes discussed above that they must be regarded as a separate and different group. They also differ markedly from the succulent halophytes which belong to the halophytes (Sect. 5.2.5); these should be distinguished as "halosucculents".

The leaves or the axial organs serve as storage organs for water. Accordingly, a distinction is made between leaf succulents including Crassulaceae, Bromeliaceae, epiphytic orchids, *Agave* and *Aloe* species, and stem succulents, including Cactaceae, succulent *Euphorbia* spp. as well as stapelias among the Ascepiadaceae. There are, furthermore, species in which only the underground organs serve as water reservoirs (Hager 1984).

Finally we must mention the morphologically unusual group of cushion plants. This habit is certainly an adaptation related to water economy, but even more to the effect of wind, since Patagonia, where there are strong winds throughout the year, is characterized by a great wealth of cushion plants; they are found also in many windy high mountain regions.

Such adaptations have enabled the homeohydric angiosperms to colonize the entire geobiosphere, except for places where the soil contains large quantities of readily soluble salts ($NaCl$, Na_2SO_4, $MgSO_4$) and has a very low water activity (hydrature). Examples are the kavirs in Iran, the Tsaidam desert in central Asia, the Tuz-Golü in Anatolia, the Great Salt Desert in Utah, and many salt pans in arid areas. Nevertheless, both the edges of these vegetationless areas and also less extremely saline soil have been settled by another ecological group, the halophytes. We must therefore consider these plants in more detail.

5.2.5 The Halophytes

Areas where the soil contains high concentrations of readily soluble salts support a quite distinct vegetation, usually poorer in species than the neighbouring salt-poor areas; these plants are known as halophytes. A halophytic habitat is commonly defined as an area where there is at least 0.5% $NaCl$ dry weight in the soil. This is not, however, very meaningful; a more significant factor is the concentration of salts in the soil water,

and this can fluctuate considerably with changes in the water content of the soil. A better method of characterizing saline soils is by estimating the quantity of free sodium ions relative to the total free cations. Determination of the electrical conductivity of the soil water can also give a clearer indication of the degree of salinity than a value based on dry weight. If Na^+ ions are in excess of 15% of the total free cations, or if the conductivity of the saturated soil solution exceeds $4 m\Omega^{-1} cm^{-1}$, then the soil can be regarded as a saline soil. The latter value is the conductivity of a 0.2% solution of common salt ($NaCl$) (= $40 mM$ $NaCl$).

There are several angiosperm groups in which halophytes have arisen; among the other cormophytes there are no true halophytes (vide 3.8).

Salt generally has a toxic effect on terrestrial plants. True halophytes, or euhalophytes, actually require a certain salt content in order to thrive. These are highly specialized species or genera, in widely divergent families. We are probably dealing with a very late adaptation of specialized taxa of very particular groups of forms. Halophytic species are especially numerous among the Chenopodiaceae, the Aizoaceae (*Mesembryanthemum* s.l.), the Zygophyllaceae, the Tamaricaceae, the Frankeniaceae, the Plumbaginaceae (*Limonium, Limoniastrum*), among mangroves in the Rhizophoraceae (*Rhizophora, Brugiera, Ceriops*), the Sonneratiaceae (*Sonneratia*), the Verbenaceae (*Avicennia*), the Myrsinaceae (*Aegiceras*), the Combretaceae (*Laguncularia, Lumnitzera*), and others, as well as in some grasses, the Juncaceae and the Cyperaceae. The monocotyledons, however, generally use different mechanisms and adaptations from those of the dicotyledons.

The colonization of saline soils by plants is difficult, since as well as satisfying requirements for their water economy, it is necessary to regulate salt concentration. In order to absorb water from saline solution in the soil with a low osmotic potential, the osmotic potential of the cell sap in the vacuoles must be even lower. The plant achieves this by first absorbing salt along with water until the osmotic forces of the soil water and the cell sap are in equilibrium. Too much salt may not be taken up, however, as it will

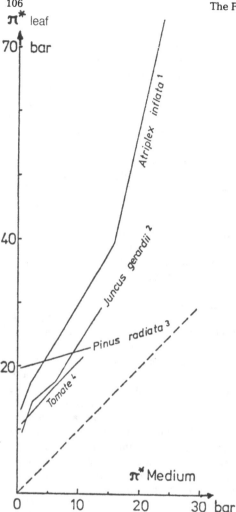

Fig. 55. Osmotic adaptation of some halophytes and non-halophytes. *Ordinate* potential osmotic pressure of cell sap. *Abscissa* potential osmotic pressure of the nutrient medium. *1* Ashby and Beadle 1957; *2* Rozema 1976; *3* Sands and Clarke 1977; *4* Slayter 1961. (All quoted in Albert 1982)

have a toxic effect on the cells. Older roots are generally almost impermeable to salts. Almost pure water is absorbed from the saline soil water, and this is under very high cohesive tension in the conducting vessels (Scholander 1968).

Through uptake of salt, osmotic potentials in the cells of −50 bar or lower are often achieved, in other words, values hardly ever encountered in non-halophytic species, and would, in fact, represent a marked fall in the hydrature. In halophytes, however, the

effect is very favourable, for the chloride ions in the salt solution cause swelling of the proteins in the protoplasts. Chloride ions, at not too high a concentration, thus have a stimulating effect on the development of euhalophytes and bring about a hypertrophy of the cells, which results in the condition of halosucculence. The higher the chloride content of the leaves, the more marked the halosucculence. This process is controlled by hormonal feed-back cycles but the primary effect is elicited by the action of chloride ions on the cytoplasm. This can be demonstrated particularly clearly by the localized action of salt on leaves. Along sea coasts, salt dissolved in fine droplets of seawater falls on the leaves on the windward side of shrubs, while leaves in the wind shadow receive no salt. The seawater penetrates into the leaves through fine cracks in the cuticle and these leaves gradually become halosucculent, while the leaves sheltered from the wind are not affected. In nine species investigated (e.g. *Quercus virginiana, Iva imbricata, Ilex vomitoria*) the salt content of leaves on the windward side was 5–10 times greater than in those in the wind shadow, and the degree of succulence 2–5 times greater (Boyce 1954). Along the sea coast, under the constant action of sea spray, the short stalks of the xeromorphic *Casuarina equisetifolia*, which normally have very reduced, scale-like leaves and are reminiscent of horse tails, assume a succulent appearance and come to resemble *Salicornia* (Schnell 1963). The details of the way in which salt uptake is regulated in the plant as a whole have not yet been completely elucidated. In most cases there is a highly developed osmotic adaptation which maintains a gradient in absorption tension necessary for water uptake (Fig. 55).

It is possible to distinguish different types of plants on the basis of their reaction to a high salt concentration in the soil. The *nonhalophytes* or *halophobes*[3] die on saline soils

---

3 The term glycophyte is often used, although neither the cell sap nor the soil on which the plants grow is "sweet". The term has been derived from the German word Süßwasser, which, while meaning literally sweet water in English, in fact means fresh water. Thus when used internationally, glycophyte is liable to be misunderstood.

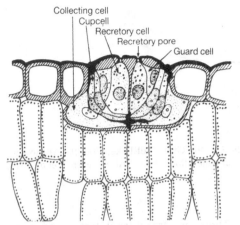

Fig.56. Salt gland in the leaf epidermis of *Limonium gmelinii*, as an example of a very complex salt gland. The cutinized cell walls in black. (Ruhland 1915)

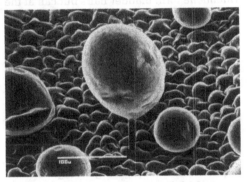

Fig.57. Stereoscan electromicrograph of the upper epidermis of *Atriplex hortensis* showing bladder hairs (photo P Gerstberger; Schirmer and Breckle 1982)

either as a result of water shortage, since the osmotic gradient necessary for water uptake cannot be reestablished, or because the salt taken up has a strongly toxic effect.

Facultative halophytes (also known as pseudo-halophytes) are to a certain extent able to adapt the water uptake system of their roots osmotically, by a secondary salt absorption mechanism. This seems typically to involve retention of salt in the xylem-parenchyma of the roots, as a result of some ion exchange processes, so that the salt concentration of the shoot is kept low. In general, ion exchange processes between the vascular system and the surrounding par-

enchyma probably play a very important role in the development of salt tolerance and the creation of specific ionic conditions. This even finds structural expression in some cases, such as *Puccinellia* (Stelzer and Läuchli 1978), in that the endodermis of the roots is covered by an inner, suberinized, cortical layer with porous cells. This "double endodermis" may be responsible for enhanced potassium and inhibited sodium transport in the stele.

In *euhalophytes* the roots act as an ultra-filter which allows only minimal amounts of salts into the transport vessels. These salts accumulate in the organs of transpiration and give rise to their halosucculence: this is to be seen in the leaf succulence of *Suaeda* and *Salsola* and stem succulence of *Salicornia, Arthrocnemum* and other genera.

The *salt-recreting halophytes* are species which take up somewhat more salt than others and continuously eliminate a large part of this.[4] This is effected by salt glands (Fig.56) or by bladder hairs (Fig.57), and also by discarding whole organs, as, for example, old leaves in which salt has accumulated. The latter phenomenon is also known in many facultative halophytes, in *Juncus* for example, and is seen in premature yellowing of the leaves (Steiner 1934).

Salt glands are found in such halophytes as *Avicennia, Tamarix, Limonium, Frankenia* and *Glaux* and in salt grasses such as *Aeluropus, Distichlis* (Lipschitz and Waisel 1982). Bladder hairs occur in trichohalophytes such as *Atriplex, Obione, Halimione, Chenopodium* and *Salsola sclerantha* (Schirmer and Breckle 1982) and possibly in some Aizoaceae (vide also Osmond et al. 1980).

Salt glands regulate the salt economy in *Glaux* but their effectiveness is limited; with 150 mM NaCl in the nutrient solution, Na^+ recretion was 20 times, Cl^- 5 times that of the salt-free control; at 300 mM NaCl, however, there was only a tenfold increase in Na^+ recretion and a twofold increase in Cl^- recretion (Rozema 1975). Potassium was recreted in traces only. *Glaux* is thus a salt-recreting, facultative halophyte. Its growth is not stimulated by NaCl, but is also not

---

4 "Recretions" are defined by Frey-Wissling (1935) as plant excreta which are eliminated in the same form in which they were taken up (e.g. NaCl).

Normal haloseries of the Great Salt Lake, Utah

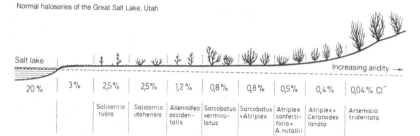

Fig.58. Diagramatic profile of the haloseries on the shore of the Great Salt Lake, Utah (USA) showing Cl⁻ content as percentage of dry soil wt. in the individual vegetational belts. (After Kearney et al. 1914, from Breckle 1976)

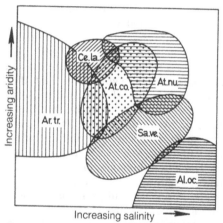

Increasing salinity ➤

Fig.59. Aridity-salinity ecogram for dominant species of the haloseries on the Great Salt Lake, Utah (USA). *Art.tr: Artemisia tridentata; Ce.la: Ceratoides lanata; At.co: Atriplex confertifolia; At.nu: Atriplex nutallii; Sa.ve: Sarcobatus vermiculatus; Al.oc: Allenrolfea occidentalis.* (Breckle 1976)

There are transitional forms and some overlapping between the different halophytic types. Thus there are species which show more than one type of adaptation: for example, some *Atriplex* species and *Halimione* form bladder hairs but can at the same time become succulent.

Apart from the purely osmotic effect, the ion-specific toxic effects on halophytes must be distinguished from one another; these are caused by chloride ions, sodium ions and other salt ions (HCO_3^-, SO_4^-, Mg^{2+}, Ca^{2+}, $H_2BO_3^-$) which can accumulate in large amounts in the soil.

In deserts, saline soils are located mainly in depressions; here desert halophytes are found immediately adjacent to non-halophytic desert xerophytes growing on small areas of higher, salt-free ground. The two types must be carefully distinguished one from another, as applies also to succulent halosucculents and halophobic eusucculents. Unfortunately, this distinction is usually not made, even in ecological investigations, and despite the fact that these are quite different ecological types. Intermediate forms between the two types have, up to now, been found only in the Aizoaceae (Mesembryanthemeae) (see p 109).

Cell-sap concentration in halophytes is always high (low osmotic potential), while in desert xerophytes it is relatively low, and in eusucculents particularly low (high osmotic potential).

Examination of the zonation of different halophytic types along a typical salt gradient near an inland salt lake shows that where the salinity is highest, stem-succulent euhalophytes predominate, to be replaced

inhibited by concentrations up to 150 mM. In the genus *Juncus*, this applies only to *J. gerardi* and for concentrations up to 60 mM NaCl, although both this species and *J. maritimus* can survive at concentrations up to 300 mM NaCl. *Juncus alpino-articulatus* and *Juncus bufonius* are better described as halophobic (Rozema 1976). Succulence has not been observed in any *Juncus* species, nor has an increase in the content of organic acids, such as oxalic acid, been found. In the latter two species the osmotic concentration of the cell sap is over-compensated when they grow on saline soils, but in the former two species it is suitably adjusted.

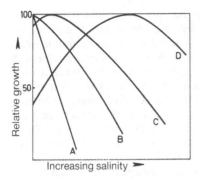

Fig. 60. Schematic representation of the dependence of growth rate on the salt content of the soil of various plants: *A* halophobes; *B* poorly salt-tolerant non-halophytes; *C* facultative halophytes; *D* euhalophytes. (Kreeb 1974)

by leaf succulents further along the gradient. In the next zone, not infrequently, several salt-recreting halophytes are found, with salt glands, bladder hairs, or both. Facultative halophytes also start to appear in this zone, to become dominant further along the gradient and finally form a gradual transition to the salt-free habitats. This applies fairly generally both to zonation near sea coasts and also inland, near extensive areas of salt accumulations, as in Utah, USA (Figs. 58 and 59), in Iran, Afghanistan and Australia.

In addition to the various adaptive types so far mentioned, a series of further criteria have been recognized during the past few decades for the classification of halophytic plants and saline biotopes.

The source of the salt can be important. Some coastal plants, like *Glaucium flavum*, are insensitive to salt dust and sea spray, but are sensitive to salt in the soil. Euhalophytes, on the other hand, are insensitive to soil salt. Aerohaline and hydrohaline salt differ in their subsequent effects (Stocker 1928).

The dependence of growth on salt supply is a further possible criterion for differentiating between types of halophytes (Kreeb 1964, Fig. 60). Chemical characterization on the basis of elements found in the ash is also important (Duvigneaud and Denayer-de Smet 1968), and this is also a source of information on the ecology of mineral metabolism, not only in halophytes. This can be seen, for example, in the fact that

potassium and sodium can be interchanged (Harmer et al. 1953; Weissenböck 1969). On similar types of saline soils different plant species behave quite differently and accumulate in their cell sap their own particular mixture of ions at different concentrations. Especially remarkable in this connection are the physiotypes (Albert and Kinzel 1973), which describe certain peculiarities of mineral metabolism in particular taxa. The Chenopodiaceae, for example, can be characterized as a plant family which has a particular preference for alkaline ions, especially sodium. Only a few species prefer potassium (e.g. *Salsola kali*). There are, however, also species which prefer the anions of strong mineral acids.

Depending on the particular anion in the cell sap (Walter 1968), distinction can be made between chloride- and sulphate-halophytes, as well as the alkali-halophytes in which oxalate predominates. When litter from alkali-halophytes decomposes, Na_2CO_3 is formed, so the soil becomes alkaline. In the Chenopodiaceae all three types are known. Most halophytic forms in this family, however, have large accumulations of Cl^-. In ruderal habitats this is accompanied by an above-average accumulation of nitrate and phosphate. In general, the Chenopodiaceae are characterized by a very high capacity for electrolyte storage compared with the other plant groups which include halophytic forms, such as the Zygophyllaceae, Compositae and Poaceae. These differences are shown in the most varied habitats. The species-specific ionic composition of the cell sap is usually influenced very little by soil factors. Even when the salt content of soils differs by the power of ten, the ionic milieu remains astonishingly constant (Breckle 1976). On the other hand, there are species which reflect the edaphic ionic conditions much more strongly.

The mesembryanthemums must be regarded as a distinct type. They behave as eusucculents, but are really halosucculent euhalophytes. The salt content of their cell sap can be very high. Certain species can live for 2 years without taking up water, but can nevertheless still flower.

The grasses (Poaceae) represent a quite different physiotype. Here NaCl concentrations in the organs seldom even approach

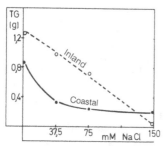

Fig. 61. Comparison of the effect of salinity on growth of an inland and a coastal ecotype of *Juncus bufonius* ssp. *bufonius. TG(g)* dry wt. in g. (Rozema 1978)

the average values found in Chenopodiaceae. A possible exception is *Spinifex hirsutus* on the fore-dunes of the coast of southwestern Australia. This is the only known example of a somewhat succulent grass which is strongly halophytic, but the salt concentration in the succulent leaves has not yet been estimated. The potassium concentration is, however, usually higher than that of sodium. The halophytic Plumbaginaceae, Zygophyllaceae, Tamaricaceae, Brassicaceae and Plantaginaceae are not infrequently sulphate-containing physiotypes.

Albert (1982) has compiled a bibliography of the literature dealing with obligatory and facultative halophytes, and shows how differently these terms have been used up to now. Misunderstandings have arisen because no sharp distinction has been made between the ecological and the physiological behaviour of particular species. Especially in unusual habitats, physiological and ecological optima are often widely different. The reason for this is competition (vide Sect. 7).

The growth of the different developmental stages of a species is always decisive for its competitive strength. Various competitive factors involved have been demonstrated experimentally in artificial mixed cultivations (Freijsen 1971; Barbour 1970). Rozema (1978) describes as an example the growth of two ecotypes of *Juncus bufonius* (Fig. 61). The coastal ecotype is superior to the inland ecotype in dry weight production only when cultivated in conditions of higher salt content, and this is, of course, only a relative superiority.

In those halophytes in which growth is promoted by salt, physiological and ecological optima are closer. In its natural habitat *Salicornia* often has its roots in seawater, but in many experimental cultivations it has shown optimal development at only half this salt concentration. As in all mesophytes and many halophytes, the root system of *Salicornia* can act to a certain degree as an ultrafilter, allowing the uptake of water, but entry of only small quantities of salts.[5]

Identical physiological and ecological optima have been observed in only a very few species. Albert (1982) thus differentiates between the following four groups on the basis of ecological distribution:

1. obligate halophytes,
2. facultative halophytes,
3. habitat-indifferent halophytes,
4. non-halophytes (halophobes).

Finally, we must consider the terms salt tolerance and salt resistance. By salt resistance is understood the ability of a plant to tolerate an excess of salt in its habitat, without any significant impairment of its vital functions. Salt resistance also involves a characteristic of the protoplasm—namely, salt tolerance.

Studies on the cell physiology of salt tolerance made by Repp (1939), involved putting leaf sections in salt solutions of increasing concentration and determining the concentration at which the cells were damaged. Salt-tolerant protoplasts survived for 24 h in 4–8% NaCl, while salt-sensitive cells died, even in solutions of 1.5% NaCl. Such experiments cannot be evaluated ecologically. In a plant the salts are differently distributed

---

5 von Willert (1968) found daily fluctuations of up to 20% in the chloride content of sap extracted from *Salicornia* plants growing on the sea coast but not inundated at high tide; he suggested that this amount of salt was recreted and reabsorbed by the roots each day, since there were apparently no appreciable fluctuations in water content. However, his Figs. 3a and 3b show the water content to be about 92% or 93% of the fresh weight with a daily fluctuation of 2%. A change from 91% to 93% wet weight implies a fluctuation of 20% in dry weight, which corresponds to the fluctuations of 20% in salt content. It is therefore not necessary to assume that there is a daily recretion and reabsorption of salts through the roots.

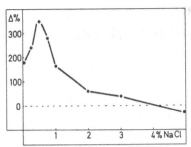

Fig. 62. Relative increase in dry weight in 9 days of *Suaeda maritima* suspension cell cultures in nutrient medium of different salinities. *Ordinate* dry wt. increase in percent. (After v. Hedenström and Breckle 1974)

between different tissues, even between different organelles of a cell; the maintenance of this marked compartmentalization of the salts would seem to be particularly important. The greatest amount of the salts is stored in the vacuole, while the cytoplasm is far poorer in salts. This compartmentalization of salts is effected and maintained by active transport mechanisms, which involve the use of energy, by ionic pumps in the plasma and organelle membranes. The osmotic imbalance which results is usually compensated by an accumulation of organic compounds in the cytoplasm. These are known today as compatible solutes and examples which may be mentioned are proline, glycine, betaine and other quaternary compounds, cyclite, etc. They have now been found in all halophytes and appear to be partly taxa-specific.

According to investigations by Scholander (1966, 1968), in natural conditions there are only small quantities of salt in the water of the transpiration stream in the conducting vessels and in the intermicellar spaces in the cell walls of leaf cells; this is in a state of high cohesive tension even in a water-saturated condition. The protoplast is thus bathed in almost pure water and not in a salt solution as in the experiments of Repp; at the same time, the protoplast is separated by the tonoplasts from the more saline cell sap in the vacuole.

The concept salt resistance must be applied to the plant as a whole; it is less meaningful for individual cells or tissues. Thus cell cultures of *Salicornia* and *Suaeda* (Fig. 62) and those of the salt-sensitive bean *Phaseolus* react almost identically to the addition of salt to the nutrient medium, namely, initially by enhanced growth of the cells. The effect of salt on intact plants of the two types is different (Hedestrom and Breckle 1974). Callus cultures of *Suaeda* and *Atriplex* show at 250 mM NaCl (approximately 1.45%) the same reduction of growth as does *Phaseolus* (Smith and McComb 1981). Salt resistance is thus a characteristic of whole plants. It is comprised of a series of interrelated processes and adaptations, so that in the different halophytic types sometimes one, sometimes another aspect is more important.

In saline habitats, however, the effects not only of sodium and chloride, but also those of other ions, e.g. hydrogen carbonate, sulphate, boron, magnesium and heavy metals, must also be taken into account (Örtli and Kohl 1961; Breckle 1976). The ecological behaviour of halophytes on different saline soils will be considered in more detail both in the discussion of arid areas and in the discussion of the pedobiomes in several of the other zonobiomes.

6 Ecological Aspects of Assimilation and Primary Production in Humid and Arid Areas

6.1 Photosynthesis in C_3 and C_4 Plants and Diurnal Acid Metabolism

Photosynthesis by autotrophic green plants —the producers—is the essential prerequisite for all life on earth and initiates the material cycle and energy flow within every ecosystem. In this process, atmospheric CO_2 is taken up by the chloroplasts and, with the aid of absorbed light energy, is converted to organic compounds, mostly carbohydrates; during this process storage of chemical energy takes place. Intermediate compounds with three carbon atoms may be formed; this is therefore known as C_3-photosynthesis. In C_3 plants fixation of CO_2 and absorption of the light necessary for the synthesis take place in the same chloroplast. It has been found that the situation is different in some tropical plants in which CO_2 fixation takes place in the chloroplasts of the mesophyll, with the formation of malate as an intermediate product. This is then transported to the bundle-sheath cells, in the chloroplasts of which the actual synthesis takes place. Since a tetra-carbon acid is formed as an intermediate compound, this is known as C_4 photosynthesis. From an ecological point of view it is important that in C_3 photosynthesis not all the available CO_2 can be incorporated into the photosynthetic pathway, while in C_4 photosynthesis almost complete assimilation occurs, and this may be of some advantage. This depends on differing affinities to CO_2 of the enzymes involved.

It was formerly believed that C_4 photosynthesis was particularly characteristic of species growing in hot deserts, but as increasing numbers of species have been examined, it has become ever more difficult to associate C_4 photosynthesis with any particular ecological or taxonomic group.

Single genera, such as *Atriplex, Euphorbia, Heliotropium, Kochia* and *Zygophyllum,* include both species with C_3 photosynthesis and others with C_4 photosynthesis.

Further detailed investigations under natural conditions will be necessary to determine whether C_4 photosynthesis in fact confers any significant ecological advantages over C_3 photosynthesis.

It has now been generally established that the high optimum temperature of photosynthesis and the usually more efficient use of CO_2 (even of that produced by photorespiration with almost closed stomata) enhances the growth of C_4 plants in warm and dry areas. In fact, in the USA their proportion within certain vegetation groups shows a clear increase from north to south. This is especially marked in plant communities on saline soils. There are few C_4 plants which are adapted to cooler climates; these include a few small shrubs of the genus *Atriplex,* such as *Atriplex confertifolia* in North America, the grass *Spartina townsendii* on the coasts of Europe and the annuals *Crypsis aculeata* and *Camphorosma annua* which grow east of the Lake of Neusiedl near the Austro-Hungarian border during the short period when the saline lakes dry out. The main area of distribution of C_4 plants is in grasslands (ZB II), where tropical grasses, like Panicoideae, form a large part of the vegetation.

Another ecologically important photosynthetic mechanism, involving a diurnal acid rhythm, is found in succulents. In 1907, Livingstone found that in the Sonoran desert the relative rate of transpiration of cacti is not lower at night but, on the contrary, reaches its maximum during the night. Shreve (1915) provided indirect confirmation of this through the observation that the stem segments of *Opuntia versicolor* make periodic movements: at night they droop, as a result of fall in turgescence and during the day are raised once more. Water balance was found to be negative during the night, when the stomata were opened, and positive during the day, with closed stomata. Uptake of CO_2 must thus occur during the night, while photosynthesis itself occurs during the day under the influence of light. Spoehr had already shown in 1913 that in this same species of *Opuntia* the acid content of the cell sap is highest at sunrise, when 2.45 ml of 0.1 N KOH was required to neutralize 1 ml cell sap, whereas in the afternoon only 0.31 ml KOH was required. It was thus clear that the CO_2 taken up during the night must be stored as an organic acid. The precise details of this "diurnal acid cycle" were worked out later by investigations on succulent Crassulaceae and today it is still known as Crassulacean Acid Metabolism (CAM) (Nuernbergk 1960; Kluge and Ting 1978). When water is plentiful, these succulents behave as typical C_3 plants, but with a diminishing water supply, opening of the stomata and uptake of CO_2 occur increasingly at night. As in C_4 plants, malate is formed, but photosynthesis takes place in the same chloroplasts during the day. There is thus no spatial separation between the site of CO_2 fixation and photosynthesis, the processes are only separated in time. The ecological advantage is clear. With stomata open only at night, the rate of transpiration is greatly reduced, with the result that succulents are able to survive a long period of dryness —large cacti more than a year—using water stored during periods of rain. The diurnal acid metabolism has been observed in all cacti and in succulent epiphytes (Orchidaceae, Bromeliaceae, Cactaceae) (Coutinho 1964, 1965, 1969; Medina 1974; Kluge et al. 1978). Lange and Zuber (1977) have shown that in *Frerea*, a succulent stapeliad from India, C_3 photosynthesis occurs in the normal leaves which are formed during the rainy season, while diurnal acid metabolism takes place in the leafless plants during the dry season. Succulent halophytes do not show a diurnal acid rhythm, with the exception of halophylic Mesembryanthemum subfamily. In these it has been observed when the salt content of the soil increased. (Literature references can be found in the summary of Kluge and Ting 1978.) Since this diurnal acid rhythm is by

no means limited to the Crassulaceae, it should be designated not as CAM but as DAM ("diurnal acid metabolism"). The oldest name for it is "de Saussure-Effect".

Recent laboratory investigations (Lange and Zuber 1980) have shown that DAM is very temperature-dependent; this effect should, however, be more closely investigated in natural conditions. In the Sonoran desert the low night temperatures may thus be of importance (see Vol 2).

6.2 Measurement of Photosynthesis in the Field

Precise field measurements of the rate of photosynthesis of plants growing in their normal habitats was, until recently, very difficult. To measure CO_2 uptake in an assimilating leaf, it has to be enclosed in a transparent container through which air is passed. The transparent container, however, acts as a "heat trap". After only 1 min in direct sunlight, the leaf becomes overheated; this inhibits photosynthesis and increases the rate of respiration, with a resulting increase rather than decrease in the CO_2 content of the air passing over the leaf. Only when measurements are made in relatively low light intensity are meaningful results obtained.

In strong light reliable values can be obtained only by using a complex apparatus with air-conditioned chambers in which both temperature and humidity are automatically adjusted to those of the ambient air. CO_2 measurements are made with an infrared absorption apparatus, Uras or Binos (vide Lange et al. 1969).

The effect of wind, which is important for CO_2 uptake in leaves is, of course, eliminated in an assimilation chamber. The magnitude of the resultant error is difficult to assess.

The rate of photosynthesis is usually expressed as mg CO_2 dm^{-2} h^{-1}, where the surface area of only one side of the leaf is measured. Measuring photosynthesis in this way involves very expensive equipment and requires either a central base or a mobile laboratory. Since it is possible to use such methods only in projects for which adequate financial support is available, the question arises as to whether simpler and less expensive methods will not suffice for large-scale ecological investigations. Great precision is, in fact, a lesser priority in ecological research than the number of measurements made; indeed, because of the high degree of variation in the results obtained, average values of many measurements on different plants are usually more meaningful than very precise measurements on a single plant (vide Sect. 4.2.1). Simple methods have in fact been used for ecological field investigations in areas where the rate of photosynthesis is limited by natural conditions as, for example, the dim light within caves, in forest shade and in the polar regions, or in winter at low temperatures around 0°C. These methods are described here briefly; for more details either the original works should be consulted, or Walter (1960b).

1. Measurement of the light compensation point with the Kauko colorimetric method of CO_2 estimation. This is very simple and can be rapidly repeated on many different objects. It has been used and described in detail by Lieth (1960). (See also pp 116, 121; also fog forests p 218.)

2. The method of Ålvik can be used for estimating extremely low rates of photosynthesis. This is based on the same principle as method (1), and is very sensitive. It was used by Zeller (1951) to discover whether CO_2 assimilation occurs at about 0°C, or whether, at low temperatures, CO_2 is produced as a result of respiration (cf. Sect. 6.3.2). Sieb (unpublished results) used the same method to determine whether the net daily turnover of CO_2 in plants growing in deep forest shade is positive or negative (vide p 123).

3. The increase in dry weight or the increase in the size of plants can be used as an indication of photosynthetic assimilation. If, for example, growth can be demonstrated in lichens in a rainless desert with frequent fog or dewfall at night, this shows that they receive enough water at certain times of the day for a sufficiently high rate of photosynthetic CO_2-assimilation to offset any losses through respiration; as a result, there is the net assimilative gain needed for growth.

It is, of course, always desirable to have exact quantitative data such as those obtained by Lange et al. (1970). Using an air-conditioned chamber and all the associated expensive equipment, they were able to measure photosynthesis of the lichen *Ramalina maciformis* after a night during which dew fell in the Negev desert (Fig. 63). As the lichen imbibes moisture during the night, it respires and gives off CO_2. After sunrise the following morning, photosynthesis starts at a high rate; this soon falls sharply, however, because the lichen dries out. It remains in a state of anabiosis or "suspended animation" for the rest of the day; that is, gaseous exchange comes virtually to a standstill. The following values show the balance over a 24 h period:

CO_2-uptake in the early hours of the morning	$1.32\,\mathrm{mg\ CO_2\ g^{-1}\ d^{-1}}$
CO_2-output during respiration over 24 h	$0.78\,\mathrm{mg\ CO_2\ g^{-1}\ d^{-1}}$
Net CO_2-uptake in 24 h	$0.54\,\mathrm{mg\ CO_2\ g^{-1}\ d^{-1}}$

During nights when there is no dew, air humidity rises to such an extent that some imbibition by the lichens takes place, but net uptake of CO_2 is then only $0.11\,\mathrm{mg\ CO_2\ g^{-1}\ d^{-1}}$. If the total number of dewless nights in a year (about 200) is taken into account, it can be estimated that the increase in dry weight of a lichen should be about 5–10% per annum; in other words, they must grow very slowly: this is well corroborated by direct observations.

6.3 The Dependence of CO_2-Assimilation on External Factors

The most important factors affecting photosynthesis are: light intensity, temperature, wind and the CO_2 concentration of the atmosphere. Here we will consider only those conditions to which C_3 plants are naturally exposed. The initiative for such experimental ecological investigations was taken by Lundegårdh (1921, 1924, 1925, 1957), who made measurements in the nature reserve on the island of Hallands Vaderö in southwest Sweden. Very few investigations have been made of C_4 plants in their natural habitat.

In ecological research CO_2 assimilation is understood to mean *net assimilation*, also sometimes referred to as *apparent assimilation*; this is the difference between the total

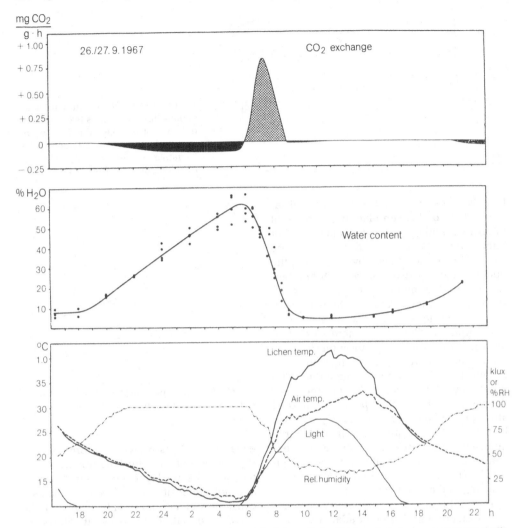

Fig. 63. a Gaseous exchange of CO_2 in *Ramalina* on being wetted with dew; **b** water content of the thalli; **c** environmental conditions in the course of 24 h. (Lange et al. 1970)

quantity of CO_2 taken up from the air and the amount released at the same time in respiration. The latter cannot be very satisfactorily estimated, for while it is possible to measure respiration rate in the dark, this is not the same as the rate of respiration in the light, as can be shown by measuring respiration in the dark both before and after a long period of illumination. Immediately after illumination, the rate of respiration is higher, probably an after-effect of a higher rate of respiration in the light. An exact estimate can be made only of net CO_2 uptake, and it is this which is of importance for production

in a leaf. Since in the field light intensity fluctuates constantly, Lundegårdh expressed illumination as a fraction or a percentage of full daylight.

6.3.1 The Effect of Light Intensity on CO_2 Assimilation

Lundegårdh (1925) represented graphically the relationship between light intensity and the rate of assimilation and obtained a logarithmic curve. If the assimiliation curves of a light- and a shade plant are compared

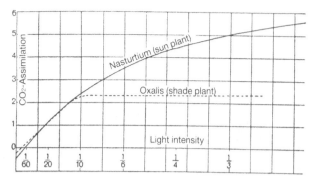

Fig. 64. CO_2 assimilation curves at various light intensities (expressed as fractions of full daylight) for *Nasturtium palustre* and *Oxalis acetosella*. Negative values = CO_2 released. (Lundegårdh 1921, from Walter 1960b)

(Fig. 64), it can be seen that the best utilization of light by the sun plant is at high light intensities, and by the shade plant in weaker light (below $\frac{1}{20}$ of daylight). From these curves it can also be seen by comparing the quantities of CO_2 given off when light intensity is zero that the rate of respiration in the shade plant is lower than in the sun plant. At very low light intensities there is still a net output of CO_2. With increasing light intensity the rate of photosynthesis rises, until a point is reached at which the CO_2 output from respiration is equal to CO_2 uptake by photosynthesis. This is known as the *light compensation point*. At normal atmospheric concentration of CO_2 this will be the point on the graph at which the curve intersects the abscissa; in the above example this is at a light intensity below $\frac{1}{60}$; it was $\frac{1}{70}$ (1.4%) of full daylight for *Nasturtium palustre,* and at 0.7% for *Oxalis*.

The value of the compensation point shows clearly the degree of adaptation of a plant to a greater or lower light intensity. Lieth (1960), for example, found that for 18 shade plants from the deciduous forests of Europe the compensation point during the summer months was always below 250 lx. It rises rapidly as the leaves turn yellow. In plants growing in sunny habitats, on the other hand, this point is reached only between 800 and 2000 lx.

The compensation point is not, however, constant for any one species; it depends on the amount of light falling on the leaves. Figure 65 shows compensation points for sun- and shade leaves of the beach *Fagus silvatica*. The values obtained for sun leaves are always higher. The same observation was made by Lundegårdh (1925) with *Sambucus nigra, Pinus silvestris* and *Picea*

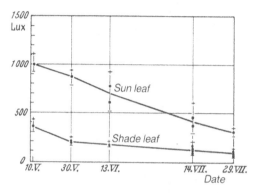

Fig. 65. The light compensation point of leaves of *Fagus silvatica* in summer. (Lieth 1960)

excelsa. In plants growing in forest shade, the compensation point falls rapidly as the foliage of the tree layer becomes more dense and light intensity decreases, and attains a constant summer level only when all the leaves are out. Conversely, it shows an immediate rise in meadow plants after the grass has been mown (Lieth and Vogt 1959). Light is, however, not the only determining factor, as Harder (1923) has shown for mosses and algae. The fall in the compensation point in forest shade plants is paralleled by a decrease in the rate of respiration which appears to be endogenously determined. Photosynthetic capacity remains in this case more or less unaffected (Lieth 1960).

The effect of frost is different (Pavletić and Lieth 1958) as it causes a rapid but temporary rise in the compensation point (Fig. 66).

In tropical greenhouse plants, which "catch cold" even at temperatures above 0°C, the compensation point rises in plants kept in the dark at 3°C, while in controls kept warm and also in the dark it falls. As an

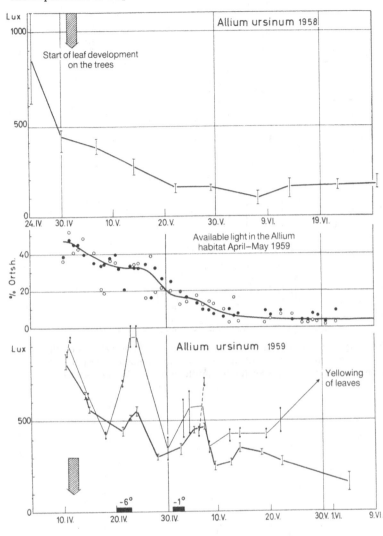

Fig. 66. Change in light compensation point of *Allium ursinum. Above* measurements made in 1958; *below* 1959, in which there were two frost periods. ○ ●, measurements of available light made with two different light meters 6 m apart in a stand of *Allium.* In the *bottom figure* the *heavy line* represents measurements on plants in shady habitats, the *fine line* on those transplanted to the open, where they were more exposed to frost and turned yellow earlier. (Lieth 1960)

example we cite the following experiment with *Gloxinia* (Önal, unpublished results).

Gloxinia leaves	Compensation point in lx: start of experiment 9 May. Sample taken			
	9 May	13 May	17 May	23 May
Plants kept in dark at 3°C	210	280	300	850[a]
Controls at 18°C	210	110	104	90

[a] Leaves already showing brown flecks

Since the compensation point is lower in shade plants than in sun plants, the former can assimilate CO_2 even at low light intensities. The minimum light intensity at which a plant can still just continue to survive lies, of course, above the compensation point. Photosynthesis must provide not only for the simultaneous losses through respiration, but must also cover losses in organic material during respiration at night and in other organs which are not green. A further quota of organic material is required for actual growth.

If the minimal light intensity (L_{min}) at which plants can still grow in forest shade is compared with the magnitude of the com-

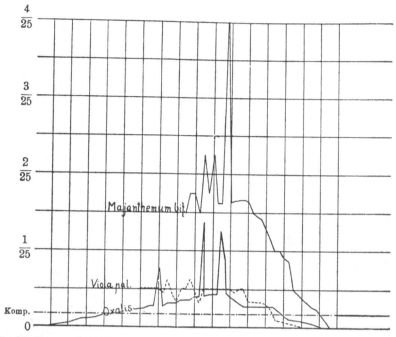

Fig. 67. Changes in the intensity of daylight from before dawn to dusk in three habitats on the forest floor. *Abscissa* time in hours; *ordinate* relative light intensity. *Comp.* compensation point of *Oxalis acetosella*. The sharp maxima are due to light flecks. (Lundegårdh 1921, from Walter 1960b)

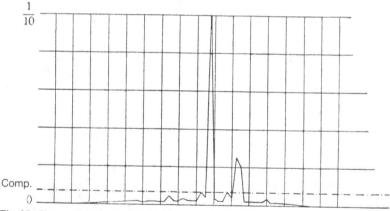

Fig. 68. Curve showing the light intensity in an *Oxalis acetosella* habitat under ferns in the forest. Units as in Fig. 66. (Lundegårdh 1921, from Walter 1960b)

pensation point of such plants, it is found that at L_{min} the plants are in fact on the point of starvation. For *Oxalis acetosella* L_{min} is 1.4% of full daylight; the compensation point is 0.7%. Figures 67 and 68 show the changes in light intensity in shady habitats. If the CO_2 turnover of *Oxalis* plants is calculated with the help of these curves and the assimilation curve shown in Fig. 64, it is found that a net uptake of CO_2 occurs during only 9 h of each day, and reaches high values only for brief periods under the influence of a sun fleck. Nevertheless, with the normal CO_2 content of the air, this would be insufficient to balance the CO_2 lost in respiration during the remaining 15 h; Lundegårdh (1921) in fact estimated that the CO_2 balance would be -0.94 mg CO_2. If, however, the higher CO_2 content of the air in a forest is taken into account, a small positive balance of $0.12\,\mathrm{mg}$ CO_2 is arrived at.

In conditions such as those represented in Fig. 67, the compensation point is surpassed only by the light intensity of the two light flecks. These last for too brief a period to result in a positive CO_2 balance. The plants are able to survive in such habitats only by making use of reserves built up during the spring and autumn. Furthermore, they usually remain sterile.

It is thus clear that plants growing in deep forest shade, including the seedlings of the forest trees, are constantly threatened with starvation. Any slight increase in light intensity, as well as flecks of light moving across the forest floor, will result in increased CO_2 assimilation and this is of critical importance for the continued existence of these plants.

A light intensity of 1% of full daylight is probably the lower limit for the existence of higher green plants. Below this the "dead forest shade" begins, in which only heterotrophic flowering plants and fungi are to be found.

Light conditions in a tropical rain forest do not differ significantly from those in temperate forests. The canopy formed of the upper tree storey is generally subject to greater disturbance and is less closed. More light thus penetrates into the lower storeys of the forest; most of this light is, however, absorbed by the lower tree storey, epiphytes and lianes. A major difference from temperate forests is that in the tropics the herbaceous layer reaches a height of 2m and more. People will move about in and below this layer, and easily gain the impression of a very low light intensity, for this herbaceous layer allows very little light to penetrate to the ground; as a result, the latter is usually covered with a thin layer of dead litter. In the forests of southern Nigeria the light intensity immediately above the ground is about 0.5% to 1% of daylight. In tropical rainforests at higher altitudes, for example, at a height of 2400m on Mount Kilimanjaro, the herbaceous layer is by no means so luxuriant, and here far more light penetrates to the ground. As a result, the forest floor is covered with a brilliant green *Selaginella* carpet, just as the floor of a temperate forest is covered with mosses.

Bünning (1947) found that in the darkest forests on Sumatra the intensity of green light is very much weaker than in temperate forests; only light of longer wavelengths, red and infra-red, penetrate to the forest floor. One cannot therefore speak of "green shade", but rather of a "grey twilight". Here only 0.1% of daylight reaches the forest floor, and there is no vegetation at all. Even mosses form only protonemal threads, although the leafy mosses grow higher up on the buttresses of trees and on dead stems. At light intensities between 0.5% and 0.2% Bünning found, besides mosses, Hymenophyllaceae, ferns, Lycopodiae and Selaginellae, and amongst the flowering plants, begonias, Commelinaceae, Zingiberaceae, Rubiaceae and *Impatiens.*

These tropical shade plants are characterized by a very low rate of respiration (Geiger 1928), which makes it possible for them to survive with very little light. It is for this reason that many are popular house plants as they survive even in the darkest corner, while, because of the existence of water-holding tissue in their leaves, they do not immediately die if watering is forgotten. Examples are leaf begonias, *Tradescantia zebrina, Chlorophytum* and *Aspidistra.*

Our discussion of light minima has been limited thus far to the angiosperms. Lower plants, which consist almost entirely of chlorophyll-containing cells, do not need to synthesize organic matter for the growth and respiration of chlorophyll-free organs; they have, therefore, lower light requirements. This applies to mosses and, to an even greater degree, to their protonemal threads, to simple green algae and to Cyanophyceae.

The low light requirements of algae explain why Cyanophyceae are found 3–5 mm deep in the wet sandy layers of shoal mud (Wadden Sea) under reclamation. In the soil profile a layer of white sand covers the green algal layer, and beneath the latter lies a violet-red layer of purple bacteria while lower still is a layer of blackened sand containing iron sulphide. According to Hoffmann (1949) as much as 6–10% of the daylight reaches the upper algal layer. Endolithic algae, such as *Gloeocapsa* and *Trentepohlia,* and lichens manage to penetrate even the finest cracks in limestone.

Here, at a depth of 4–8 mm, the light intensity cannot be more than 0.1% of daylight. The compensation point for these organisms must thus be very low indeed. Vogel (1955) has described the "window algae" which grow beneath transparent quartz stones in the desert; here the light is adequate and there is a longer humid period after rain or dewfall. Other algae in the upper layers of fog-desert soils are also very important in those particular places where these layers are often wetted by dew.

The magnitude of the light compensation point, that is, the light intensity at which green leaves neither take up nor give off CO_2, depends not only on the rate of photosynthetic assimilation, but also, and to a similar degree, on the rate of respiration. Since at low light intensities the gradient of the assimilation curve is almost the same for sun- and shade-plants, it could even be said that the compensation point is mainly determined by the rate of respiration. This is shown by the following data (estimated by Boysen-Jensen 1949):

Type of plant	Respiratory rate ($mg CO_2$ 50 cm^{-2} h^{-1})	Compensation point (rel. light intensity)	Ratio of compensation to respiration
Sinapis alba	0.87	1.8%	2.1
Ash (sun-leaf)	0.6	1.4%	2.3
Beach (sun-leaf)	0.5	1.0%	2.0
Oats	0.37	0.6%	1.6
Ash (shade-leaf)	0.2	0.4%	2.0
Beach (shade-leaf)	0.1	0.3%	3.0
Marchantia	0.06	0.2%	3.3

It can be seen that in the shade leaves the respiratory rate is significantly lower than in the sun leaves and the compensation point is correspondingly lower. The lower light requirements of shade leaves or plants are thus not due only to a better utilization of the available light.

The factor limiting the rate of photosynthesis at higher light intensities may be the rate of diffusion of CO_2 into the leaves,

for shade leaves have far fewer stomata than do sun leaves.[6] While shade leaves reach maximum assimilation rate at 10% of daylight, the CO_2-assimilation curve for sun leaves becomes more nearly horizontal only at 20–40% daylight, and maximum assimilation rate is reached at 45% of the highest daily light intensity. Overcast days are thus often just as productive for these plants as are sunny ones (Tranquillini 1955).

Gams (1927) investigated the distribution of plants in caves in the Vallais, in Switzerland, and found the following zonation with increasing distance from the mouth of the cave:

flowering plants → ferns → mosses → algae

Lieth (1960) gives the following description of the distribution of plants in caves in the Swabian Alb in southern Germany and near Hallstadt at the foot of the Dachstein mountains in the Alps (Fig. 69).

The light compensation point is very dependant on the state of activity of the plant. When plants are producing new leaves and shoots, the rate of respiration is very high and photosynthesis low, so that the light compensation point is correspondingly high. If, during later development, the respiratory rate falls while photosynthesis increases, the light compensation point falls (Fig. 70).

Spring geophytes start turning yellow when the forest trees come into leaf, at which stage the compensation point rises sharply, because their chlorophyll is broken down and photosynthesis is coming to an end.

The light compensation point thus reflects very accurately the balance which exists between photosynthetic activity (gross assimilation) and respiration.

The capacity for adaptation to different light conditions was investigated in some forest shade plants (Lieth 1960). This is reflected in changes in the magnitude of the light compensation point: before the trees come into leaf this is about 900 lx, but as the leaves come out it falls rapidly to 200 lx, and subsequently remains at this value. This fall in the compensation point is due to a fall in the respiratory rate (Fig. 70); the productive

6 This is contradicted by the findings of Harvey (1980). Vide p 97.

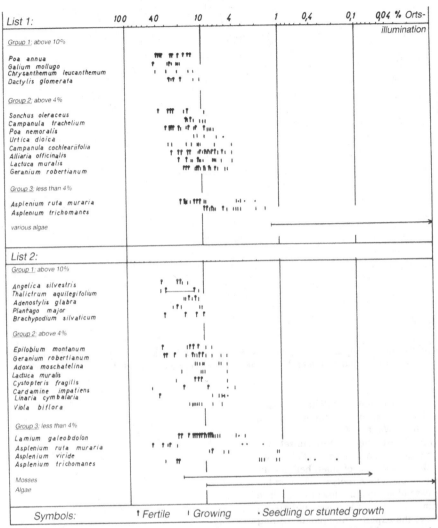

Fig. 69. The occurrence of plant species in relationship to available light. *List 1* from caves in Rosenstein in the Swabian Alb; *list 2* from caves around the Hallstadter Lake in Austria. (Lieth 1960)

capacity of the photosynthetic mechanisms remains, however, fairly constant. If late frosts occur in the spring, the compensation point does rise again for a few days, but this is without the respiratory rate being affected (Fig. 66).

Kaben (1959) investigated photosynthesis in *Lamium galeobdolon* during the course of a whole year. He found that in summer the plants were in a state of starvation. Highest productivity was in the autumn after leaf-fall, while in winter the balance was still positive.

The colorimetric method makes it possible to measure the daily production balance in shade plants on the forest floor, in very unfavourable light conditions.

Principle of the Method. Round-bottomed, 500-ml flasks with a side piece were used; 2–3 ml of a cresol red solution (containing 87 mg NaHCO$_3$, 7.43 g KCl and 10 mg cresol red l^{-1}) are placed in each flask. The open flasks are then left for 1 h in normal air, so that the NaHCO$_3$-solution can come into equilibrium with the CO$_2$ of the air and the colour of the cresol red changes accordingly (pH value, depending on temperature, 7.9–8.1). To read off the values, the solution is transferred to the side

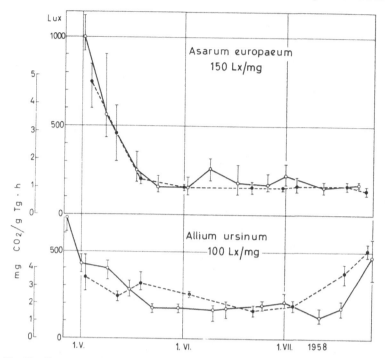

Fig. 70. Changes of light compensation point in lx *(solid line)* and dark respiration *(broken line)* in mg $CO_2 \, g^{-1}$ dry wt. h^{-1} of *Allium ursinum* and *Asarum europaeum* in a deciduous forest. (Lieth 1960)

arm-piece of the flask and its colour compared with standard indicators in test tubes showing different pH values (Walter 1960b).

Making the Measurement. After sunset, the plants to be investigated are placed in the round-bottomed flasks: in the case of small herbaceous plants, the whole plant with roots is placed in the flask; in the case of shrubs, the lower twigs. The flask containing an indicator is closed and placed in the habitat of the experimental plants. During the night the plants respire, the CO_2 content of the flasks rises and the pH value of the indicator solution falls, for example, to 7.5, when the indicator turns yellow. After sunrise, photosynthesis starts, the CO_2 content of the flasks falls and the pH value rises. When the starting point of the previous evening, before the plants were introduced into the flasks, is reached again, the *daily balance point* (DBP) has been reached. That is, the loss of organic material through respiration during the night has been compensated; all subsequent assimilation of CO_2 represents a net gain to the plant. The earlier the DBP is reached, the greater the gain in organic mass for the day as a whole. Under very unfavourable conditions, the DBP is not reached at all; on such dull days the plant suffers losses in organic matter. Such experiments can be conducted only in deep forest shade, since in direct sunlight the flasks become overheated.

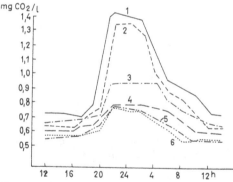

Fig. 71. Daily fluctuations in CO_2 content of the air in a stand of pine (mean for June–July), 70–80 years old, closed canopy 0.7. *1* At the soil surface; *2* 15 cm above it; *3* 1.3 m above it; *4* at the base of the crown; *5* within the crowns; *6* 2–3 m above the crowns. (After Kobak 1964)

This method allows even the smallest turnover to be detected but it cannot be used for large plants, for then the CO_2 content of the flasks would become unnaturally high at night; in natural conditions on the forest floor only a small increase in CO_2 concentration of the air of course occurs (Fig. 71).

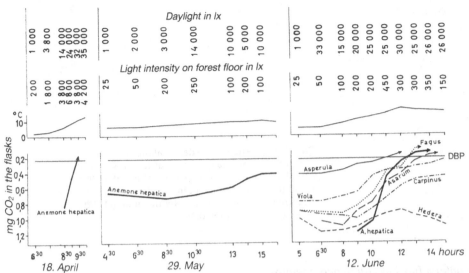

Fig. 72. Experimental study of the daily balance point (DBP) of various forest plants. The DBP is reached when all CO_2 formed during the night has been re-assimilated and the CO_2 concentration of the air in the flasks is the same as it was the previous evening. *Anemone hepatica* showed a high rate of respiration during the night of 18th April, but the DBP was reached as early as 09.30 h. After a period of bad weather, the rate of respiration during the night of 29th May was very low, but the DBP was not reached because of the poor level of illumination. On 12th June, after a long period of good weather, the rate of nocturnal respiration was again very high; the DBP was reached at 12.30 h; this was true also of young *Fagus* plants. On the same day, the DBP in the slowly respiring *Asperula* was reached at 10.40 h, in *Asarum* at 13.30 h, in *Viola* at 15 h, while the balance remained negative in young *Carpinus* plants and in *Hedera*. Temperature measurements were made in glass flasks on the forest floor (cf. Figs. 73, 74). (Walter 1960b)

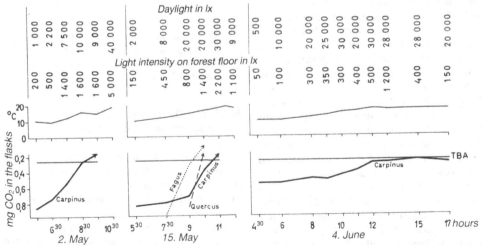

Fig. 73. Measurements of the daily balance point in tree saplings on 3 different days. (Walter 1960b)

Figures 72–74 show some results obtained with this very simple field method (Sieb, unpublished results).

In central European forests the plants of the forest floor adapt to very poor light conditions by greatly reducing the rate of respiration, so that the CO_2 turnover remains very small. There are only two plants, *Hedera helix* and *Fragaria*, in which this does not occur; it is particularly marked

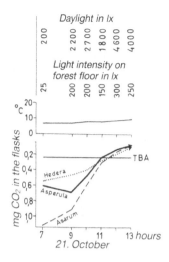

Fig. 74. Estimation of the daily balance point on an overcast autumn day, when leaf fall was almost complete. The DBP was reached at about 11.00 h by all three species, including *Hedera*, which was respiring at a very low rate by this time (cf. 12th June, Fig. 72). (Walter 1960b)

in *Dryopteris filixmas, Anemone nemorosa, A. hepatica, Viola silvatica, Asperula odorata, Asarum europaeum* and *Convallaria majalis* and in the seedlings of the trees *Fagus, Carpinus, Quercus petraea* and *Abies alba*. It has also been found that even a period of bad weather before the trees came into leaf causes a reduction in respiration, the rate of which increases again during a subsequent sunny period (Fig. 72, *Anemone hepatica*). The impression is thus gained that reduction in respiratory rate is a direct response to a state of starvation and thus to a low level of carbohydrate reserves. Figures 73 and 74 show some examples from the large number of measurements made. We wish to draw attention to the following conclusions.

Productivity of forest floor plants is not determined primarily by relative light intensity, but by the total amount of light reaching the forest floor in the course of a day. Thus on a very overcast day, even before all the leaves are out on the trees, the daily balance (that is, reassimilation of CO_2 produced at night) cannot be met by photosynthesis. Similarly, once the trees are in full leaf, those plants which are adapted to very low light intensities will achieve a compensation of the daily balance only on bright days, not on rainy days. In *Hedera* and *Fragaria* in which there is no reduction of respiratory rate, compensation is rarely achieved in summer. This becomes possible only in the autumn, after leaf-fall commences (Fig. 74).

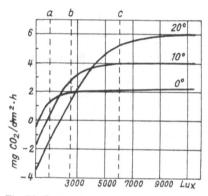

Fig. 75. Dependence of net assimilation on light intensity and temperature in air of normal CO_2 content. For further details see text. (Müller 1928, from Walter 1960b)

6.3.2 The Dependence of Net Assimilation on Temperature

Since net CO_2-assimilation is the difference between gross photosynthesis and respiration, and since the rate of respiration rises more rapidly with increasing temperature than does photosynthesis, the dependence of net assimilation on temperature is very complicated. The quotient Q_{10} is about 2.5 for respiration, but for photosynthesis only about 1.2.[7] From Fig. 75 it can be seen that at very low light intensities net assimilation is higher at 0°C (a) than it is at 10°C or 20°C; at moderate light intensities net assimilation is

---

[7] Q_{10} = increase in rate of reaction for a 10°C rise in temperature.

highest at 10°C (b) and lowest at 20°C; at relatively high light intensities it is highest at 20°C and lowest at 0°C. This is ecologically important because low temperatures are generally associated with low light intensity, both during the colder seasons of the year and also in the morning and evening. In polar regions this latter applies also to the light polar nights. Polar plants can thus use these nights for assimilation even when, in July and August, the temperature falls to below 5°C. This continuous productivity over 24h is important to the plants of this region, since the period of vegetational growth is only 2–3 months long; the plants develop very rapidly in this time. In 1950, for example, it was observed that at the end of July *Epilobium angustifolium* was still at the bud stage in central Sweden, while at the polar tree line it had already flowered.

Stålfelt (1938) showed that, for mosses, the ratio of net assimilation to dark respiration fell with rising temperature, and was most favourable at 0°C. If net productivity is calculated taking into account normal day length in summer and in winter, it is found that the largest gains in summer occur at high temperatures, while in winter they are achieved at low temperatures. In other words, the CO_2 metabolism of mosses is so adapted that at all seasons of the year the optimal temperature approximates to the temperatures which actually prevail. Of course, CO_2 assimilation is affected also by hydrature conditions, which in summer are very unfavourable for mosses. Drying out not only inhibits photosynthesis, but long periods of dryness also inactivate the assimilation mechanisms. The longer such conditions continue, the longer it takes for normal CO_2 assimilation to be resumed once the level of hydrature has become more favourable. Mosses thus show their greatest productivity in spring and autumn (Romose 1940).

A very similar situation is found in lichens, even though these are symbiotic organisms. Temperature adaptation to changing seasons of the year takes place in a similar manner. Productivity is no higher in these plants in summer than it is in winter (Stålfelt 1938).

It is thus clear that photosynthesis is still possible at temperatures at which growth

has practically ceased. In these circumstances the assimilates are not used as building materials but are stored, partly as sugar (Andersson 1944). It is possible that the marked formation of anthocyanins at low temperatures in spring is related to this (anthocyanins are glycosides).

This raises the question as to the lower temperature limit to photosynthesis by plants which remain green during the winter; that is, whether on cold but sunny winter days, they are able to use the light for CO_2-assimilation.

Zeller (1951) investigated this problem during the cold winter of 1946/47 and the milder winter 1947/48, using the method of Ålvik. She examined herbaceous evergreen species such as winter grain, winter spinach (*Spinacea oleracea*) and corn salad *(Valerianella olitoria)*], spruce and the less hardy evergreen *Prunus laurocerasus*. In the winter of 1946/47 there were three cold periods at Hohenheim (near Stuttgart) with a series of freezing days with minimum temperatures down to −20°C. These occurred in mid-December, the beginning of January and from the end of January to the beginning of March. We will first describe the behaviour of winter wheat (Fig. 76).

The wheat germinated on 20.10.1946 and passed the whole winter under natural conditions. On only two of the days on which experiments were conducted was there no gaseous exchange at all (28.1.47 and 31.1.47). On 20.12.46 and 17.2.47 there was no respiration at night, but a small assimilation during the day could be measured. Night respiration was quite considerable up to the first period of cold; thereafter, it remained very low for the rest of the winter. During some very cold nights it ceased altogether, but began to rise again slowly in March. The rate of assimilation was high up to the middle of December, particularly on 30.11.; thereafter it fell, but remained higher than the rate of respiration and in March, with rising temperature, it rose very rapidly. The time it took for the daily balance point (DBP) to be reached, calculating from sunrise, depended on the rate of respiration and that of photosynthesis. The DBP was reached latest on cold days when, during the long night, there had been respiration, although at a low rate, and less intense assimilation during the short day. Nevertheless, over an entire winter there was no negative daily balance, as can be seen clearly in the graph showing percentage increase in dry weight (Fig. 76). There was a sharp increase in

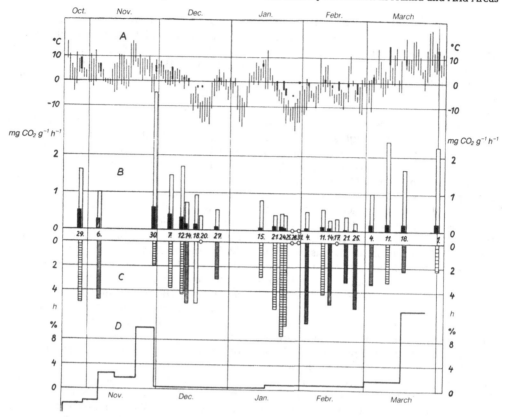

Fig. 76. Assimilation and respiration of winter wheat during the cold winter of 1946/47. **A** Temperature fluctuations: fine lines, in the open; *heavy lines* in the flasks. **B** Rate of assimilation *(white bars)* and rate of respiration *(black bars)* (the numbers are approximations only). **C** Daily compensation time (hours after sunrise). The level of illumination during the assimilation period up to the daily balance point (DBP) is indicated by the *density* of the horizontal striations; the *closer* the striations, the brighter the light. **D** Percentage daily increase in dry wt. (material gain). Up to early November this was still negative in most seedlings which drew nutrients from the endosperm. (Zeller 1951, from Walter 1960b; likewise Figs. 77–80)

productivity in March after the end of the cold period, and the DBP was reached 2 h after sunrise. Light intensity played a very minor role. Between 24.2.47 and 25.2.47 the ground was covered with snow. The experimental flasks, too, had a light covering of snow, but there was no noticeable inhibition of CO_2 assimilation.

Winter wheat generally has no dormant period during the winter. Only after several very cold days may a certain degree of inactivation occur. Thus the plants showed no sign of assimilation on 31.1.47, although the temperature in the experimental flasks rose to 0°C. If wheat plants are transferred in winter to a warm room, photosynthesis and particularly respiration both increase rapidly, but they never reach the summer

rates; this is probably due to the plants being in a winter-hardened state. The minimal temperature for assimilation was found to be −3°C, and that for respiration −7°C. The fact that while on some days CO_2 assimilation could be registered there was no respiration can be attributed to the far lower temperatures prevailing at night. This, too, accounts for the almost continuously positive productivity balance in the winter months.

In the winter of 1947/48 there was but a single short period of cold from the 17.2.48 to the 26.2.48. The rates of respiration and assimilation were thus higher during this winter than in the previous year. There was no day on which gaseous exchange ceased and material production was much greater than in the previous year. While the average daily increase in dry weight for December 1946 and the first half of January 1947 was only 0.4%, it was 2% for the same period in 1947/48.

Conditions for winter barley were found to be similar in all details to those for winter wheat. Those for winter spinach *Spinacia oleracea* and corn salad *Valerianella olitoria* were not very different. The relationship of CO_2 assimilation to respiration was somewhat more favourable and the time taken to reach the daily balance point was thus a little shorter. These two species are, however, more sensitive to frost than wheat and barley, and lose some of their leaves.

The spruce, *Picea excelsa*, reacts in very different manner from these herbaceous winter greens. This is illustrated in Fig. 77: the large number of days on which there was no gaseous exchange (marked with a circle) is immediately striking. On December 18, 1946 respiration could still be recorded at night (T = −6° to −12°C), as could assimilation during the day (T = −2° to

−6°C), albeit at extremely low rates. The minimal temperatures for respiration and assimilation are, as in wheat, around −7°C and −3°C respectively. During the long frost period from 22.1.47 to 3.3.47, gaseous exchange ceases altogether, although sometimes the temperature rose to freezing point or above, as on 21.2.47. The spruce has thus a definite period of winter dormancy, even if this is not an absolute condition (cf. 27.12.46, 15.1.47, 21.1.47, 11.2.47, and 14.2.47). Normal gaseous exchange was not reestablished even in March, although the temperature had already risen. In the milder winter of 1947/48 no such winter dormancy was observed (Fig. 78). Respiration ceased only on frosty days (18.1.48, 18.2.48, and 20.2.48) and on 29.2. On one day (19.2.48), no assimilation was recorded, although the temperature was only 0°C. After this mild winter, normal autumn values were reached again as early as 6.3.48. Winter dormancy in the spruce is thus entirely dependent on temperature and may well be a state of absolute dormancy in its natural area of distribution. On 25.2.1947, when outside there was no

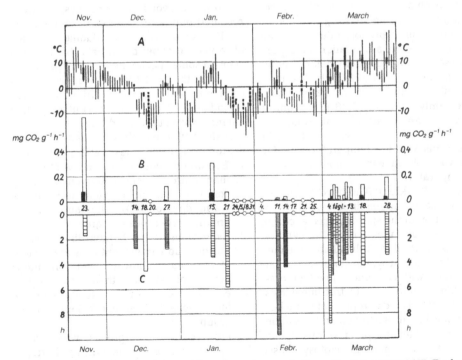

Fig. 77. Assimilation and respiration of branches of spruce in the cold winter of 1946/47. Explanation as in Fig. 76

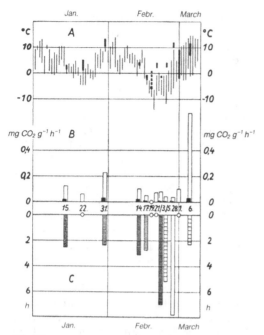

Fig. 78. Assimilation and respiration of branches of spruce in the mild winter 1947/48. Explanation as in Fig. 76

gaseous exchange, branches of spruce, brought into a warm room, showed an enormous increase in respiratory rate during the night, and the rate of assimilation during the day was so high that the daily balance point had been reached before 14.00 h. The ratio of assimilation to dark respiration was, however, only 4.6 on 23.2.48 in a room at 12–16°C, whereas in the open, at a temperature of −1.5° to −3°C, it was 27. Thus, in relative terms, very low temperatures are far more favourable in winter months than is room temperature.

Pisek and Winkler (1958) have examined in greater detail CO_2-assimilation capacity and respiration in the spruce *(Picea excelsa)* from different altitudes and in the cembra pine *(Pinus cembra)* from the alpine treeline. They confirmed that winter dormancy is far more marked with continuous low temperatures at high altitudes. At the tree line it lasts for 4–5 months. The chlorophyll of branches in sunlight is in part destroyed and has to be regenerated when the plant warms up. In intermittent warm periods photosynthesis did not reach summer levels, and the rate of respiration also remained low. Only when

winter dormancy comes to an end in spring do the assimilation mechanisms once more function normally. Annual productivity at the tree line is relatively low; for, even in summer, photosynthetic capacity is lower than at lower altitudes, partly as a result of a higher rate of respiration.

The cherry laurel *Prunus laurocerasus*, is indigenous to areas with very mild winters and often suffers frost damage in central Europe. It belongs to a third type of plant, the evergreen angiosperms. During the cold period from January to March 1947, gaseous exchange in this plant ceased altogether, until finally, a negative balance was recorded, since respiration continued through the night, but there was no photosynthesis during the day (Fig. 79). On a few occasions CO_2 output was observed during the day. The most probable explanation is that under the influence of light, respiration was resumed and exceeded gross photosynthesis. Branches brought indoors also showed a rapid increase in respiratory rate. In some cases the daily balance point was not reached. A net CO_2-output could also be recorded at low light intensity.

Normal gaseous exchange was, however, resumed only after the rise in temperature on 18.3.47 (Fig. 79). Frost damage was not observed on experimental shrubs, but was seen on some specimens in sunny positions.

During the mild 1947/48 winter, gaseous exchange only ceased during the cold period from 18.2.48–25.2.48, but a negative balance was not recorded (Fig. 80).

A distinction can thus be made between the following three types of evergreen plant:
1. Evergreens which have no winter dormancy, are entirely passive on extremely cold days, but immediately utilize every warm day for CO_2-assimilation. These are the herbaceous evergreens.
2. Evergreens which behave like group 1 during mild winters, but pass into a state of dormancy following long continuous periods of cold, and then show no gaseous exchange, even at temperatures a little above 0°C. Spruce belongs to this group.
3. Frost-sensitive evergreens, which show no gaseous exchange during short periods of cold, but suffer material losses during longer cold periods, either as

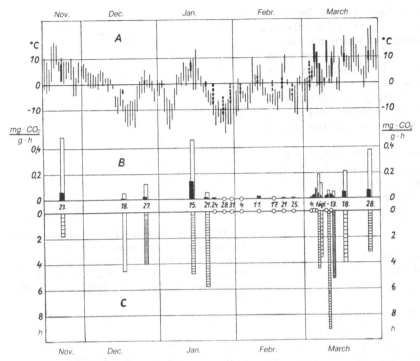

Fig. 79. Assimilation and respiration of the cherry laurel *(Prunus laurocerasus)* during the cold winter 1946/47 with, in part, a negative balance in organic material. Explanation as in Fig. 76

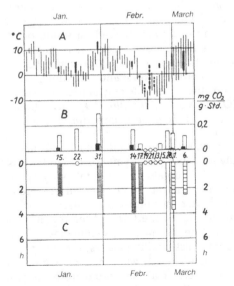

Fig. 80. Respiration and assimilation of cherry laurel in the mild winter 1947/48. Explanation as in Fig. 76

a result of night respiration which is not compensated for during the day, or through CO₂-output in the light. Examples are species like *Prunus laurocerasus.*

The exotic plants cultivated in places with a climate like that of central Europe belong either to 2 or 3, according to their degree of winter hardiness.

The absolute temperature minimum for CO₂-assimilation appears to be about −2° to −3°C, and for respiration at −6° to −7°C. Older observations, according to which photosynthesis is still possible at −20°C, even −40°C, must be attributed to methodological inaccuracies. Pisek and Rehner (1958) made a broad investigation of temperature minima by determining the dependence of the compensation point on temperature in atmospheres of normal CO₂ content and at 3300 lx. They, too, obtained values between −2° and −4°C. Only the mediterranean genera *Arbutus, Olea* and

Laurus showed a CO_2-uptake at temperatures as low as $-6°C$, and this is only just above the temperature at which these plants freeze and suffer damage as a result. Frost-resistant genera, such as *Picea* and *Pinus*, cease assimilating at lower temperatures, but they survive freezing and low temperatures without damage, because they reach a stage of "hardening" before the frost period begins.

Apart from measurements of the compensation point in montane spruce forests, few comparable investigations have been undertaken in the high mountain ranges of the tropics (cf. Vol 2, 8.3).

6.3.3 The Effect of CO_2 Content of the Air and of Wind on Net Photosynthesis

Not only do light and temperature affect photosynthesis, but the CO_2 concentration in the atmosphere also has an important influence on CO_2-assimilation. As is well known, the mean CO_2 content of air is 0.03 vol%, or 0.57 mg l^{-1}. As Lundegårdh (1924) emphasized, this is not, however, as constant as was previously assumed. We must therefore examine the effect of CO_2 concentration on photosynthesis.

Figure 81 shows the relationship between rate of assimilation and the CO_2 content of the air at a light intensity equivalent to daylight. It can be seen that at normal CO_2 content of the air of 0.03% the rate of assimila-

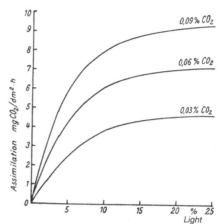

Fig. 82. Dependence of the CO_2-assimilation of the shade-plant *Oxalis acetosella* on light (as % of maximal daylight) at three different CO_2 contents of the air. The curves should really start a little below zero. (Lundegårdh 1921, from Walter 1960b)

tion is relatively low, but that it rises rapidly with increasing CO_2 concentration.

At lower light intensities also, such as those to which shade plants are exposed, increase in the CO_2 content of the air always results in an increased rate of assimilation (Fig. 82); this can be observed at extremely low light intensities and also at 10% of maximum daylight, which is equal to the intensity often measured in moving flecks of light on the forest floor. Indeed, in shade plants the maximum assimilation rate is reached at 10% light intensity, if the CO_2 content of the air is normal; exploitation of higher light intensities is possible only at higher levels of CO_2. *This shows that the CO_2 concentration of the air is a very important ecological factor.*

There are other factors which also markedly influence photosynthesis by land plants. Since the uptake of CO_2 takes place by diffusion, the rate of uptake will be greater the greater the diffusion gradient. In completely still air the CO_2 concentration must fall in the layers of air immediately surrounding assimilating leaves. A light breeze (less than 2 m s^{-1}) will cause movement of these layers of air and bring fresh air of a normal CO_2 content into contact with the leaves, substantially increasing the uptake of CO_2 (Deneke 1931).

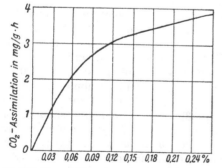

Fig. 81. Dependence of CO_2-assimilation by pine needles on the CO_2-content of the air (in vol.%). (Stålfelt, from Walter 1960b)

The uptake of CO_2 by a leaf with un-altered stomatal aperture is a purely physical process. The physical characteristics of transpiration can be measured by substituting wet, green filter paper for a leaf and the rate of evaporation from the paper can be measured as in a standard Piche evaporimeter. Environmental factors affecting the physical aspects of CO_2 uptake can be assessed in a similar way by substituting for the leaf a dish containing a CO_2-absorbing potassium hydroxide solution. The assumption is that CO_2 absorption by the potash solution will be as dependent on air movements as Deneke (1931) found assimilating leaves to be. Shallow, open dishes filled with KOH were found to be unsuitable for this purpose (Walter 1952). Any effect of the wind is too small to measure and the procedure is difficult to use in the field. Instead, 200 cm³ bottles, 11.8 cm high, with an opening of 3.5 cm diameter and containing 10 ml of a 0.5 N KOH solution were found to be very satisfactory (Fig. 83). This may be seen in the following comparisons.

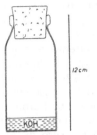

Fig. 83. Standard flask for absorption of CO_2 from the air. (Walter 1960b)

K_2CO_3 formed converted to $KHCO_3$. When the solution has just lost its colour, methyl orange is added as indicator and the titration continued until the colour changes to pink. Twice the number of cm³ of 0.1 N HCl used in the titration with methyl orange is equivalent to the total quantity of K_2CO_3 present in the flask. This number multiplied by 2.2 gives the quantity of CO_2 in milligrams absorbed by the KOH solution during the experiment. A control, with a closed flask, should also be made, since the potash solution usually contains some potassium carbonate.

	Wind speed		
	No wind	1 m s⁻¹	3 m s⁻¹
Relative rate of CO₂ uptake			
Standard flask	100	277	315
Open dish	100	137	135

A wind speed of 2–3 m s⁻¹ results in a three-fold increase in the rate of CO_2-uptake by an assimilating leaf. A similar increase is found using the standard bottles, but with the open dishes maximum absorption is reached by 1 m s⁻¹ and the increase is only 30–40%. The standard bottles have the added advantage that they can be filled in the laboratory, closed with a cork, transported to the experimental site, opened there and, at the end of the experiment, closed again and returned to the laboratory. The amount of CO_2 absorbed can be estimated by titration in the flask itself. For this purpose, either 0.1 N or 0.25 N HCl can be used, initially with phenol-phthaline as indicator: in this procedure excess KOH is neutralized and the

Example: 10 cm³ 0.5 N KOH is equivalent to 50 cm³ 0.1 N HCl

Quantity used for titration with phenol-phthaline	39.62 cm³ 0.1 N HCl
Quantity used for titration with methyl orange	10.36 cm³ 0.1 N HCl
Less control with methyl orange	0.86 cm³ 0.1 N HCl
Difference (methyl orange)	9.5 cm³ 0.1 N HCl

CO_2 absorbed: $9.5 \times 2 = 19$ cm³ 0.1 N H_2CO_3
$= 19 \times 2.2 = 41.8$ mg CO_2.

While the degree of accuracy of the titration is very high, problems in setting up the experiment in the field lead to a certain scatter in the results. Using ten bottles, the following average values were obtained:

On a balcony	24.89 ± 0.22 mg CO₂
On open ground	19.27 ± 0.35 mg CO₂

It is advisable to use not less than three bottles in one place, but five is preferable.

The degree of sensitivity of this method is shown by the observation that, over 3-h periods, the quantity of CO_2 absorbed in an empty room was 4.6 mg CO_2, while in a full lecture hall it was 43.4 mg CO_2.

Several absorption measurements made in different plant stands and at different heights above the ground may be described here as an example (Walter and Zimmermann 1952). It must constantly be borne in mind that CO_2 absorption depends not only on the CO_2 concentration but also on air movement. Determination only of the CO_2 content of the air therefore does not provide a complete picture of the conditions affecting CO_2 uptake by assimilating leaves.

Absorption measurements made in the open, 8 m above the ground, between November and October 1951, while showing marked fluctuations from day to day, did not display any seasonal pattern. Particularly high values were admittedly obtained in March and April, but these can be attributed to the rise in temperature at that time. Furthermore, the daily fluctuations showed no clear relationship to particular weather conditions, although at times there was a close correlation with wind strength and at others with temperature and light intensity. The rate of CO_2 uptake at night was not very different from that during the day. The frequently low values obtained at night are to be explained by the generally smaller air movements at night. It is very striking that CO_2 absorption over a bare field is always higher at night than during the day; this is due to greater soil respiration (cf. p 138).

More important from an ecological viewpoint are conditions at different heights above the soil within plant stands. Figures 84 and 85 show a layout of absorption flasks in a wheat field and in a potato field respectively. The results from the measurements made in the wheat field are shown in Fig. 86. During the day, CO_2 absorption increases with increasing height above the soil. The difference between absorption at the height of the plant tops and that above the plants is very small. The lower values nearer the soil arise because there is very little air movement at this level. Night values have little bearing on photosynthesis. It is, however, striking that the highest values at night were at the soil surface. This must be due to accumulation of carbon dioxide at the soil surface because of lack of turbulence (cf.

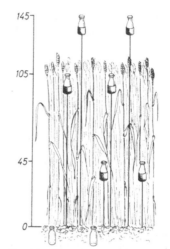

Fig. 84. CO_2 absorption experiment in a wheat field: arrangement of the flasks. (Walter 1960b)

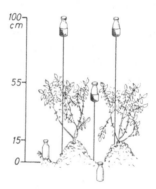

Fig. 85. Arrangement of absorption flasks for an experiment in a potato field. (Walter 1960b)

CO_2 content of the air in a stand of pines on p 122, and Fig. 71). Results obtained in a number of different stands are summarized in Table 15.

These results show clearly that not only light conditions but also the conditions for CO_2 uptake by leaves become more favourable at increasing height in this plant stand; this is due to greater air movement near the tops of the plants. The same conditions were found in a deciduous forest with a shrub undergrowth and a herbaceous layer. The relative absorption in the latter was:

Height above soil (in cm)	0	40	80	110
Absorption during the day	32	46	91	100
Absorption at night	87	44	83	100

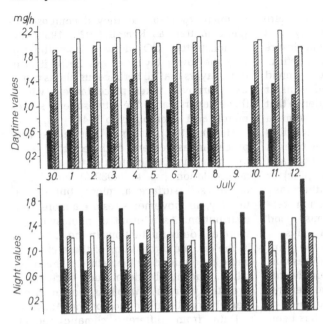

Fig. 86. CO_2 absorption of wheat from 30th June to 12th July: *above* during the day; *below* at night. *Black columns* at the soil surface; *double cross-hatched columns* at 45 cm above the ground; *single cross-hatched* 105 cm height; *open columns* above the stand at 145 cm height. (Walter 1960b)

Table 15. Relative CO_2 absorption values in different crop stands. The opening of the flasks was at soil surface (1), in the stand (2), at the upper surface of the stand (3) and above the stand (4); for maize 3 and 4 were both within the stand. (Walter 1960b)

Stand	Height above soil (cm)				Absorption during the day				Absorption at night			
	1	2	3	4	1	2	3	4	1	2	3	4
Wheat	0	45	105	145	34	65	96	100	127	55	88	100
Rape	0	40	105	165	37	64	76	100	89	49	52	100
Sugar beet	0	15	40	85	44	60	100	111	87	80	100	
Oats and peas	0	40	80	110	36	69	94	100	156	76	90	100
Potatoes	0	15	55	100	34	49	83	100	100	69	71	100
Clover	0	–	40	100	40	–	69	100	110	–	64	100
Pasture	0	15	40	65	34	50	86	100	201	141	71	100
Pasture (dense)	0	15	–	75	48	–	100	220	155	–	100	
Maize (2 m tall)	0	40	110	150	55	80	94	100	188	101	92	100

Diffusion of CO_2 from the air into the inside of a leaf takes place through the stomata, since direct diffusion through the cuticle and epidermis is not possible. Stålfelt (1935) showed that CO_2 uptake increases as the stomata open more widely. When the stomata are not fully open and the light bright, the limiting factor to assimilation is the stomata; when the stomata are fully open and the light weak, the light is limiting; with fully open stomata and strong light or with almost closed stomata and weak light, both stomatal aperture and light intensity are of equal importance. In lichens the limiting factor is the permeability of the cortex of the fungal hyphae (Ried 1960).

Since stomatal aperture is influenced by various external factors, such as humidity of the air and wind strength, these must in turn have an indirect effect on assimilation (Stocker 1937). Hydrature has a direct effect: Stocker (1954) found that the rate of assimilation in desert plants in southern Algeria falls in the middle of the day. This was believed to be due not to partial closure of the stomata, but to increasing water deficit.

In experiments conducted under humid conditions in the Alps, Pisek and Winkler (1956) could, however, find no evidence that a fall in water content with open stomata had any effect on photosynthesis. According to Stålfelt (1924) the stomata of pine trees in Sweden open at 07.00h on sunny days, and close again at 12.00h (in Scotch pine) or 13.30h (in spruce). In bad weather they remained open longer. Assimilation rate therefore falls during continuous good weather, and is maximal after rainy days.

Finally, there are factors within the plant itself which affect photosynthesis; these include assimilate accumulation in the leaves (Kostytschew et al. 1926) and the general conditions under which the development of the plant has occurred. The latter has been thoroughly investigated in lichens and mosses, but in higher plants, too, development under dry or wet conditions, as well as the availability of nitrogen and mineral nutrients all affect CO_2 assimilation (Müller and Larsen 1935). It is not possible, however, to make any definitive statement about the ecological implications of these physiological investigations. We have already mentioned the effects of winter hardening, of the onset of a degree of winter dormancy and the detrimental effect of chilling injury on subsequent CO_2-assimilation.

Naturally the age and condition of a plant also affect CO_2-uptake and other aspects of assimilation (vide Drautz 1935). With increasing age in pines, the needles assume more and more the character of shade needles (Stålfelt 1924).

6.3.4 Carbon Dioxide Content of the Atmosphere and Soil Respiration

The carbon dioxide of the air is the sole source of the carbon which forms the basis of the nutrition of green plants. It is thus important to consider the limits within which CO_2 content of the air may fluctuate and whether such changes are of any ecological significance. The most important source of CO_2 is the soil where it is released by the activities

of microorganisms as they decompose organic matter; as Romell (1922, 1932) has demonstrated, CO_2 diffuses from the soil into the lower layers of air. Since it is in these layers that all herbaceous plants grow, CO_2 turnover takes place largely within these layers, and the atmosphere serves merely as a reservoir (cf. p 3–4).

Huber (1952), using a Uras, made very precise measurements of the effect of vegetation on the CO_2 content of the air. Some accumulation (up to 0.039%) may occur at the soil surface at night, but during the day, air movements restore a more even distribution. As a result of photosynthesis there are daily fluctuations in the CO_2 level in the air, the dimensions of these changes being determined by the nature of the vegetation. The amplitude of such fluctuations over the year as a whole is, however, extremely small (Huber and Pommer 1954).

Data from different climatic regions reveal no substantial differences in the CO_2 content of the atmosphere. Thus the concentration measured in the tropics (on Java) by Stocker (1931) was 0.59mg l^{-1} and that obtained on Greenland by Müller (1928) was about 0.55mg l^{-1}. With increasing altitude above sea level the CO_2 content of the atmosphere does decrease, in accordance with the fall in air pressure. In the Alps, at an altitude of 2600m, 0.367–0.395mg l^{-1} was measured (Cartellieri 1940), on the Pamir highlands in Central Asia, at 4000m, it was only 0.25mg l^{-1} (Blagowestschenski 1935). Recently, some increase in atmospheric CO_2 has resulted from the burning of coal and oil in industrial areas and from deforestation.

Ecologists have often pointed out the significance of an increased CO_2 level in the air on the forest floor for photosynthesis by shade plants. This factor should, however, not be overestimated. Normally, CO_2 which has accumulated during the night is carried away to higher layers by air turbulence which commences in the early morning and continues throughout the day. CO_2 absorption measurements (p 133) have shown that even slight air movements at soil level during the day outweigh the possible advantage of CO_2 accumulation from soil respiration. The CO_2 content of air in pores in the soil is always much higher than that of the air immediately above the soil surface; it

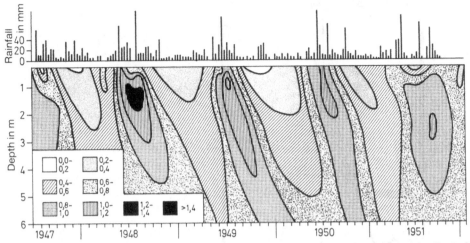

Fig. 87. Annual fluctuations in the CO_2 content of the air in pores in the steppe soil of the Central Black Earth Reserve. CO_2 content in vol.%: *1* 0.0–0.2%; *2* 0.2–0.4%; *3* 0.4–0.6%; *4* 0.6–0.8%; *5* 0.8–1.0%; *6* 1.0–1.2%; *7* 1.2–1.4%; *8* > 1.4%. (Fedorov and Gilmanov 1980)

increases with increasing depth but is subject to constant fluctuation. There is always a marked increase after heavy rain. Figure 87 shows the results of measurements made in the soil of the grass steppe in the Central Black Earth Nature Reserve.

Continuous soil respiration is dependent on a constant supply of organic matter. The source of this, in turn, is photosynthesis in the plant cover. When, in natural conditions, formation of organic matter by green plants and its destruction in the soil are in balance, then soil respiration is always equal to the difference between the total assimilated CO_2 and the quantity (in the form of assimilate) which is utilized by the plants in respiration and by the organisms (epiphytes, parasites, animals) which live above the soil. This carbon cycle in a plant community is shown in Fig. 88. The production of green plants which is not utilized by the aboveground parts in respiration sooner or later ends up in the soil as:

1. dead plant parts (litter);
2. assimilates which have been passed to plant organs below the soil surface. These may (a) be utilized in root respiration, (b) form structural material for underground organs, which in time also die, (c) form organic excretions from the roots, or (d) be incorporated into reserve substances which are transferred to the aerial organs once more in the spring.

There is some loss of CO_2 because a certain amount dissolves in soil water which percolates through to the ground water. Organic matter is also lost in this way. Furthermore, part of the CO_2 formed by respiration of the roots is removed from the soil in the transpiration stream (see below). This CO_2 is, however, usually reassimilated immediately. Additional CO_2 is released when acids dissolve any $CaCO_3$ in the soil, but this is of minor significance.

It is thus clear that in natural conditions soil respiration is determined not by the soil, or rather its living component, but by the photosynthetic activity of the particular plant cover.

Romell (1932) was thus correct in pointing out that there is no difference in the "respiration" of raw humus and mull soils.

Here we wish rather to treat of the ecological problems and to draw attention to the work of Haber (1958, 1959), who reviews the problem as a whole and provides an extensive bibliography. Soil respiration in the tropics is discussed in Volume 2.

For ecological purposes the most suitable of various methods for measuring soil respiration is the *absorption method*. It is simple to apply and gives integrated values for 12 h periods of the day or night. We describe the method here in the form devised by us (Walter 1952) and improved by Haber (1958).

CO_2 diffusing out of the soil was absorbed by a KOH solution under a bell jar. The absorption

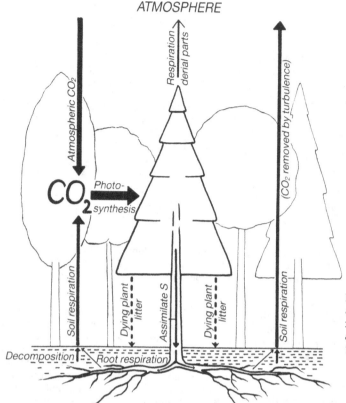

Fig. 88. Relationship between assimilated CO_2 and soil respiration in a plant community in a state of equilibrium. (Walter 1960b)

vessels used were 200-ml bottles (Fig. 83), the absorption medium was 10 ml KOH. The strength of the potassium hydroxide solution in the range $0.5N$ to $2N$ had little effect on the results, but it should not fall below 80–90%. For the determination of absorbed CO_2, see p 131. The absorption vessel was covered with a tin, 25 cm high, 22 cm internal diameter and 380 cm^2 in cross section. The tins were pushed 5 cm deep into the ground. Beneath each, a bottle with potassium hydroxide stood on a tripod, 3 cm above the ground. The bottles were removed after 12 h and the quantity of CO_2 absorbed was determined. During an experiment, the tins should be shielded from direct sunlight, while the area of soil beneath the tin should be cleared of all plant parts before the start of the experiment.

As Haber was able to show in preliminary experiments, the amount of CO_2 absorbed by the KOH was equivalent to $75.0 \pm 4.65\%$ of the actual quantity of CO_2 released from the soil; this proportion was unaffected by the intensity of soil respiration. The values obtained by titration should thus be multiplied by ⁴/₃.

Since soil is always very heterogeneous in composition and the distribution of respiring roots

very uneven, the rate of soil respiration varies greatly even over very small distances. It is thus to be recommended that 5 units always be used and the average be taken of these five values. With continuous measurements it is expedient to alter the positions of the tins from day to day. Only where the soil shows no plant growth can use be made of a two-part cover, the lower part of which can be left permanently in the soil.

The soil horizons with greatest respiration are those which contain the most organic matter; these are the upper horizons, and this applies also to raw humus soils, as can be seen from the following relative values for soil respiration:

Unaltered soil profile 100
After removal of A_0
(10–15 cm raw humus) 49
After further removal of A_1
(6–9 cm, damp and containing
many roots) 21

In some places soil respiration shows great fluctuations. These depend on environmental factors, in particular, temperature and humidity.

The sudden increase after a fall of rain may be due to displacement of soil air by rainwater. Apart from this, it is the activity of microorganisms that is paramount in CO_2 formation, and this is influenced by temperature and hydrature. Prusinkiewicz (1959) found that the soil bacteria become inactive below 96% hy.

Haber (1959) investigated the relationship between soil respiration and the numbers of bacteria; the latter were estimated using the Strugger fluorescent microscopic direct method. A high level of correlation was found in open fields, but in forest soils this was not the case, probably because fungi predominate in these soils. It has not as yet been possible quantitatively to assess the fungi in the soil.

If the soil respiration of different plant stands is to be compared, an adequate number of measurements must be made to ensure statistical validation of the mean values. No such data are available. Table 16 shows the maximal values for approximately 800 measurements obtained with the method described here.

Romell (1932) had found earlier that all 900 values for soil respiration obtained in Europe, in an area from 46°N to 63°N, lay between 200 and 700 mg CO_2 m^{-2} h^{-1}. Values outside this range obtained by Lundegårdh, although often cited, cannot be regarded as reliable (Walter and Haber 1957). Of the values shown in the table, some do fall outside these limits. They include two measurements made in winter, another two from very humus-poor, sandy soils and one from river-bank soil (from the Drau flood plain); all these lie below 200 mg, while two measurements made on dry turf on a southerly slope (near Klagenfurt, Austria) were above 700 mg. These latter are unexpected, for dry turf certainly does not belong among the communities which have a high rate of material production. It must be remembered, however, that these are maximal values of short duration, obtained when the ground was thoroughly wetted and the temperature high. The mean annual values would probably have been very low.

The observation that there is a higher rate of soil respiration at night (see p 133)

has been confirmed by Haber. This is more noticeable in forest than in an open field. It is very marked after sunny days but when the weather is overcast there is hardly any difference between day and night values. The precise causes of this phenomenon are not known. We were of the opinion that after sunny days there was increased root respiration as a result of the greater supply of assimilates, but Haber has suggested another possibility. On clear days transpiration is very great, in particular in forest, and CO_2, which would otherwise have been measured as soil respiration, is carried in the xylem sap to the leaves. This must lead to a lowering of the daytime values for soil respiration, while at night, transpiration as a factor can be ignored.

Experiments conducted at the University of Hohenheim (West Germany), using containers filled with soil, have shown that with free-standing containers, which cooled down considerably at night, higher values for soil respiration were obtained during the day, whereas with containers sunk into the ground a higher rate of soil respiration was observed at night. Thus fluctuations in temperature also seem to have an effect. These experiments were further designed to determine the part played by roots in soil respiration. Two types of soil were used: chalk-free, sandy field soil in some and poor quartz sand in others. Mineral fertilizer was added to each. Some of the containers were sown around the edge either with oats or with lupins; others were left as controls. A covered absorption flask was placed in the centre of each container. The exposed soil was covered so that all the CO_2 emitted from the soil would reach the absorption unit. The difference between the quantity of CO_2 absorbed in the containers with plants and that absorbed in the controls was considered to represent root respiration.

Before this, respiration by the roots had been thought to make up one third of total soil respiration. These experiments showed, however, that this cannot be put at any constant fraction; in conditions of optimal humidity, its value depends on the plant species, the density of the roots, the amount of organic matter in the soil and on temperature. Although there was little difference in dry weight between the roots of the oats and

Table 16. Maximum soil respiration in different plant communities, measured with the simplified absorption method. The figures represent the mean of 12-h measurements, expressed as mg CO_2 $m^{-2} h^{-1}$. (Walter and Haber 1957)

I. Northwest German Plain

    a) Oak-hornbeam-ash forest, without soil vegetation on heavy alluvial clay (near Datteln)
        April–June 1954 ⁻ 240
        October–November 1954 275

    b) Botanical Gardens, Münster, clay soil (cast-up earth)
        1. Arboretum, April–July 1955 512
           Arboretum, November–December 1955 148
        2. Meadow (*Arrenatherum-Dactylis* type) April–July 1955 623
           Meadow (*Arrenatherum-Dactylis* type) November–December 1955 156

    c) Nature reserve Heiliges Meer, diluvial sandy soil (July–October 1955)
        1. Black alder forest, fen peat 415
        2. Birch bog forest, intermediate fen peat-podzolic .216
        3. Pine forest, humus-podzolic soil 203
        4. Poor *Agrostis* meadow, humus-podzolic soil 315
        5. *Calluna* heath, humus-podzolic soil 228
        6. *Erica tetralix* heath, peaty soil 264
        7. Sugar beet fields 419

II. Southwest Germany

    a) Near Waiblingen. Loess-clay. April–May 1952
        1. Beech forest 407
        2. Spruce forest 220

    b) Sand-dune area near Schwetzingen, Upper Rhine plain, July 1952
        1. Vegetationless sandy soils 89
        2. *Pinus-Robinia* forest 157

    c) Spruce forests in the southeastern Black Forest, altitude 900–1000m, August–September 1954
        1. On Triassic limestone (Rendizina or brown clay) .431
        2. On Triassic sandstone (podzolic brown forest soils) 360
        3. The same (peaty gley soils, heavy raw-humus layer) 264

III. South Kärnten (Summer 1956)

    a) Flood plain of the River Drau, south of Klagenfurt. Fine sandy river deposits, various stages of development of vegetation
        1. Meadow shrub in the flood plain 410
        2. Transition to grey alder wood 235
        3. Grey alder wood with *Asarum* and *Struthiopteris* 150
        4. Areas trampled by animals in grey alder forest 370

    b) Nock Mountains north of Klagenfurt, 500–600m altitude, clay soils, in part gravel (5–8 and 9–12 represent vegetational development stages which belong together)
        1. Xerobrometum, 30° southern slope 780
        2. Golden oats meadow, below the latter 866
        3. *Trisetum* meadow, wet (valley) 551
        4. Ash-alder-bog forest, adjacent to the above 251
        5. Black alder forest, peaty soil, transition to spruce forest 348
        6. Black alder-spruce forest 241
        7. Spruce forest with many herbaceous plants 228
        8. Clearing in spruce forest (7), with nitrophilous herbs 308
        9. 30-year-old clear-felled area, 30° southern slope. *Calluna* heath, with young pines . .392
      10. Pine forest with *Calluna*, southern slope 255
      11. Pine-spruce forest with *Vacc. myrtillus* and *Pleurozium schreberi* 304
      12. As 11, but with *Rhytidiadelphus triquetrus* 299

Table 17. Daily and seasonal changes in soil respiration of a pseudo-podzolic soil under a many-layered Piceetum near Moorema, Estonian Peoples' Republic, during 1971. (Lutsar and Pork, from Fedorov and Gilmanov 1980)

	Soil respiration at different times of the day (g CO$_2$ m^{-2} per day)						Daily total of soil respiration (g CO$_2$ m^{-2})	Monthly total of soil respiration (g CO$_2$ m^{-2})
	00	04	08	12	16	20h		
10. VI	0.556	0.489	0.366	0.317	0.375	0.461	10.34	
20. VI	0.604	0.508	0.451	0.508	0.651	0.594	13.28	283.8
30. VI	0.746	0.670	0.747	0.740	0.680	0.737	17.23	
10. VII	0.747	0.680	0.604	0.566	0.651	0.718	15.71	
20. VII	0.528	0.528	0.347	0.394	0.451	0.565	11.22	445.2
30. VII	0.594	0.651	0.661	0.813	0.556	0.594	15.43	
9. VIII	0.622	0.594	0.603	0.622	0.585	0.613	14.58	
19. VIII	0.489	0.489	0.394	0.366	0.461	0.537	10.94	403.5
29. VIII	0.632	0.518	0.432	0.480	0.499	0.622	12.91	
8. IX	0.391	0.366	0.328	0.300	0.204	0.373	8.03	
18. IX	0.475	0.440	0.329	0.359	0.367	0.443	9.52	271.7
28. IX	0.396	0.352	0.242	0.233	0.398	0.433	8.19	
8. X	0.377	0.409	0.339	0.366	0.344	0.338	8.69	
17. X	0.305	0.386	0.288	0.224	0.347	0.352	7.58	236.8
27. X	0.265	0.262	0.311	0.231	0.226	0.255	6.24	
5. XI	0.378	0.368	0.413	0.332	0.346	0.428	9.02	(47.7)
Mean	0.506	0.482	0.426	0.429	0.450	0.504	11.18	

lupins, the respiration rate of the lupin roots was many times higher. This can be attributed to respiration by the N-fixing nodule bacteria. The contribution of root respiration to total soil respiration increased as development of the plants proceeded and reached a maximum at the time of flowering. There was wide scatter of the values obtained for oats growing on field soil: on average, root respiration made up 20% of soil respiration, but with a maximum of 65%. The average contribution of lupin roots was 82%, with a maximum of 91%. In the poorly developed plants growing on sand there was less difference: the value for oats averaged 81.5% and for lupins 93%. These very high values can be attributed both to the greater density of roots in the containers than in the field; furthermore, soil respiration of the controls with sand and no plants was very small, because the sand was very poor in organic matter.

A detailed study was made of the daily fluctuations in soil respiration in a dense pine forest, growing on brown pseudo-podzolic forest soil in Estonia SSR, by Lutsar and Pork (1971, cited in Fedorov and Gil-manov 1982). As can be clearly seen in Table 17, there was a daily maximum between midnight and 04.00 h, and frequently a minimum just before noon.

A different situation was found on dry, high mountain steppe on desert burosem soil, where the daily maximum was found to occur around noon. The following data, obtained by Mamytov et al. (1962), are quoted by Federov and Gilmanov (1980):

Time of day							
3	6	9	12	15	18	21	24h

mg CO$_2$ m^{-2} h^{-1}							
50	75	75	220	220	75	75	50

As far as annual fluctuations are concerned, soil respiration on the steppe was found to be greatest during the wet spring to early summer period.

Date					
15.5.	7.6.	28.6.	1.8.	1.9.	22.10.

g CO$_2$ m^{-2} d^{-1}					
2.6	2.6	2.5	1.7	0.6	0.1

This relatively low value of about $3-4 t ha^{-1}$ a^{-1} is correlated to the small annual primary production of the dry steppe.

6.4 Assimilate Economy (Carbohydrate Allocation)

There is no fixed relationship between the rate of photosynthesis by the assimilating organs of a plant and productivity, that is, formation of dry weight, during the period of vegetational growth. This must be especially emphasized.

Photosynthetic rate relates to only a brief phase in the CO_2-assimilation processes of a leaf. Even when the daily pattern of photosynthesis is known, it is not possible to calculate the increase in dry weight of the whole plant over a 24-h period; this is because a large part of the leaf assimilates are used up in respiration during the day and night by the non-photosynthetic parts of shoot and root and likewise by the leaves at night. If the leaf surface of a plant is small and light conditions poor, it is quite possible for the 24-h production of the plant as a whole to be on balance negative. The rate of photosynthesis of the leaves is even poorer as an indicator of total annual productivity of a plant or the vegetation as a whole. Annual production is the result of long-term photosynthesis and of mineral nutrient assimilation, not only carbohydrate assimilation. During this time, different plant species develop in quite different ways: some use assimilates (carbohydrates and proteins) continuously to increase the size of the productive organs, that is, the leaf surface; others form supporting tissues from the CO_2 assimilates. Annual production thus depends to a very great extent on the total assimilate economy of the plant; that is, on the way in which the assimilation products are used. It does not depend directly on the rate of photosynthesis, that is, the quantity of CO_2 assimilated per unit area and per unit time.

A good illustration of this is provided by comparing the first year's productivity of seedlings of the European beech Fagus sylvatica with those of the sunflower Helianthus annuus. Nutrient reserves in the cotyledons of the seeds of Fagus and Helianthus are almost the same: in both cases the substances are mostly oils, so that the starting conditions for the seedling are approximately alike. The plants were grown in the botanical garden of the University of Hohenheim (West Germany); that is, in climatic conditions favourable to Fagus but relatively unfavourable to Helianthus. In autumn, at the end of the period of vegetative growth, the Fagus seedlings had only a few, small leaves, but had formed a small woody stem; the mean dry weight of the plants was 1.5 g. The Helianthus plants had, however, used the assimilates continuously to increase the number of their leaves and then in late summer to produce flowers. The dry weight of a plant was, on average, 600 g; that is, 400 times that of a Fagus seedling. Admittedly, the rate of photosynthesis of Helianthus was twice that of Fagus, but this cannot explain a 400-fold greater annual production. The critical factor was the use to which the assimilation products of the leaves were put, that is, the assimilate economy. This is sometimes called carbohydrate allocation, because the assimilates are mainly carbohydrates but, as we will see, assimilation of nitrogen from the soil and also of water are very important in assimilation economy. (For a discussion of the literature see Schulze 1982.)

6.5 The Assimilate Economy of Different Life-Forms

The assimilate economy of a plant is primarily determined by the genetic life-form of the species. The longevity of a species is important in this regard.

Annuals

These species start their development with the nutrient reserves of their seeds. Each year the CO_2-assimilating leaf surface has to be formed anew after germination, before flowers and as many seeds as possible can be produced for reproduction. The productivity of such a plant during its life-span cannot be very great. It will be particularly low if the life-span is very brief as in Erophila verna, where it extends over only a few weeks, but it is higher in others, such as

Chenopodium album, which has a life-span of many months. Thus there is considerable variation in productivity, depending on the period of growth. In four species of winter annuals on sandy soil (I) and four others on a rocky substratum (II) in the Mohave Desert (California) the proportion of total dry substance production which is constituted by the roots and by the inflorescence was determined (Bell et al. 1979). In I the roots made up 4–10%, in II this was 6–13%. In I the inflorescence constituted 28–37%, in II 16–50%. The average dry weight in I was 0.12–2.83 g per plant, in II it was 0.11–0.9 g per plant. The proportion made up by the roots decreased in the course of development, while the relative proportion of the inflorescence rose, being particularly high in Compositae and *Plantago*. In general, the rate of photosynthesis is relatively high in annuals.

Biennial Species

In the first year of growth these behave as annual species, except that they do not use assimilates to produce flowers, but at the end of that year store as much as possible as nutrient reserves for the next year and often form a large rosette of leaves which will last through the winter. This gives them an advantage over the annuals since, in the second year, they can immediately form a rapidly developing shoot with leaves and then many flowers to produce a large number of seeds. They thus soon replace the annuals in secondarily open habitats such as forest clearings, open fields and ruderal places.

Perennial Herbaceous Shrubs

These behave like biennials in the first year, and in cold climates usually pass the winter as hemicryptophytes; that is, they have storage organs (roots, rhizomes, tubers or bulbs) beneath the soil surface, with the buds at the soil surface. In the second year they either do not yet flower or, if they do, only poorly; their storage organs are, however, continuously enlarged. Development of these plants improves from year to year because they are able to form a large leaf surface in a shorter space of time and productivity increases. In Slovakia, ruderal

shrub stands (Atriplicetum nitentes, Tanaceto-artemisietum and Sambucetum ebuli) grew to a height of about 2 m and the dry weight of the aboveground phytomass was on average 2.22 g m^{-2} (up to 2.50 g m^{-2}) or 25 t ha^{-1}; these values are equivalent to those for dense reed beds (Elias 1978). Reproduction is partly sexual, but also vegetative. In the tropics the development of herbaceous shrubs, like bananas and the Zingiberaceae, continues uninterruptedly. Since the bulk of these giant herbs consists of green, productive plant parts, they have the highest annual primary production. On the east-Asian island of Sachalin (Japanese Karafuto) annual values up to 40 t ha^{-1} were obtained over small areas for giant herbs growing on river banks, *Filipendula camtschatica* and particularly *Polygonum sachalinense* (vide Walter 1981).

Woody Plants

Their development, particularly of the long-lived woody species, the trees, is quite different and we will limit this discussion to these latter. Their assimilate economy is adapted to a long time-scale of development; in each succeeding year, the buds of a growing tree are formed at higher points above the ground and the leaves thus enjoy increasingly favourable light conditions compared with the herbaceous competitors. Immediately following germination, the shoot axis is formed, which becomes more and more lignified; this means that in the first year the assimilate is used unproductively. As a result, during their first very critical years, tree seedlings are not very strong competitors with herbaceous species. If they are able to survive these years and grow above the herbaceous layer, they become dominant in the stand and have the highest productivity. In this regard a difference can be detected between the softwood species and the longer-lived hardwood trees. The wood of the softwood (in central Europe, for example *Populus tremula*, *Betula pendula* and others) consists of thin-walled cells. These species thus use only a small proportion of their assimilate as structural material for the trunk, while a larger part of the assimilate can be used for formation of the productive leaf mass than is the case in hardwoods. These trees increase rapidly in height in forest clearings, but their life span is shorter than that of hardwoods, so that they are replaced by the latter and are unable to propagate themselves in the shade of the hardwoods.

There is a very great diversity in the detailed aspects of assimilate economy and each species has its own special features. Precise investigation of assimilate allocation is required in each individual case.

6.6 Changes in Assimilate Economy in Response to External Factors

In the above discussion we have dealt exclusively with genetically determined types of assimilate economy. In one and the same species, however, assimilate economy can be greatly altered by external factors. We cite examples from Walter (1960b).

When plants of the same species are grown under different *hydrature conditions*, those with an adequate water supply form large, thin, hygromorphic leaves, while those growing in dry conditions form thicker, xeromorphic leaves. This means that for the same expenditure of dry substance, a larger leaf surface is formed by plants with hygromorphic leaves, and this, in turn, makes possible a larger production of assimilates. Even when the rate of photosynthesis dm^{-2} is lower in the hygromorphic form, its total productivity exceeds that of the xeromorphic form.

An investigation made by Simonis (1941, 1947) provides an example of this. *Trifolium incarnatum* was cultivated in containers under different conditions: so-called "wet plants" were grown on soil at 80% water capacity, and "dry plants" on the same soil at 40% water capacity.

The rate of CO_2 assimilation per unit area of leaf was found to be 11–25% higher in the dry plants than in the wet plants; the results were the same whether estimated on the basis of CO_2 uptake, assimilate formation or daily increase in dry weight. Nevertheless, the wet plants had a 50% higher yield. This apparent contradiction becomes com-

prehensible when the assimilate economy is examined. The dry plants used more assimilate for extension of the root system, while in the wet plants it was used for the formation of more leaf surface. Average values for the relative amounts of dry substance constituted by the different parts of the plants are as follows:

Plant type	Percentage of total dry weight made up of:		
	Roots	Stalks	Leaves
Dry plants	44.4%	22.7%	32.9%
Wet plants	32.6%	24.8%	42.6%

In absolute terms, the leaf surface area in wet plants was twice as large as that in dry plants. *It is this greater development of leaf surface which, despite the lower rate of photosynthesis, accounts for the greater yield in wet plants.*

The use of nitrate fertilizers greatly increases yield. Nitrogen shortage causes the rate of CO_2 assimilation to fall by one half in *Sinapis alba*, while the total yield of dry substance is reduced eightfold (Müller 1932). This discrepancy is again to be explained by a reduction in growth of leaf surface area in favour of greater root development, which is evoked by lack of nitrogen, as by dryness. In plants with an inadequate nitrogen supply, the ratio of shoot plus root dry substance to that of the leaves is three times greater than it is when nitrates are added to the soil. According to Maiwald (1930), the increase in leaf area occurs parallel to N-uptake. Thus the assimilates formed as a result of photosynthesis can only be used for the growth of the leaves when nitrogen is available at the same time.

The effect of nitrate fertilization on yield is thus not direct, but arises from a change in the assimilate economy. Potassium, on the other hand, does not influence the formation of leaf surface area.

Assimilate economy is also markedly affected by day length. The following data, due to Rasumov, have been taken from Boysen-Jensen (1932).

Plant type	Day length in h	Percentage of total dry weight			
		Leaves	Roots	Stalks	Ears
Long-day plants	18	12	11	55	22
(Barley)	12	30	32	37	1
Short-day plants	12	19	10	26	45
(Millet)	18	25	17	40	18

Date	Mean total dry weight per plant		
	Plant type	Covered	Exposed
18.1.	Winter wheat	0.914	0.575
13.2.	Winter rye	2.221	1.043

By weekly sampling of a fixed number of plants the average increase in dry weight was determined as well as mean leaf surface, and expressed as quantity $m^{-2} d^{-1}$. The rate of assimilate formation calculated on this basis was as follows:

It can be seen that although the long-day plants flower when exposed to long days, their leaves and roots are three times more strongly developed when the days are short. The total yield of dry weight is, however, greater with a shorter day, but grain yield is less. In short-day plants the situation is reversed.

The greater productivity of shade plants in light of low intensity can be explained as the result of a phenomenon described by Blackman and Wilson (1951). They found that if sun- and shade plants were first grown in full daylight and then transferred to shade, the leaves of the sun plants showed very little change, while the small leaves of the shade plants increased in size very rapidly.

Finally, we will consider the effect of temperature on assimilate economy. Favourable temperature conditions result in a high yield of dry substance. Can this be explained as a result of a higher rate of CO_2 assimilation? The following experiment, conducted at the University of Hohenheim, was designed to answer this. On 28.9.1949, seeds of winter wheat (Karsten V) and winter rye (Petkuser) were sown in four similar glass-covered seed-beds, two for each type of plant. After germination on 6.10.49, one of each pair was left covered with glass, the other exposed. The temperature of the upper soil layer and the air in the covered containers was on average 2°C, and in January 3°C, higher than in the uncovered containers. Growth rate was correspondingly higher and the yield of dry substance at the end of the experiment twice that of the exposed plants. This can be seen from the following figures.

Average temperature (°C) (weather station)	Time between samples (1949–50)	Rate of assimilation in g ($g\ m^{-2}\ d^{-1}$)			
		Winter wheat		Winter rye	
		Covered	Exposed	Covered	Exposed
13.0	13.10.–19.10.	4.5	4.45	3.62	5.05
13.4	20.10.–26.10.	4.05	4.40	3.32	3.06
2.9	27.10.– 2.12.	2.94	3.70	3.32	3.06
2.1	30.11.–15.12.	1.25	1.61	1.28	1.06
2.6	16.12.– 6. 7.	1.51	1.22	1.17	1.33
3.2	7. 1.–16. 1.	1.49	2.58	1.23	1.66

It can be seen from this table that the rate of assimilation in each replication was higher in October than in December and January, reflecting the prevailing temperature conditions. If, however, the values for the covered and exposed plants are compared, it will be seen that they are lower in the covered plants, although these were better protected from the cold. This paradoxical effect can be explained as follows: it was mostly during the night that there was a temperature difference between the covered and uncovered cases, since the former had to be opened for airing during the day. The higher night temperatures increased the rate of respiration and the gain in assimilate over a 24 h period was thus reduced.

The higher yield of the covered plants was thus not a result of a higher rate of assimilation, but, as in the other cases mentioned, due to a greater development of leaf area. The roots were proportionally larger in the uncovered plants, in wheat by an average of 35% and in rye by 43%, than the roots of covered plants. It can be seen that *yield, that is, the production of organic matter, is indirectly affected by external factors which bring about changes in assimilate economy.*

The time of onset of the reproductive phase also affects the production of dry material. With the formation of flowers or inflorescence at the tip of shoot, all increase in leaf surface ceases, and the assimilates are used for the production of reproductive organs.

It would be wrong to overestimate the ecological importance of dry weight production. The reproduction process is decisive for continuation of the species. There is a certain antagonism between vegetative growth and the formation of reproductive organs, as the effect of day-length shows. Vernalization, too, promotes flowering and leads to a reduction in the production of dry substance. The factors which actually initiate flowering are not yet clearly understood. The first visible sign is in the apical meristem of the shoot, when the vegetative cone no longer increases in length, but broadens out at its tip. It is known that development of flower primordia is promoted by a certain degree of water deficit, resulting in reduction in the hydrature, in other words, an increase in cell sap concentration and subsequent loss of water from the protoplasm, including that of the meristem cells. In trees flower buds develop on those branches which are exposed to sunlight, transpire at a higher rate and have a low osmotic potential (low hydrature), and also in short twigs to which the supply of water is restricted. Plants in dry, sunny positions flower similarly more profusely. Ephemerals growing in the desert in conditions of water shortage develop dwarf forms which flower early. In column cacti in the Sonoran desert, flowers form on the southwest side of the plant; this is the warmest side with the lowest osmotic potential, and they do not form at all on the north side where the osmotic potential is highest (Walter 1931; Walter and Kreeb 1970). Similarly, low N availability, which inhibits growth, promotes flowering.

In horticultural and agricultural undertakings in which the flowers, fruit or seeds are harvested, production of dry substance is of less importance than promotion of flowering.

6.7 Leaf Area Index and Productivity in Plant Stands

The productivity of a plant stand depends to a large extent on the Leaf Area Index (LAI). This is the ratio of the total leaf area of a stand (taking account of one side of the leaves only) to the ground area covered by it; it is expressed in $m^2\,m^{-2}$. The greater the LAI, the greater the productivity of the stand, up to a certain limit. The upper leaves always receive more light than do those lower down. If a stand is so dense that the light intensity at the lower layers is permanently below the light compensation point, the balance of production of the lower leaves will be negative; they will turn yellow and be shed. This implies that the LAI can never exceed a certain maximum.

In a leaf which is perpendicular to the incident light, increasing light intensity evokes at first a steep, linear rise in the rate of photosynthesis; soon, however, this becomes less pronounced, until finally, at the point of "light saturation", the curve becomes horizontal (vide Fig. 64 and curve a in Fig. 89).

As Boysen-Jensen (1949) has shown, these relationships do not apply to plant stands. Here the rate of photosynthesis continues to rise linearly to the point of full sunlight (Fig. 89, curves b and c). This is possible because even when the point of light saturation has been reached for the uppermost leaves, the rate of photosynthesis continues to rise in the lower, partially shaded leaves. Furthermore, if the leaves in the upper canopy of a beech wood are observed —and this applies to other plant stands as well—it will be seen that the upper sun leaves are not held horizontally, but hang at an oblique angle. As a result, when the sun is at its zenith around noon, the rays of sunlight fall at an angle onto the leaves; these are thus not optimally illuminated and allow more light to pass through to the lower leaf

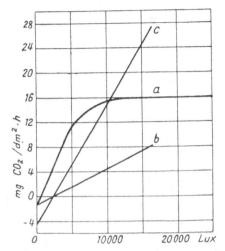

Fig. 89. Relationship between the rate of CO_2 assimilation and light intensity in *Sinapis alba:* *a* leaves perpendicular to the incident light, *b* in a stand in which the leaves were variously oriented to the light (expressed as mg CO_2 per dm^2 leaf surface), *c* the same, but per dm^2 of soil surface (with a leaf surface index of 3.4). (After Boysen-Jensen 1932, from Walter 1960)

layers, increasing their productivity. Only the leaves of the lower layers, which receive but a fraction of the light falling on the whole stand, are orientated perpendicularly to the incident light. In the middle of the crown, these lower leaves are thus arranged horizontally, but at the periphery, where the incident light is oblique, the lower leaves are held obliquely and are thus perpendicular to the light rays.

The LAI of a stand depends not only on light conditions, but is influenced by other factors as well, in particular the water relations. A high leaf area index makes possible high productivity in a stand, but a high rate of photosynthetic activity is always associated with high water loss through transpiration. If there is insufficient soil water to meet the demands of the stand, *the leaf area index is limited by the water economy*. It is immediately noticeable that forest stands on moist river banks are always very dense; that is, they have a high LAI, while forest stands in dry biotopes are, in comparison, much sparser. It is for this reason, too, that even when other conditions are the same, the vegetation of humid areas is always much denser than that of arid areas. This is

particularly striking as desert areas are approached: the plant cover becomes increasingly sparse with decreasing rainfall, until finally the landscape is no longer characterized by the vegetation, but by the exposed rock. An observer not personally acquainted with desert conditions is astonished that in such conditions there should be any vegetation at all, and assumes that the plants possess some special protoplasmic resistance to drying out, the capacity to take up water from the air, or some other special physiological characteristics. The matter is far simpler, however. Such unproven hypotheses are not necessary, as the following will show.

6.8 Vegetation Density and Rainfall in Deserts

Diffuse Vegetation

If the relationship between annual rainfall and plant growth is to be established, it is essential to bear in mind that rainfall, expressed in millimeters, means the quantity of water, in litres, which falls on one square meter of soil surface. Thus the aim should be to establish the relationship between these quantities of water and the plant mass, also expressed as quantity per square meter of soil surface.

To establish the degree of dependence of vegetation density on annual rainfall, it is necessary to ensure that other factors such as temperature or soil type remain constant. Furthermore, the comparison must be limited to similar types of vegetation.

If these conditions are fulfilled, it is possible to show that the plant mass, and thus also the transpiring surface, decrease proportionally to the decline in the amount of rain which falls. This was demonstrated for the first time in Namibia in a comparison of *ungrazed* areas of grass on sandy soil (Walter 1939, 1964). Rain falls in this area in the summer and the magnitude of the rainfall declines from over 500 mm in the wet northeastern part of the country, to almost none on the coast (Fig. 90). At the same time, the grass becomes shorter and the soil covering of plants more sparse, up to that point at

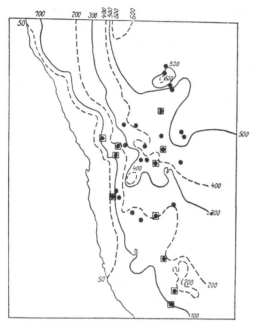

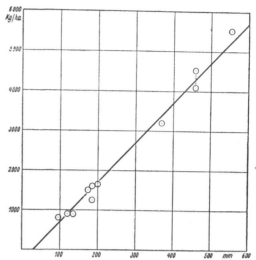

Fig. 91. Relationship between productivity of grassland and rainfall in Namibia. *Ordinate* production of dry substance in kg ha^{-1}; *abscissa* mean annual rainfall. (Walter 1939)

Fig. 90. Mean annual rainfall (in mm) in Namibia. □ Test areas for estimation of productivity of grassland; ● places where samples were taken for analysis of grazing plants. (Walter 1939)

which the annual rainfall is less than 100 mm per year, where the extreme desert begins. Figure 91 shows the quantitative relationship between the aboveground production of dry substance and annual rainfall. It can be seen that annual production of plant mass indeed decreases in direct proportion to the decline in rainfall. At the same time, the larger grass species of the wet areas are replaced in arid parts by smaller species with a short growth period.

Since this comparison has been limited to a single plant type, namely, grasses, one might reasonably assume that this relationship applies not only to the quantity of dry substance produced, but also to the transpiring surface of the grasses; that is, that this also decreases in proportion to the decrease in rainfall. This would imply that the quantity of water available per unit of transpiring surface is the same in humid and in arid areas, that is, the water supply to the individual plant is not very much worse in arid than it is in humid areas. Certainly the air in the desert is dryer, so that leaf structure in desert plants is more xeromorphic.

This conclusion was confirmed by investigations in the Mohave desert in California. Here it has been shown that there is a very close correlation between the density of *Larrea divaricata* stands and the quantity of rainfall within the range 50–250 mm (Woodell et al. 1968).

Le Houerou (1959) found the same type of relation in olive plantations in Tunisia. He found that the yield per hectare decreased in proportion to decreasing rainfall. The yield per tree remained more or less constant, but the number of trees per hectare, that is, the average density of the plantations, declined in proportion to the decrease in rainfall.

We are able to show that this rule applies to other types of plant as well, by comparing *Eucalyptus* forests in Australia (Walter and Stadelmann 1974). In this case, not only the annual production of leaf mass (leaf litter) but also its leaf area was directly determined. It was, however, necessary to compare separately old forest with undergrowth and young plantations which, devoid of any undergrowth, had greater leaf production.

In Fig. 92 rainfall is recorded on the abscissa, relative quantities of litter or of litter area (with maximum values set at 100) on the ordinate. Apart from a few irregularities, unavoidable in such investigations, the pro-

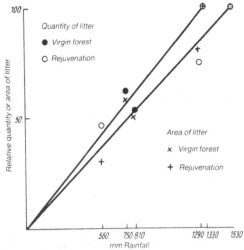

Fig. 92. Dependence on rainfall of the quantity and surface area of leaf litter in Australian *Eucalyptus* forests. (Data supplied by OW Loneragan, Forestry Department of W Australia)

portionality is shown fairly clearly. Furthermore, these results relate to a winter rainfall area and one in which the rainfall is far greater, ranging from about 500 mm to over 1500 mm.

It follows from these observations that *the water utilization per unit area of leaf remains almost unaltered in humid and arid conditions*. The only precondition is that the plant has the capacity to take up the required amount of water through its root system. This, indeed, is the case. Reduction in leaf surface in arid areas is accompanied by extension of the root system, not in depth, but horizontally. In other words, the ratio of aboveground to belowground parts of a plant decreases with increasing aridity. The only exception are the succulents, which are not comparable to other plant types.

6.9 Restricted or "Contracted" Vegetation

Another feature which must be borne in mind when considering an arid area with an annual rainfall of about 100 mm is that soil water is not evenly distributed; this, in turn, is reflected in the distribution of the vegetation. The absence of closed plant cover

results in very heavy surface run-off. The soil surface becomes covered with a crust or vesicular strata are formed.[8] Thus the quantity of water seeping into the soil on higher ground is very small and less than the rainfall would suggest, but is more in drainage gullys and depressions. As a result, the vegetation is not evenly distributed over the whole area, but is concentrated where there are larger reserves of water in the soil; in such places the water supply is, in fact, often no worse than it is in humid areas (Fig. 93).

Assuming, for example, that in an area with an annual rainfall of 25 mm, 40% of the rainwater flows into a depression which has an area of one fiftieth of the total, then the quantity of water available to plants in this depression is equivalent to a rainfall of 500 mm.

The different patterns of the vegetation can be seen particularly clearly from an aeroplane. In humid areas the vegetation is spread over the whole surface and depressions are marked only by denser growth; this dispersed *"diffuse vegetation"* is rapidly replaced by an increasingly restricted *"contracted vegetation"* with increasing aridity of desert areas (Monod 1954). In other words, most of the surface is completely devoid of plants and the vegetation is limited to valleys and depressions. The water which runs off into the depressions penetrates relatively deeply into the soil; here it is protected from evaporation and can be stored over many years. The plants in such depressions have deep roots which penetrate as far down as there is any water, often several meters.

Results of investigations in the Wadi Hoff, near Heluan, south of Cairo, provide a good example of this. In this area mean annual rainfall is 31 mm, but even after a fall of only 10 mm of rain, water flows off the bare limestone plateau, first into shallow erosion gulleys which unite, becoming deeper, and finally lead the water into the broad, dry valley or wadi. Only in rainy years is the bed of this water-course actually filled with water for any significant time. As it flows off, water seeps continuously into the sand, until

---

8 These arise when rain falls on dried, silty soil, the air is unable to escape from such soil and remains trapped as fine blisters in the surface layers (vide Evenari et al. 1974, Volk et al. 1970). Rainwater penetrates only poorly into such soil.

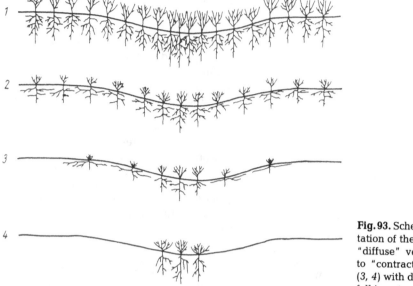

Fig. 93. Schematic representation of the transition from "diffuse" vegetation (*1*, *2*) to "contracted" vegetation (*3*, *4*) with decrease in rainfall in extremely arid areas

Fig. 94. The Wadi Hoff near Heluan, showing the area investigated by Batanouny (cf. Fig. 95). (Photo by O Stocker, from Walter 1973)

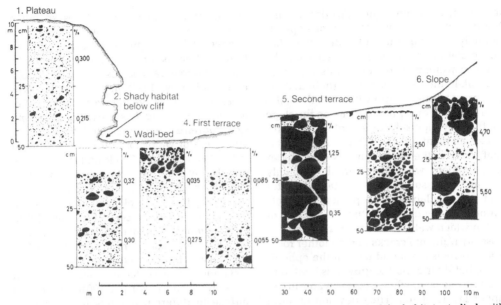

Fig. 95. Diagonal section through the Wadi Hoff. The soil profiles are those of the habitats studied, with depth of profile (cm) on the left and total salt concentration from 0–25 cm and 25–50 cm depth at the right (as % dry wt.). Two horizontal scales are used: that to the right, from the second terrace *(5)* to the slope *(6)*, is one fifth of that to the left, which extends from the plateau *(1)* to the first terrace *(4)*. *Vertical scale* to the left relates to the left side only: on the *right* the vertical scale is 1 mm = 10 cm. (After Batanouny 1963, from Walter 1973)

Table 18. Lowest water content recorded during the drought period in the Wadi Hoff, expressed as a percentage of the dry weight of the soil. (Abd-el-Rahman and Batanouny 1965)

Habitat (see text)	1	2	3	4	5	6
Minimal water content	1.0%	2.0%	2.3%	1.4%	0.8%	2.5%
Degree of plant cover	12%	50%	0%	20%	1%	0%

finally the flood ceases. In this way some underground reserve water accumulates in the soil, enabling plants to grow there. There is a continuous increase in vegetation density from the shallow rain furrows of the plateau to the wadi. Even trees (*Acacia, Tamarix*) are able to grow in the wadi. The plant cover would be even more dense were it not for its continual destruction by the grazing camels of nomads and by the chopping down of firewood.

The water economy of this vegetation has been examined over a period of a whole year (Abd-el-Rahman and Batanouny 1965). An aerial view of Wadi Hoff is shown in Fig. 94. A section across this wadi (Fig. 95) shows six distinct habitats: (1) the plateau with shallow rain furrows, (2) a steep, over-hanging cliff, at the foot of which is a shaded area where the plants receive direct sunlight early in the morning only, (3) the actual bed of the wadi, (4) the first low terrace, (5) a second, higher terrace, (6) the scree of the Hamada area. The soils are raw desert soils consisting, between the stones, mainly of coarse sand, with some fine sand and only very little silt and clay. The quantity of clay exceeds 10% only in habitats 2 and 4. The water capacity of the soil is not high; the wilting point is below 1%. In habitats 1 to 4 the soil contains very little salt, in 5 it is above 1–2.5%, and in 6 even above 5%.

The water content of the upper 50 cm of soil was determined once a month; the results are shown in Table 18. The upper layers are wet only during the rainy season in winter, in summer they are dry. During the drought period water content depends not only on the quantity of water which has accumulated in the soil, but also on the water consumption of the plants rooted in the soil. In the bed of the wadi (habitat 3) no perennial plants can grow because the soil is overturned at every flood, while on the slope (6) the very high salt content of the soil precludes all plant growth. What is important is that even in the extremely dry year under investigation, when the rainfall was only 19.5 mm, the soil always contained water that was available to the plants. The roots probably extend deeper than the 50 cm of soil in which water content was measured— possibly right into cracks in the mother rock. Shallow roots are found only in the ephemerals, but their period of growth is limited to the rainy season, when there is water in the upper layers of soil. They develop in large numbers only in years with good rainfall.

Water Turnover of the Plants. The rate of transpiration over the course of a day was determined monthly for the most important perennial species, namely *Pennisetum dichotomum*, *Zilla spinosa* and *Zygophyllum coccineum*. The quantity of water lost in this way confirms that the plants are able to take up water from the soil even during the drought, and that there is always some period of the day when the stomata are kept open, so that photosynthesis can take place. Nevertheless, water loss is reduced as the drought proceeds. This can be seen as the curve for daily transpiration falls transiently around midday, or, alternatively, a maximum occurs at 10.00 h. Summer transpiration values are higher than those in the winter, but they do not increase at the same rate as evaporation measured simultaneously. If the plants are watered in summer, the rate of transpiration doubles in both *Pennisetum* and *Zilla*—an indication that it had been limited by shortage of water. *Zygophyllum* has, however, succulent leaves, so transpiration values calculated relative to wet weight are much lower for this plant than for the other species. The water economy of this species is very closely controlled. Despite the far higher rate of evaporation in the summer, its transpiration rate is no greater than in winter. Similarly low values are shown by the annual *Zygophyllum simplex*

which, at the same time, stores so much water in its leaves that occasionally it even survives the drought. The highest rate of transpiration is encountered in the thin-leaved ephemeral species in which a maximum value of $14-25 \text{ mg g}^{-1} \text{ min}^{-1}$ was measured. Similar values were obtained for the soft-leaved *Stachys aegyptiaca*, *Artemisia fragrantissima* and *Diplotaxis harra*. Limitation of water loss during the drought is achieved not only by closing the stomata, but also by other means, such as casting off leaves, or the reduction of the transpiring surface by formation of small leaves or even thorns; this is to be seen, for example, in *Zilla*, *Pennisetum*, *Lavandula coronopifolia*, *Iphiona*, *Diplotaxis harra*, *Pituranthos tortuosus* and *Stachys aegyptiaca*.

Batanouny also calculated water loss from the entire plant cover per 100 m^2 of the soil surface in different habitats in March and June. This was calculated on the basis of total wet weight and corresponding transpiration rate. The results are shown in Table 19.

The approximate water loss per year in mm, calculated from these figures, was found to be 180 mm for habitats 1 and 2, 400 mm for habitat 4 and about 30 mm for habitat 5. To this must, of course, be added both the water lost by the ephemerals during the winter months which varies from year to year, and that lost through evapora-

Table 19. Phytomass (wet weight) and water loss per 100 m^2 soil surface per day in the Wadi Hoff. (Abd-el-Rahman and Batanouny 1965)

Habitat	Month	Wet weight of perennial species (kg)	Water loss (kg)
1. Plateau	March	15.2	44.8
	June	15.3	51.0
2. Below the cliff	March	47.4	45.9
	June	42.8	48.7
4. First terrace	March	46.8	114.2
	June	44.3	120.0
5. Second terrace	March	6.1	8.3
	June	5.5	8.0

Table 20. Lowest and highest potential osmotic pressure (in atm) in species in the Wadi Hoff. (Abd-el-Rahman and Batanouny 1965)

Plant species	Min. π^* (atm)	Month	Max. π^* (atm)	Month
1. Perennial species				
Farsetia aegyptiaca	18.2	Jan.	23.0	Oct.
Achillea fragrantissima	17.0	Jan.	22.7	Oct.
Lycium arabicum	19.1	March	21.3	Nov.
Asteriscus graveolens	16.3	Feb.	19.0	Sept.
Stachys aegyptiaca	13.8	April	19.3	Nov.
Zilla spinosa:				
Habitat 1 (plateau)	14.2	Feb.	16.6	Sept.
Habitat 2 (below cliff)	12.1	Feb.	14.4	Sept.
Habitat 4 (first terrace)	14.2	Feb.	14.9	Nov.
Iphiona mucronata:				
Habitat 1 (plateau)	13.1	March	23.3	Dec.
Habitat 2 (below cliff)	11.6	Feb.	16.0	Dec.
Habitat 4 (first terrace)	13.6	April	17.8	Nov.
Pennisetum dichotomum	11.2	March	14.8	Dec.
Ochradenus baccatus	19.7			
Diplotaxis harra	22.0			
Retama retam	14.9		18.2	
Pituranthos tortuosus	13.5		17.0	
Heliotropium arabianense	14.3		15.1	
Pergularia tomentosa	16.3			
Fagonia kahirina	12.2			
Lavandula coronopifolia	11.2			
Scrophularia deserti	12.1			
Centaurea aegyptiaca (biennial)	13.1		14.4	
2. Ephemeral species				
Diplotaxis acris	13.6			
Trigonella stellata	13.2			
Plantago ovata	11.6			
3. Halophytic species				
Nitraria retusa	25.3	Jan.	33.0	July
Zygophyllum simplex (annual)	28.9			
Zygophyllum coccineum:				
Habitat 1 (plateau)	29.1	April	35.6	Sept.
Habitat 2 (below cliff)	20.6	Feb.	29.3	Oct.
Habitat 4 (first terrace)	23.6	April	32.1	Oct.

tion from the soil surface. It is clear that the calculation made on p 147 for "contracted" vegetation is completely applicable to habitat 4. The low water loss relative to the plant mass in habitat 2 is due partly to the dominance here of *Zygophyllum coccineum*, which has a low rate of transpiration, and partly to the lower rate of water loss from the plants in this shaded habitat.

It may be said in conclusion that the water supply of desert plants in this extremely arid area is by no means poor. Favourable water conditions are reflected also in the concentration of the cell sap. Except in the halophytes, this is relatively low and there are no great differences between summer and winter levels (see Table 20).

The contribution of chloride to the potential osmotic pressure was not investigated, but it is always high in Zygophyllum spp. Nitraria is likewise known to be limited to saline habitats.

The daily fluctuations in the potential osmotic pressure so far investigated have all been very small. This was particularly so in Pennisetum; in Zilla and Farsetia there was a fluctuation of 3 atm. Water balance is also very steady. Even after irrigation, the potential osmotic pressure of Pennisetum fell by only 2.3 atm and in Zilla by only 1.7 atm—an indication that any deficit of water saturation is not great, even in summer. It is thus clear that desert plants have no need to develop any particular protoplasmic drought resistance. Their adaptations are more of a structural nature, such as the development of an extensive root system and the reduction of the transpiring surface area.

Conditions in central Egypt are even more interesting. Near Asyut the mean annual temperature is 21.7°C (maximum 49.5°, minimum −0.4°C). Mean annual rainfall is given at 7 mm, but all that this means is that there is a single heavy downpour every 10–20 years. Potential evaporation may be as much as 4000–5000 mm per year. Nevertheless, plants are able to grow in the Wadi El-Asyut, 370 km south of Cairo. The water economy of the small tree-like asclepiadacean Leptadenia pyrotechnica has been studied over the period of a year. The soil of the experimental site was sandy, with a low salt content and a pH = 7.5–9.1. It was found that this species transpires at quite a high rate throughout the year. Calculated on the basis of wet weight, the maximum rate was 2–3 mg g^{-1} min^{-1}, while maximal evaporation was 1.5 mm h^{-1}. In July, the maximum rate of transpiration was 5 mg g^{-1} min^{-1}, with a maximum rate of evaporation at 2 mm h^{-1}. In summer, the transpiration rate is generally only half that of evaporation, with a maximum value at about 10.00 h. The saturation deficit of the plants is very small (up to 16.8%), the potential osmotic pressure as a measure of the hydrature of the protoplasm is very low. The upper and lower values recorded during the year were

11.6 and 15.1 bar; by comparison, in other habitats values of 12.3 to 16.9 were obtained. The daily fluctuation was, on average, 1 bar. The values obtained for other species in the neighbourhood were: Calligonum comosum (psammophyte) 12.3 to 14.8 bar, Pulicaria crispa (malacophyllous species) 18.7 and 20.9. It was higher only in halophytes, such as Zygophyllum coccineum at 38.2 to 42.8 bar, and Anabasis setifera at 36 bar (Migahid et al. 1972).

Leptadenia can be regarded as a stenohydric xerophyte, as seems to be true of all plants with milk sap. The very steady water economy of this species arises in the following way. The total transpiring surface of the plant is small, because it is a species of broom, and these plants form short-lived leaves only at the tips of the youngest shoots. The root system is, by contrast, very extensive. For example, the roots of a 1.6 m high bush extended 11 m in depth and horizontally over a radius of 10 m. Although the upper 75 cm of the soil was completely dry, the lower layers contained water derived from the very rare but torrential rain, which penetrates deeply into the soil, especially in the wadis. Since water consumption by the sparse vegetation is very low per unit area of the soil surface, the water reserves in the soil can last for several decades.

Several further investigations have been carried out at the University of Cairo (Batanouny and Batanouny 1968, 1969; Batanouny and Zaki 1969).

The distribution of vegetation associated with wadis arises when water runs off the land in streams, that is, when there is gully erosion. In flat or very gently sloping terrain the water flows off broadly as sheet flow. Where such sheet flow is dammed up, the water seeps into the ground, and the plant cover is thicker. In this way strips of vegetation arise which run at right angles to the direction of the slope; sometimes they even form a regular pattern, which can be recognized on aerial photographs. In Somalia dense strips of Andropogon alternate with strips almost entirely free of vegetation. The former collect almost twice as much water as the latter from the annual rainfall of 150 mm. After a rainy year, the water content of the soil on which the Andropogon was growing was found to contain 10% water, that of the intermediate strip only 6–7% (Hodge 1962; Hemming 1965). Formation of these strips usually starts with single plants, around each of which is deposited soil carried by wind; this results in damming up of the sheet flow and there is further deposition of soil at the sides. In this way a bow-shaped strip finally arises (White 1971). Even in the tropics, striped patterns of vegetation have been observed when termite mounds form areas denuded of plant growth. From these rainwater runs off as sheet flow, to the benefit of vegetation on the perimeter of the area (White 1970).

The effect of run-off is very clearly seen along the edges of motorways which pass through arid areas. Water which flows off the road gives rise to a strong growth of plants which, in flat, extreme deserts, form often the only recognizable vegetation.

The principle underlying contracted vegetation was exploited even in ancient times for *run-off farming* in arid areas. In southern Tunisia, in the dry, hilly area around Matmata on the edge of the Sahara, isolated date palms are to be seen scattered over the slopes. Closer inspection shows that the Berber population has provided each small erosion gulley with a dam and then piled up soil above the dam wall. After rain, any water which runs down the gulley wets this soil sufficiently to enable one or more date palms to be grown on it or even barley or field beans *(Vicia faba)* to be cultivated.

In the Egyptian coastal region are similar systems which date from Roman times and were then used for viticulture. It is today still possible in this region with only 100mm rainfall to grow olives by diverting the sheet flow from a stony ridge to the olive gardens. The catchment area must be at least four times the size of the area under cultivation, and there should be 100% run-off. Farming is possible whenever these conditions are fulfilled.

Old systems of ditches and dams have been discovered in the Negev desert; these were used in pre-Arabian times to channel the flow-off after rain from higher surfaces to more low-lying fields (Tadmor et al. 1960, 1961). Recently, some of these systems have been renewed and good harvests have been achieved, even with the very small annual rainfall.

Results obtained on the experimental farm Avdat in the Negev (Evenari et al. 1971) are summarized in Table 21; they give some indication of the quantities involved in the run-off into a wadi during the winter rain period. On a small plot of $250\,m^2$ about $15\,m^3$ of drainage water accumulated; this was equivalent to 60mm of rain. The quantity of the run-off depends very much on whether the rain falls on dry or wet soil (cf. 15.iv.71 in Table 21).

The statements made here relate to perennial desert species. Only these are adapted to the typical climatic conditions of such regions. In unfavourable years they may lose a large part of the shoot mass because it dries out or in some places they may die altogether. In good, rainy years they grow more luxuriantly and may colonize new areas with their seedlings. Besides these perennials, there are, however, in all arid areas, many species which belong to the "ephemeral vegetation"; these are entirely dependent on the amount of rain

Table 21. Quantity of run-off measured on a test area in the Wadi Maschasch (Negev desert) during the rainy season of 1970/1971. (Evenari et al. 1971)

	Day of rain												
	8. xii	10. xii	13. xii	10. i	12. i	16. i	14. ii	13. iii	3. iv	13. iv	14. iv	15. iv	24. iv
Rainfall (mm)	7.0	3.0	17.0	18.5	15.5	1.0	4.5	6.0	11.0	23.0	13.0	4.0	10.0
Run-off (mm)	0.4	1.3	7.2	9.1	8.8	0.3	1.0	1.5	3.5	10.3	5.8	3.8	6.8
Percentage run-off	5	43	42	49	56	30	22	25	32	45	45	95	68
Overall run-off (%)	42.8												

which falls in any particular year. They serve as a sort of "buffer" in the vegetation, balancing the fluctuations in rainfall. In favourable years the perennials cannot utilize all the available water; excess is available for the ephemerals. They germinate in their thousands and cover the entire desert first with a green, and later, during the flowering, with a colourful veil. They are not strictly limited to a particular habitat, but grow wherever the ground remains moist for some time. If, during the drought years, this is restricted to the depressions, then they grow there only. If soil moisture content is low, the cell sap concentration of these annuals rises rapidly, and dwarf forms develop which flower early, bearing few fruits and seeds. If the rainy period lasts longer, the plants flourish in the favourable hydrature conditions and at the end of the period of vegetational growth they form large numbers of flowers and fruits, so that the seed reserve in the desert soil is again replenished. Thus the density of the ephemeral vegetation depends on the prevailing rainfall. The latter, however, also influences floristic composition, for the following reasons.

1. Each species of annual germinates within a very narrow range of temperature, some at higher temperatures (the summer ephemerals), others at low temperatures (winter ephemerals). Which species germinate and which remain dormant thus depends on the temperature conditions following the rain.
2. Among ephemerals there is considerable variation in growth rate and water requirements. There are hygromorphic, soft-leaved species which grow rapidly but which require the soil to be thoroughly moist and the evaporation not too high. Others are naturally xeromorphic, frequently very hairy or succulent, and have a well-developed root system; these plants are able to survive a short period of drought within the rainy season better than the other species, but they grow more slowly.
3. During a short, favourable season, competition is just as significant among ephemerals as it is in any other plant community. If the rainy season is particularly wet, and the soil and air remain humid for a long time, the hygromorphic and

fast-growing species will be at an advantage, and will displace the more xeromorphic species. When, however, the rainfall is irregular, so that the soil dries out in the intervals between periods of rain, the xeromorphic species, having the capacity to endure a marked drop in hydrature, gain the upper hand.

It is thus easy to understand why, unlike the perennials, ephemeral vegetation shows a very variable picture from year to year. When, in 1934, in Namibia, following three years of very bad drought, a quite exceptional period of rain set in and lasted for several months, farmers maintained that there appeared entirely new species, which they had never seen before. Closer investigation revealed, however, that these were species which preferred moist habitats; thus in normal rainy years they appear in modest numbers and flower only in sheltered places between rocks and in erosion gulleys; they had thus not previously been noticed by farmers. An example is one of the Compositae, *Nidorella*. As a result of the torrential rain, the soil was everywhere so wet that *Nidorella* was soon dominant and in a short time had covered such vast areas with its yellow flowers, that it could not fail to be noticed.

The geophytes can also be regarded as ephemeral in that they frequently have a very brief period of vegetational growth and die back at the onset of drought.

The geophytes or *ephemeroids* (see p 56) also show a marked dependence of their growth on rainfall. They usually have storage organs, bulbs or tubers, which often lie deep in the soil. According to local observers, if insufficient rain falls, the geophytes may not appear at all for many years, but then flower abundantly in years when there is very heavy rain. We have been unable to check such data personally, since this would require an observation period extending over many years; it nevertheless seems to us very likely. The soil in which the bulbs or tubers lie dries out completely during the periods of drought. Only when rain is so plentiful that it moistens the deeper soil layers, making it possible for any newly formed roots of the storage organs to take up sufficient water, will a flower stalk and leaves develop. If this is not the case, the bulbs and tubers remain dormant. In the soil they are so well protected from water loss that they do not themselves dry out, even after many years.

In some years the geophytes may start to shoot when soil temperature rises, but before the onset of rain. This is an indication that there is still a certain amount of reserve water in the deeper soil layers following a good rain in some previous year. This is usually the case on overgrazed land. No investigation has yet been made to determine whether some geophytes store so much water in their bulbs or tubers that they are able to form a flower stalk without taking up water, but this appears unlikely.

There is almost no competition between perennials and ephemerals. As has already been said, the perennials are unable to utilize all available soil water during a good rainy season. Their roots seldom occupy the upper soil layers. Furthermore, the perennials are so widely spaced that their shading of ephemerals is unimportant. It has frequently been observed that the shade beneath the crowns of perennials has a favourable effect because evaporation from the surface is here reduced and more delicate ephemeral plants find protection from direct sunlight. Thus at the end of the rainy period, when the ephemeral vegetation in exposed areas has already dried out, that in the protection of the woody plants may continue for some time.

What has been said about the behaviour of ephemerals in deserts where there are more or less regular periods of rain applies also to those where rain is very erratic. The difference is that the ephemerals do not appear every year, but only in years when there is exceptionally heavy rain and in those regions of the desert where this falls. For example, the outer Namib Desert is normally almost entirely devoid of vegetation. In 1934, however, over 100 mm of rain fell and the desert was covered with a rich flora of *Mesembrianthemum* species. These we saw ourselves the following year, because these succulents stored so much water that, even without further rain, they survived for more than a year and were still able to flower (vide Vol. 2, Namib Desert).

6.10 The Importance of Soil Type for Water Supply and Vegetation in Dry Regions

The quantity of water which penetrates into the soil depends not only on the rainfall but, as we have seen, also on amount of run-off; not all of this water is available to plants. To make this clear, the highly complex conditions can best be explained with the help of the diagram in Fig. 96.

Of the total amount of water reaching the soil surface, whether directly as rain or as drainage from elsewhere, some will seep into the soil, some will immediately evaporate and some may flow off the surface. The

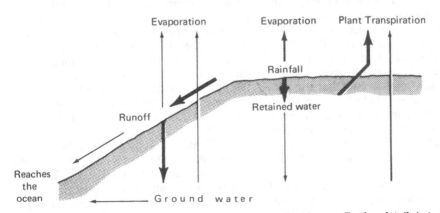

Fig. 96. Schematic representation of the water economy of arid areas. Further details in text. (Walter and Volk 1954)

water which seeps to a greater or lesser depth into the soil is then held by soil particles as capillary water. In arid areas there is usually no excess which would then percolate to the groundwater. Here the soil is moistened periodically only to a certain depth. Beneath this is the so-called "dead soil" which contains no available water for plants.

The water which flows off the surface reaches the valleys by way of erosion gullys. After very heavy rain, the flow of water may be so considerable that in coastal deserts it can even reach the sea; otherwise it flows into depressions from which there is no further drainage. After light rain, on the other hand, the water sinks rapidly into the river beds and is there stored as groundwater. Where the groundwater lies close to the surface it may be lost by evaporation or be used up by deep-rooting plants.

Apart from river valleys, groundwater reserves in arid areas are mainly found in rocky clefts or in the beds of dry rivers, that is, rivers in which water flows only periodically. There may even be no groundwater at all.

For the vegetation, it is the quantity of capillary water in the soil that is important. This forms the water reservoir on which the plants can draw. The quantity of soil water thus held depends not only on the annual rainfall but, to a very large extent, also on the nature of the soil itself.

The slope of the ground and the porosity of the soil determine how much water flows to or drains from any particular area. Generally, the run-off is greatest from clay soils, which increases their dryness; there is very little run-off from sandy soils and none at all from loose, stony ground. The run-off from a rock surface is also high, provided it is intact, but less if there are fissures in the rock. Smith (1949) pointed out that any bare surface in an arid area is further compacted by the tread of grazing animals, by the pressure of car tyres and by the impact of falling raindrops, and this results in greater run-off. This applies also to soils which are periodically flooded, particularly if they are lightly clayey. There is usually very little plant growth on flooded terraces, since, after every fall of rain, all the seeds are carried away by the sheet flow.

The nature of the soil determines also how much of the water remains in it as capillary water and how much evaporates. It is important to note that the water-retaining capacity of different types of soil in arid areas is exactly the reverse of that in humid conditions. In the latter, clay soils, with their high water-retaining capacity, are regarded as the wettest soils, sandy soils, and particularly stony ground, as the driest. In arid areas precisely the opposite is the case. The reason for this is that in arid regions the soils are never thoroughly wetted. Large reserves of water are found only in the upper layers into which water seeps during the rainy season. Assuming, for example, that 50 mm of rain falls, this will wet only the upper 10 cm of a clay soil, but will penetrate 50 cm deep in sandy soil and even deeper in ground where there is coarse stone or rock; in the latter case, water is retained only in cracks in the rocks where fine soil has collected (Fig. 97).

After rain, evaporation takes place. A clay soil soon develops cracks as the upper 5 cm rapidly dries out; this means that 50% of the rainwater which had penetrated the soil is lost again. In sandy soil the surface dries out, but the capillary connection between this surface water and that lower down is soon broken, so that more than 90% of the water remains in the soil. There is almost no evaporation from rocky ground, where storage capacity is thus maximal. Thus in an arid region it is possible to find trees growing in a rocky area, grass vegetation in sandy areas, while there is no vegetation at all where there is clay soil. This explanation by the authors of conditions in western North America was confirmed by Hillel and Tadmor (1962) for the Negev Desert. They found that in areas with the same annual rainfall, the quantity of available water for the plants varied according to the nature of the soil: where there was loess, available water was 35 mm, in a stony habitat (from which the run-off was, however, high) it was about 50 mm, on sandy soil it was 90 mm and in dry valleys (into which rainwater drained during the rainy season) it was 250–290 mm.

If there are only a few clefts in a rocky surface, rainwater may run along them and percolate through to reach the groundwater. The soil which has collected in such fissures

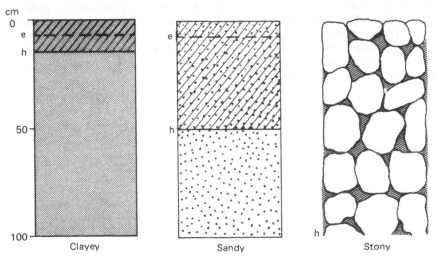

Fig. 97. Diagrammatic comparison of water retention in different soil types in arid areas after 50 mm rainfall. *h–h* Lower limit of water-saturated soil; *e–e* lower limit to which the soil dries out again. Clay soil retains 50%, sandy soil 90% and stony soil 100% of the rainfall

is consequently moistened from the surface right down to the groundwater. Plant roots can thus extend inside the fissure to the level of the groundwater, thereby ensuring a supply of water to the plant throughout the year. In such places plants with very deep tap-roots are, in fact, found, but this type of root system is very rare. The depth to which such roots grow depends entirely on the depth to which the soil is moistened and in arid areas this is normally not very deep.

We saw an interesting example of the way in which groundwater can be used for aforestation near Basra in Iran. In this region there is an inexhaustible supply of groundwater at a depth of 15 m, for it is constantly replenished from the Tigris and Euphrates via gravel strata. The scanty rainfall of 120 mm moistens only the upper layers of soil; the layer of dry soil beneath this prevents plant roots from penetrating as far as the groundwater. Thus the only vegetation is formed of ephemerals which develop after rain. Access is gained to the groundwater by wells; the water is raised, usually by primitive means, and used to water furrows in which onions, tomatoes and pumpkins are grown. At the high summer temperatures which prevail (up to 50°C), water must be led into the irrigation furrows eight times a day. As a result of rapid evaporation, the soil soon becomes saline, so that vegetables can be grown on any particular plot for 1 year only. Cuttings of *Tamarix articulata* are stuck into the ground between the vegetables. These cuttings rapidly

take root, and since the irrigated soil is moist from the surface down to the groundwater, the *Tamarix* roots can continue to grow downwards deeper and deeper until they finally reach the groundwater. Thereafter, survival of the trees is ensured. They are cut back every 25 years, but regenerate from the stumps. Thus where there had been desert, a forest has developed around the site of the abandoned watering hole.

When account is taken of all the factors affecting water supply—climate, nature of the relief, type and composition of the soil—it becomes clear that in arid areas there is an endless variety of habitat conditions; indeed, no two habitats are exactly alike. It is thus impossible to agree with the opinion frequently expressed that the multiplicity of adaptive types found amongst plants is far greater than the variety of habitats. Such a statement is based on a very rough division of habitats into a small number of different types, while at the same time according great importance to the smallest differences in the vegetation, even differences in the rate of transpiration in some species. It has perhaps a certain validity when applied to a small and seemingly uniform habitat, in which there are plants belonging to many different ecological types. In this case, however, we are usually not able to determine whether the soil does, in fact, provide identical conditions for the different types of root system. Added to this is the fact that each species of plant has a certain ecological tolerance, so that species which belong to one ecological type may be encountered in different habitats. Since the number of plant species is limited, and the number of ecological types even more so, while habitats show countless finely graded differences, the statement could in fact be reversed. Such generalizations are, however, of little scientific value.

7 The Competitive Factor and Root Competition

7.1 General

The assumption is frequently made that the distribution of plants is determined directly by environmental conditions; this is, however, incorrect, for these are of importance only indirectly in that they affect the competitiveness of the plants. Only at the absolute limits of distribution, in arid and icy deserts, at the edge of the salt desert or where the forest shade is deepest—in other words, where competition is absent—do environmental factors (usually one extreme factor) have a direct effect on plant distribution. Apart from these exceptional cases and the pioneer plants on soil that is otherwise devoid of vegetation, all plants in natural conditions are in competition with each other. We must therefore move away from the purely physiologial to a more ecological approach. A physiologist studies plants in clearly defined conditions, quite independently of the natural environment, and has a more analytical approach. An ecologist must adopt a more integrated outlook, taking into account the constantly changing conditions of the natural environment and the competitive pressure exerted by neighbouring species.

What we understand by the competitive factor is, in broad terms, the inhibitory effect which plants growing together in a limited space exercise upon each other, while excluding actual parasitism. Plants compete for light, soil water and nutrients— in other words, purely physicochemical relationships are here involved (Clements et al. 1929; Boysen-Jensen 1949). There is no clear evidence that in natural conditions small amounts of certain compounds (e.g. terpenes), released from roots, litter or aerial plant organs, have an inhibitory effect on other species and in this way affect competition.

Hitherto, none of the so-called "allelopathic" effects, repeatedly observed in the laboratory, have been found to be effective in natural conditions (Ahlgren and Aamodt 1939; Börner 1960; Cannon et al. 1962; Donald 1946; Grümmer 1955; Knapp 1967; Martin 1957; Rademacher 1959; Whittaker 1970). Grodzinski believes that such an effect probably occurs in *Salvia* in California (Muller and del Moral 1960), and in some steppe plants (Grodzinski 1965, 1968), but this view is not generally accepted.

A distinction must be made between intra-specific competition occurring between members of the same species, and inter-specific competition, which takes place between individuals of different species. These two forms of competition have quite different effects.

Intra-specific competition is important in the single-species cultivations of agriculture and forestry. If the number of plants in a field is very large, the individual plants are so weakened by the strong competition that the harvest is small. If the plants are spread so far apart that competition is virtually excluded, each plant develops well, but the number of plants is small and the harvest per unit area is low. In addition, the danger of weed infestation is very great. For this reason, an empirical calculation has been made for each crop, under different climatic conditions, of the density of planting which will give the best possible yield. Foresters prefer initially dense stands, to promote the growth of tall trees with trunks relatively free of branches; at a later stage, weakening of the stand through competition between the individual trees is prevented by thinning out.

Single-species stands also occur in nature as, for example, on burnt areas of virgin forest in the boreal zone where *Pinus* species sow themselves readily; on saline soils on areas of mud flat along sea coasts on which thousands of seedlings of *Salicornia europaea* appear; or in deserts where, after rain, the seeds of an ephemeral species are washed together and then germinate; or in the wet forests of a ravine, where one may come across masses of *Impatiens noli-tangere*. These naturally pure stands are, however, never so uniform as those of cultivated plants and selection soon takes place: the strongest individuals grow, flower and fruit, while the weaker ones die. This intra-specific competition is thus useful for the maintenance of the species. The initial high density of the stand in fact protects the species from competition from other species.

It has been found in the east European steppe, when strips of forest were laid out, that if acorns were set out singly, the young seedlings were suffocated by the grass vegetation: they succumbed in competition. If, however, the acorns were planted very close to one another, only those seedlings on the outer edge, exposed to competition with the grasses, were overgrown. These protected the other members of the species in the middle of the clump, so that these then had time to grow above the grass, to develop strong root systems and eventually grow into trees (Orlenko 1955).

The effect of inter-specific competition is different in that while some species gain the upper hand, others are suppressed. In a plant community of long standing, however, a certain equilibrium between the various species is established in the course of time. Usually several similar species in a characteristic numerical ratio to one another make up a particular stratum; examples are to be found in mixed deciduous forest with several tree species, in forests with both deciduous and evergreen species together, and in grassland or steppe where there may be several different species of tall grasses. This results whenever intra-specific competition is stronger than inter-specific competition (vide p 163).

That these species are always in strong competition is shown by the fact that the slightest change in habitat brings about an alteration in the numerical ratios between the different species, and also by the fact that none of the plants grow as luxuriantly as when it is alone in a garden. Within a community the plants suffer from a certain under-nourishment. Thus a plant commu-

nity exploits its habitat to the utmost, and *no room remains for foreign intruders*, which explains the great stability of an undisturbed plant community (Ellenberg 1954).

Just how decisive competition is for the distribution of plants has been demonstrated by an experiment involving groundwater levels undertaken at the University of Hohenheim in West Germany.

Alopecurus pratensis, Arrhenatherum elatius and *Bromus erectus* are three important meadow grasses in central Europe. The first is dominant on wet meadows, the second on moist and the third on dry grassland. If, however, they are sown individually on plots of ground where the groundwater level varies in depth from 0–150 cm (Fig. 98), all three show a growth optimum at about a middle value of groundwater level. Only when sown together do differences in ecological behaviour become apparent; these are shown diagrammatically in Fig. 99. *Arrhenatherum*, competitively the strongest species, occupies the optimal habitats in the middle range of the groundwater levels, displacing *Alopecurus* to the wetter parts and *Bromus*, especially in nitrogen-poor soil, to the driest habitats. In time, weeds also develop, but the seeds of these are distributed indiscriminately over the whole area. In competition with the grasses and with each other, however, some weeds could establish themselves only in the wet part of the plots, others on the moist and yet others on the dry areas, a distribution corresponding to their occurrence in meadows (Lieth 1958). A few are indifferent to groundwater level.

The pH of the soil also has an indirect effect on plants. This explains why a plant species which is found only in a very narrow range of pH in conditions of strong competition loses all its indicator value and becomes rather indifferent to soil characteristics when competition disappears (Ellenberg 1950, 1968).

It is thus always necessary to distinguish between a *physiological optimum* in the absence of competition, and an *ecological optimum* when the plant is in competition with other species. The ecological optimum corresponds to the conditions prevailing in habitats where a particular plant species grows most prolifically in nature. Since this

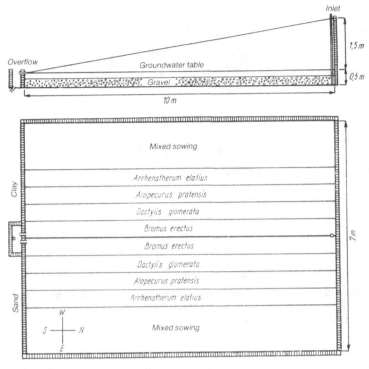

Fig. 98. Groundwater experiment conducted at Hohenheim (West Germany): basin with constant groundwater level. *Above* lateral view; *below* view from above showing the experimental layout in 1953

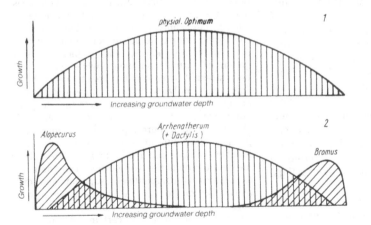

Fig. 99. Growth of meadow grasses with increasing depth of water table. Diagram **1** applies more or less to all species investigated in pure sowings; there is a growth optimum around the medium depth of groundwater; **2** shows the distribution obtained with mixed sowings: *Alopecurus pratensis* has a maximum growth on wet ground; *Bromus erectus* a maximum on dry ground. For further details, see text

ecological optimum is not, however, determined by habitat alone but, as we have just shown, often to a far greater degree by the presence of particular competitors, the occurrence and distribution of a particular species can be markedly different when other competitors are present. *Hypochoeris radicata*, for example, occurs in central Europe only in certain meadow communities. It was introduced into New Zealand and there

met with quite different competitors by which it was less strongly suppressed; the result has been that it is found in practically all plant communities, right up to the tree line in the mountains; it never actually becomes dominant, however.

If all the inhibitory influences arising from competitors are summed up as "competitive pressure", the relationships between the ecological and physiological optima under

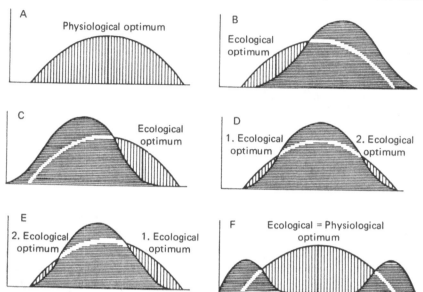

Fig. 100. Schematic representation of the growth of a plant species *(vertically striated)*. **A** In the absence of competition; **B–F** with competitive pressure of other species *(horizontally striated)*. *Ordinate* growth rate or productivity; *abscissa* changing habitat factors. (Walter 1960)

varying degrees of such competitive pressure can be represented diagrammatically as in Fig. 100.

The area in which a plant species will flourish if grown on its own, free from competitive pressure, is fairly extensive. It develops best in its physiological optimum (Fig. 100 A). As soon as other species are present, however, its range is severly limited. Its area of most frequent occurrence, that is, its ecological optimum, is displaced in one direction or the other, depending on the nature of the competitor and the degree of competitive pressure which it exerts (Figs. 100 B, C).

A species can even have two ecological optima (Figs. 100 D and E). Only when a species is especially superior to its competitors will it thrive in those conditions which are most favourable to it, so that the physiological and ecological optima coincide (Fig. 65 F), but its range of successful growth will be somewhat restricted at the periphery.

A further example may show the extent to which we need to readjust our evaluation of competition as a determining factor. In Missouri (USA) *Quercus palustris* forms whole stands on river banks, around water basins and in marshes. In dry habitats on elevated ground, however, *Quercus*

rubra is the dominant tree species. Sullivan and Levitt (1959) thus postulated that *Quercus rubra* is more resistant to drought and better able to endure water shortage. This was put to the test by comparing the physiological behaviour of young plants of the two species, some of which had been grown in normal conditions, while others had been stressed by subjecting them to water shortage. Surprisingly, both species appeared equally drought-resistant. The authors state that the only difference was that, when water was plentiful, *Q. palustris* grew more rapidly that *Q. rubra*. It is precisely this difference which seems to us of great importance and makes it possible to explain the distribution of the two species. In wet habitats *Q. palustris* is competitively superior and displaces *Q. rubra* to the drier habitats, where it is able to assert itself. It is not drought resistance, but more rapid growth in certain environmental conditions which is decisive (Zelniker 1968).

The distribution of a species thus gives little indication of its physiological requirements. The fact, for example, that in central Europe, pines are found in natural conditions only on dry, limy or sandy soils or in acidic bogs, is attributable to their displacement from more favourable habitats by stronger competitors. Knowledge of the physiological needs of a species is thus an insufficient basis either for predicting or for explaining its distribution in nature. Whether or not a species will, in fact, establish itself in a physiologically suitable habitat will depend, apart from any historical factor, on the nature of its competitors.

7.2 Analysis of the Competitive Factor

The competitive ability of any species may vary, depending on external conditions. One species may now be superior to, now inferior to another. In ecological studies it is therefore desirable to make a quantitative estimation of competitive strength. This is usually not possible, however, because the number of morphological and physiological characteristics which determine it is very large.

Productive capacity is, however, always an essential prerequisite for competitive strength. The greater a plant's productivity, the better and taller will it grow, the more fruits will it bear, the more numerous or larger the seeds it will theoretically be capable of producing. Greater productivity thus usually implies greater competitive ability. What is important is not its absolute magnitude, but productivity relative to that of competitors.

7.2.1 Vigour and Competitive Strength

The vigour (W) of a species A has been defined by Bornkamm (1961a) as the ratio of the dry substance production per individual of species A (M_A) to that per individual of species B (M_B). Thus $W_{A/B} = M_A : M_B$ or $W_{B/A} = M_B : M_A$.

Vigour has been found to be a fairly useful unit for the quantitative assessment of competitive ability, provided certain environmental conditions prevail and similar growth forms are compared (Bornkamm 1961a,b, 1963). If a comparison is made of the competitive ability of 1-year-old plants of Sinapis alba, Avena sativa, Vicia sativa and Triticum aestivum in different combinations but with a fixed density of 100 plants per m², Sinapis shows the greatest vigour and is very much superior to the others; it is followed by Avena, then Vicia and Triticum,

although the difference between the latter two is not significant. If the vigour of Triticum compared with Sinapis and that of Triticum compared with Vicia are known, the vigour of Vicia relative to Sinapis can be calculated, because $W_{T/S} = W_{T/V} \cdot W_{V/S}$ or $W_{V/S} = W_{T/S} \cdot W_{V/T}$.

Individual Sinapis plants showed poorest growth in a pure stand, that is, when the other 99 of the 100 plants on the 1 m² area were of the same species, while its development was better, the weaker the competition from the other 99 plants. The competitively weakest species, Triticum, grew best on its own, but was smaller the greater the number of strong competitors present. Thus intra-specific competition is less favourable than inter-specific only when a competitively strong plant species is involved; otherwise the reverse is the case.

It follows that a mixed stand, left without interference, will never turn into a pure stand of the competitively strongest species, but that a certain mixture will develop, as can be seen, for example, in a meadow. A field, too, becomes more and more weed-infested if the weeds are not constantly controlled (Harper 1967). Pure stands of a tree species are found only where environmental conditions are such that competitors are excluded.

Bergh (1968) has conducted some experiments on competition in grasses. It should be noted, however, that vigour can only be used as a measure of competitive ability when the plants under consideration are of comparable height, so that light is not a determining factor in competition; if one plant increases rapidly in height, while another spreads out over the surface and is shaded by the first, the taller plant is always at a competitive advantage.

7.2.2 The Role of Height in the Competitive Struggle

Iwaki (1959) has examined competition between Fagopyrum esculentum and Phaseolus viridissimus. When grown separately, the productivity of the two species was almost the same. In Phaseolus the leaves comprise 50–60% of total dry weight, which is more advantageous than in Fagopyrum, where they make up but 20–60% of dry weight. *Fagopyrum*, however, increased in height at about twice the rate of *Phaseolus*, so that its leaves are better exposed to light and the rate of assimilation is greater. In a mixed stand the leaves of *Phaseolus* receive only 16–35% of the available light and their photosynthetic capacity shows a marked decline. If *Phaseolus* is given a head-start by sowing *Fagopyrum* 3–13 days later, the development of the beans is greatly enhanced; yet *Fagopyrum* is not at a serious disadvantage, for its stalks soon grow above the beans and it always obtains sufficient light.

It is thus clear that growth in height, which is a species-specific characteristic and is dependent on productivity to a limited extent only, is of particularly great importance to competitive power (e.g. herbs in a stand of low grasses). Height is also decisive in intra-specific competition as, for example, in a dense stand of sunflowers *(Helianthus annuus)*. The seedlings from large seeds increase in height more rapidly than those from small seeds (Kuroiwa 1960). They receive more light and accordingly produce more dry substance, so that their advantage over the suppressed plants constantly increases. Since photosynthetic processes, but not respiration, are inhibited by the unfavourable light conditions, the disadvantaged plants use up a relatively large portion of their carbohydrates in respiration. While net production is about 48% of gross production in the more rapidly growing plants, it falls to only 12% in the shaded plants. In even less favourable conditions, it falls to below 0% and the plants soon die.

7.2.3 Light Conditions and Competitive Ability

The competitive power of the individual species always depends to a marked degree on environmental conditions; for the competing shoots, light is decisive; for the roots, on the other hand, water and mineral salts are the most important factors. It is usually impossible to draw a sharp distinction, however, for the growth of the roots is dependant on production by the leaves, and the

photosynthetic capacity of these is, in turn, dependent on an adequate supply of water and mineral nutrients.

Bornkamm (1961b) studied the influence of light on competition between *Sinapis alba*, *Agrostemma githago*, *Bromus secalinus* and *Anagallis arvensis*. Light intensities of 100%, 60%, 30% and 15% of full daylight were used. In each container (type "Mitscherlich" used in Germany) there were 24 plants; these were either all of the same species, or combinations of 2 (12 of each), 3 (8 of each) and 4 (6 of each) different species.

There is a logarithmic relationship between production as a whole and light intensity. The largest part (80%) of the combined yield (from all the experiments) of equal numbers of the four species was attributable to *Sinapis*, the smallest (1–2%) to *Anagallis*. In containers with only one species, reduction in light intensity from 100% to 60% of the available light had very little effect on productivity; further reduction to 15% caused production to fall to one-quarter or one-third in three of the species, but in *Bromus* to only one-half of maximum. The latter species clearly had the lowest light requirement. When two or more species were grown in combination, the situation was more complex. *Anagallis*, as the competitively weakest species, profited from the weakening of its competitors at low levels of illumination: in combination with one other species its productivity remained constant, irrespective of light intensity; in combination with two other species and at 60% and 30% of available daylight, its productivity increased fourfold, and in combination with three others, sevenfold, compared with its productivity at 100% illumination. In *Bromus*, productivity doubled when it was grown in combination with three other species and at 30% available light. It can be concluded that the competitive ability of *Anagallis* compared with the other partners increases with decreasing light intensity. Its vigour relative to *Sinapis* increased fivefold at 15% daylight, 11-fold compared with *Agrostemma* and at 30% daylight three- to fourfold relative to *Bromus*, compared with its vigour relative to these species in full daylight. A similar relationship is found between *Bromus* and *Sinapis* or *Agrostemma*, while competition between the latter two species is uninfluenced by light intensity. It should not be imagined, however, that light has a direct influence here. The poorer growth of the competitors in weaker light results not only in a reduction in the shoot, but also, and to a far greater extent, in the root system. With falling light intensity, the ratio of shoot to root increases. Thus in poor light conditions, root competition will be very much reduced, to the advantage of the weaker partner.

7.3 Root Competition

The precise way in which root competition functions in natural conditions has been elucidated by the investigations of Karpov (1961, 1962a–c, 1969a,b) in a boreal forest of spruce, Piceetum myrtillosum. Illumination in such an old forest, in which the tree canopy absorbs 80% of the available light, is very poor. The undergrowth consists mainly of bilberries *(Vaccinium myrtillus)* and mosses. The wood sorrel *Oxalis acetosella* shows very stunted growth. Light is generally regarded as the limiting factor for the development of the undergrowth, but Karpov has shown that competition from the roots of the spruce is far more important. When this was

eliminated by cutting through the roots to a depth of 50 cm around a 1 m^2 test area, the herbaceous plants showed far stronger growth in the same light conditions. It is particularly striking that *Oxalis* rapidly covered the whole area and displaced the layer of moss. After a few years, even the raspberry *Rubus idaeus* established itself and spread rapidly. Investigation of photosynthesis in *Oxalis* showed that there were no differences in the rate of CO_2 uptake per unit area of leaf between plants in the control area and those in which root competition had been eliminated. Nonetheless, productivity in individual *Oxalis* plants increased tenfold. They grew luxuriantly and developed a large leaf surface (Fig. 101). The same was true of *Majanthemum bifolium* and *Trientalis europaea*. This apparent contradiction can be explained as follows. The roots of the spruce and of the herbaceous plants are limited to the upper layers of soil—in fact, 90% of all the roots were found in the upper 20 cm of soil. The roots of the herbaceous plants thus grow in a dense network of tree roots. Water is not a limiting factor in the very moist boreal climate; the groundwater is in fact close to the surface. No difference was found in the water content of the soil of control and test sites and it was always above the wilting

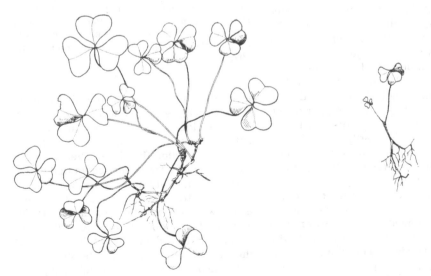

Fig. 101. Development of *Oxalis acetosella* in a spruce forest, *left* when root competition is eliminated; *right* with root competition. Half natural size. (Karpov 1962a, b, from Walter 1973)

point. The limiting factor was found to be the quantity of available nitrogen (N): while the plants on the control site showed unmistakable signs of N-shortage, namely, a yellow-green colour of the leaves, those on the test site were deep green. Analyses of *Oxalis* leaves revealed a N-content of 2.06% on the control site, but 3.62% in the absence of root competition on the test site. The acute N-shortage to which the herbaceous plants of the forest are normally exposed results in a small leaf surface relative to the non-CO_2-assimilating organs. Net photosynthesis is just adequate to cover the respiratory losses of the plant as a whole. There is no substantial gain in organic matter and therefore no significant growth. If, however, the nitrogen supply is greatly improved by eliminating root competition, a relatively large leaf surface is formed, resulting in a net gain in organic material, which in turn contributes to further increase in leaf surface. That nitrogen is indeed the decisive factor is shown by the fact that in the course of time, on test sites where not only root competition had been eliminated, but the herbs and mosses had also been removed, a series of nitrophilic species established themselves. Similar results were obtained in experiments involving transplantation of species, such as *Aegopodium podagraria* and *Ajuga reptans,* for which a high level of nutrients is essential. It must also be borne in mind that the severed roots died and decomposed within the test area; this may have had a fertilizing effect on the soil, particularly with regard to nitrates. The experiment described below involving fertilization with ^{32}P showed, however, that such an effect is not significant: the N-content of the roots is low. The poor development of the herbaceous layer in boreal spruce forests is not primarily due to lack of light, but to the shortage, arising from competition with the tree roots, of nitrogen and other mineral nutrients. Mosses are not affected by this; they obtain their nutrients from the water which drips off the trees or from the spruce needle litter. This enables them to grow luxuriantly and to cover the forest floor with their cushions (Tamm 1950).

Root competition from the old trees inhibits the growth of young trees. Thus spruce seedlings develop only very slowly in

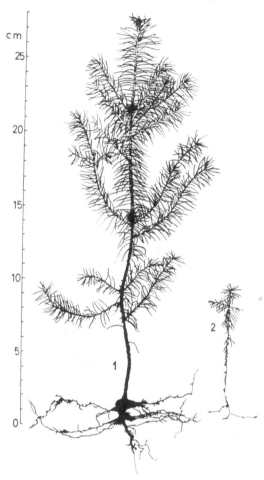

Fig. 102. Development of a spruce seedling in a stand of birch in the boreal zone: **1** when root competition of the old stand is eliminated; **2** with root competition. (Karpov 1962a, from Walter 1973)

a stand of birch, although the illumination is good. On test sites where competition from the older roots had been eliminated by cutting through them, the spruce seedlings developed better (Fig. 102). The following experiment of Karpov has shown that in this case, too, mineral nutrients are the decisive factor. Control and test areas were treated alike with a fertilizer containing radioactively labelled phosphorous (^{32}P). Subsequently, the radioactivity of the needles of the spruce seedlings was measured, and found to be five to six times higher in plants from the test area than in those from the control site (Fig. 103). In this case the additional

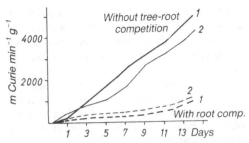

Fig. 103. Accumulation of radioactive phosphorus in the needles of spruce saplings: *1* in a birch wood with *Vaccinium myrtillus* on heavily podzolized soil; *2* in a birch wood with well developed herbaceous layer of *Aegopodium* on weakly podzolized soil. *Upper curves* without, *lower curves (broken lines)* with tree root competition. (Karpov 1962b, from Walter 1973)

phosphorus taken up clearly did not have its origin in decaying, severed roots. In no case was it possible to demonstrate any effect of inhibitory substances or excretions from the roots.

The original experiments of Karpov were repeated by Dylis and Utkin (1968) in a mixed spruce-deciduous forest with a well-developed herbaceous layer. Elimination of root competition had a noticeable effect after only 3–4 weeks. The dark green colour of the leaves of herbaceous plants on the test site was particularly striking. In August, the chlorophyll content of *Oxalis* and *Picea* seedlings was more than double that of the plants from the control area; in late autumn the green colour of the test sites stood out against the already yellowing surroundings. The water content of the herbaceous plants (*Carex pilosa, Galeobdolon, Oxalis* and *Dryopteris linnaeana*) was markedly increased. Similarly, the nitrogen content was 1.5–2 times higher, which was more or less optimal. The phytomass of the herbaceous layer showed a marked increase: after 2 to 2.5 months, it was 120–160% of that on the control site and by the next autumn 180–195%.

The species of tree involved may also be important: *Aegopodium*, for example, reacted far more markedly to elimination of root competition when under spruce than it did under oaks or aspen.

When illumination of the forest floor was especially poor, no clear effect could be detected, even after 1½ years; clearly the light was here at a minimum.

Since the test sites had a higher soil-water content to a depth of 55 cm, the possibility that this factor alone had led to the favourable development of the herbaceous plants was investigated. During June and July, some areas were watered four times, so that the plants received an extra 143 mm of water; no effect was observed, however. Other factors must thus be involved. Fertilizing at the rate of 40 kg ha^{-1} with N, 40 kg ha^{-1} with K and 29 kg ha^{-1} with P likewise had no effect; the mineral nutrients were probably taken up by the tree roots. Only when N fertilizer was applied at a rate of 75–200 kg ha^{-1} was there any deepening of the green colour of the herb leaves and thereafter stronger growth. Thus it seems very likely that after the elimination of root competition from the trees, the nitrates which would otherwise have been taken up by the trees became available to the herbs. It is not certain whether there was any significant effect of a possible additional fertilization of the soil due to decomposition of severed roots. One would have to investigate the rate at which they were mineralized and what quantities of N were involved. The rapid effects of eliminating root competition argue against this idea, since most of the nitrogen contained in the roots is initially incorporated into the cell substance of microorganisms and only gradually mineralized.

A similar investigation was carried out in the dark *Abies sibirica* forests near Krasnojarsk in Siberia; here 80% of the forest floor is covered with moss (*Hylocomium splendens*), while the herbaceous flora is not much in evidence. On test sites in which root competition had been eliminated, the number of herbaceous plants and the extent of their coverage were observed over a 3-year period (Latschinski 1968). Here the main mass of tree roots lies in the upper 20 cm, and the roots of the herbaceous plants are scattered among these, making up only 0.18% of the total root mass. After 3 years, all the herbs were more numerous: in particular, the shoot count of *Oxalis* rose from 58 to 856 per m^2 while its ground coverage increased from 18 to 60%. The greater development of the herbaceous plants resulted in repression of the mosses, the ground cover of which declined to 33%.

Generalizations should not be made, however. In beech stands on deep brown forest soils, root competition for nutrients seems to be of less importance to the herbaceous layer, since these latter have their roots in the upper humus layer, while the tree roots lie somewhat deeper. In this case shading is more important. The investigations of Burschell and Schmalz (1965a) on root competition in an old beech wood showed that in a wet year, the roots of the old trees had no influence on the development of the beech seedlings. On the other hand, Slavikova (1958, 1965, 1966) was able to show that in a deciduous forest near to its aridity distribution limit, root competition for water had a marked effect on the development of the undergrowth, which in this area is mostly very repressed.

It is very striking that at the aridity limit of forests of certain tree species there is no undergrowth, that is, no herbaceous layer at all. These are known as "nudum" communities. An example of such a "Fagetum nudum" is to be seen on the Calcidean Peninsula in Greece, but it is found also in the subcontinental climate of central Europe, for example, in Bohemia (Slavikova 1958). In other words, it occurs wherever the water supply is just adequate to meet the needs of the beech tree layer. At the same time, a massive litter layer of still largely intact beech leaves accumulates in such stands, presumably because it is so dry that their decay by saprophages and decomposers is inhibited. In pine forests on dry, sandy, shallow granite soils there is likewise no herbaceous layer, because the trees use all of the limited water reserves. Lichens do occur, but they do not take up water from the soil. Similar "nudum" *Abies* stands are met in northern Anatolia as, with increasing dryness towards the interior, the aridity limit of *Abies bornmuelleriana* is approached; yet another is in southern Anatolia at the aridity limit of *Pinus brutia* forests. In clearings within such "nudum" forests, forest herbs or dwarf shrubs do, however, occur, for here root competition is absent.

It is very striking that mycotrophic saprophytic flowering plants such as *Neottia* (Pyrolaceae) are frequently encountered on the floor of "nudum" stands; as a result of the extremely low, almost non-existent transpiration in these non-photosynthesizing species, their water requirement is insignificant.

This all shows that tree roots have a greater capacity for taking up water from the soil than do those of the herbaceous plants which have thus to depend on any residual water.

In a mixed oak and hornbeam wood, Ellenberg (1939) observed marked wilting of species in the herbaceous layer during an exceptionally dry period in late summer, and was able to demonstrate a clear relationship between the degree of wilting and the water potential of the soil. Slavikova (1965) has since shown in container experiments that the maximal suction pressure (minimal water potential) of the roots differs in different herbaceous species, being very low in species from moist forests of river banks and much higher in those from dry forests. These results are shown in Table 22.

In a further investigation (Slavikova 1966) it was shown that the maximal suction pressure (minimal water potential) of an ash stand was 37–39 bar, and thus considerably higher than for all the herbs of the same

Table 22. Maximal suction pressure ($\equiv$ minimal water potential) in the roots of different forest plants; values given in bar (Slavikova 1965). The first 8 species are flood plain forest species

Milium effusum	6.3
Asarum europaeum	10.7
Lamium galeobdolon	14.0
Viola riviniana	15.8
Aegopodium podagraria	16.1
Hepatica nobilis	19.3
Mercurialis perennis	19.8
Pulmonaria officinalis	21.0
Geum urbanum	27.2
Fragaria vesca	28.4
Carex muricata	29.0
Lamium maculatum	29.2
Poa trivialis	36.4
Cynanchum vincetoxicum	37.4
Viscaria vulgaris	44.2
Sanguisorba minor	48.9
Calamintha clinopodium	53.6

stand. This shows that the ability of tree roots to take up water is very much greater than that of the herbaceous plants, so that the latter are entirely dependent on the quantity of soil water not taken up by the trees. The higher pressure values (lower water potential) of the woody species can be related to their need for water to be raised in the conducting vessels right up to the crown. The presence of mycorrhizae is also important for better exploitation of soil water.

7.4 Life Form and Competitive Ability

The examples mentioned above dealt with competitive ability during periods of vegetative growth. Each individual species must, at all stages of the life cycle, be able to maintain its position in a plant community despite inter-specific competition. The species must be able to reproduce successfully within the community. The age at which the individual species set seed has a bearing on this. Annuals can produce seeds once only, and have to begin their development each year from the seeds; longer-lived plants are able to establish themselves and become competitively stronger over many years before they produce new seeds. This is a distinct advantage. The various developmental types will be discussed briefly (Monsi 1960).

Annuals

The period of development of these plants, from germination to the production of fruit, is very brief. The number of seeds produced by a single plant can be very great, with the result that annuals multiply rapidly in an area where they are not exposed to competition, always provided environmental conditions are otherwise favourable to growth. The ephemeral vegetation is typical of extreme deserts in a good rainy year and may be absent for many years with poor or no rain. Furthermore, they are found as complementary species (see p 154) (ephemerals, therophytes) in all plant communities of those arid regions where there is only poor cover of perennials; they make their appearance if, during a short, relatively warm period of the year, there is an excess of water in the upper layers of the soil, which cannot be fully exploited by the perennials. The same considerations apply to open areas which have arisen as a result of some catastrophe or have been created by man. Compared with the other life forms, annuals are competitively the weakest.

Biennials

During the first year, these plants behave as annuals, except that they do not form reproductive organs. Instead, a rosette is formed, which survives the unfavourable conditions of the winter or a dry season and which can begin photosynthesis as soon as conditions become more favourable. As a result, they have an advantage over annuals in the second year (Fig. 104); their productivity is greater, they develop more strongly and can produce more seeds. The precondition for their development is that there should be two successive favourable periods separated by one which is not extremely cold. This precondition is satisfied only in areas with a relatively favourable climate, but here perennial plants are also present. With these, biennials cannot compete and they are thus found only in disturbed habitats with an unstable plant community, as weeds or ruderal plants, in forest clearings, and so on.

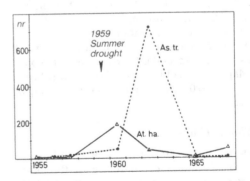

Fig. 104. In the first year (1960) after bad drought in the summer of 1959, the annual *Atriplex halimus (At. ha.)* spread over the naked soil of *Festuca rubra* salt marshes near Birdwater (England); in the following years it was displaced by the biennial *Aster tripolium (As. tr.)*. Thereafter, the perennials again became dominant. (Ranwell 1972)

Perennial Herbaceous Species

During their first years, these species show only vegetative growth. In areas with an unfavourable season they behave as geophytes or hemicryptophytes; that is, their aerial parts die. They accumulate reserves, however, in underground organs; these increase in size from year to year so that the plants show increasingly strong new annual growth and become competitively ever stronger compared with annuals and biennials. They have a life span of many years—in fact, for plants with rhizomes or bulbs, it is almost unlimited. After a few years they begin to flower and produce seeds which serve to disperse the plants more widely. The established plants reproduce vegetatively and therefore often occur in masses, which further increases their competitive ability. In a humid-tropical, permanently favourable climate they continue to grow without interruption and often reach heights of many metres (bananas, Zingiberaceae, etc.). If one moves from an area with cold winters to one where the winters are mild but where there is also a period of summer dryness, as in the mediterranean area, it will be noticed that genera which were represented exclusively by herbaceous species in the temperate zone, are now represented by woody species. Examples are *Linum suffruticosum, Lithospermum suffruticosum, Herniaria suffruticosa,* thorn cushions of *Verbascum spinosum* and *Cichorium spinosum* (on Crete) and the halophytes *Salicornia fruticosa* and *Suaeda fruticosa.* Cytogenetic investigations have shown that the herbaceous species should be regarded as recent descendants which have become adapted to a cold season of the year (Ehrendorfer 1970); this is particularly striking in *Astragalus.*

Woody Plants

During the first few years, these plants are at a disadvantage compared with the other groups, for the seedling uses the CO_2-assimilates to form a woody stem and not, as is the case in the other life forms, to develop as large a leaf area as possible. As a result, total production in the first year is very small: in beech, for example, it is only 1.5 g dry substance.

The woody stem has, however, the advantage that in successive years the new leaf buds open at a higher level and that the axillary organs do not have to be formed anew each year. Thus, in the course of time, their competitive ability relative to the herbaceous plants increases, until finally, provided they have survived the early critical years, the woody plants become dominant.

If there is no marked unfavourable season, even the leaves may be retained for several years. Sooner or later, however, they are cast off, to be replaced by new leaves, for with time they accumulate too much "slag", such as salts which are dissolved in the transpiration stream but which cannot be used.

If there is a cold season or a period of winter dryness, the leaves fall either before or during this season. If there is a period of summer dryness, the plants form leaves which are evergreen, but are small and hard. These sclerophyllous plants are in this way at least in part able to exploit such periods of dryness for material production. In climates with long, cold winters and only brief summers, the evergreen conifers are able to survive by retaining their organs of assimilation, the needles, throughout the winter; material production can thus begin again as soon as conditions become favourable. One genus of deciduous conifer, the larch *Larix,* is nevertheless able to survive in the most extreme cold at the tree line in the boreal region of Central Siberia.

In favourable conditions the competitively strongest of the woody species are the tall trees. If the climate does not allow the development of trees, smaller woody plants, shrubs and dwarf shrubs prevail. These are found otherwise only as lower, dependent layers in forests.

Many European geographers use the term "steppe" to describe any vegetation type in the tropics between forest and either almost treeless savanna or desert. Yet it is precisely here that the variety of vegetational types is so great that they cannot simply be grouped together under a single term. Their description as "steppe" is totally unsuitable: the word is derived from the Russian "stepj" and is used for the grasslands of the temperate zone, not for woody plant formations. Neither should the grassland formations of the subtropical-tropical zone be called "steppe", but simply "grass-land", to avoid misunderstandings. We will return to the concept of "savanna" in Volume 2.

The relative competitive ability of different trees depends on both their height and their age. The latter is not unlimited, for the relationship of the size of producing, assimilating leaf surface to that of the respiring axillary organs and roots becomes decreasingly favourable with time; this is so despite the fact that the oldest woody parts die completely and no longer respire, while all the axil organs which no longer carry leaves (old branches and twigs) are cast off. With time, net production of new dry substance becomes less and less in old trees; when it is no longer adequate to form a sufficiently large leaf surface, the tree dies.

In light-loving woody plants such as birch and aspen, which have relatively small leaves, these changes occur earlier than in shade trees such as beech. The latter finally achieve dominance because the deep shade they cast prevents the development of the seedlings of the light-loving species (vide p 173).

Competition between species of the same life form. A very clear example of this from Namibia may be cited. In the area around Mariental, the Fish River Sandstone is exposed. The rock is either very finely stratified or is formed of thicker shelves. Three shrub species root in the cracks: a large,

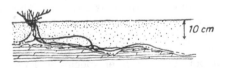

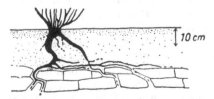

Fig. 105. Rock formation and the roots of plants. *Above* shallow-rooting *Rhigozum; below Catophractes,* broken line shows torn off roots. The relatively fine-grained soil is indicated by the stippled areas. Further explanation in text. (Walter 1939)

strong *Boscia foetida* (Capparidaceae), a medium-sized *Catophractes alexandrii* and a small *Rhigozum trichotomum*, both Bignoniaceae (Fig. 105).

On the finely stratified rock only *Rhigozum* grows, because only this species can survive on the small amount of water held in the fine cracks. On the coarsely shelved layers there are somewhat larger reserves of water in the wider cracks and *Catophractes,* which requires more water, is found here. The large *Boscia* grows only in particularly favourable places. Close examination showed the seeds of the three species to be distributed evenly over the whole area and germinate in all three places. In the most favourable positions, however, *Boscia* displaces the others, which are found only as dead plants beneath the *Boscia.* In other places were there is less water *Boscia* does not grow; here *Catophractes* is able to establish itself and suppresses *Rhigozum;* the latter grows only in rock cracks where there is too little water for the other two species.

The complex competitive relationships between grasses and woody plants in savannas is dealt with in Volume 2, Zonobiome II.

7.5 Indicator Value and Individual Plant Species

We wish to consider here the question of the ecological indicator value of individual plant species; that is, the extent to which detailed conclusions about conditions prevailing in any habitat can be based on the presence in that habitat of particular plant species. This has repeatedly been attempted as, for example, with regard to soil pH; one speaks of acid-indicators in this case, while nitrophilic species have been called nitrogen-indicators, and so on. This is permissable in this limited form and when applied to a small area with a uniform climate. It was, however, noticeable that on international excursions, whenever a local guide pointed out particular indicator plants, he would invariably be contradicted by participants from other climatic regions. The explanation for this is that plants which occur, for example, only on limestone with a high pH in humid areas, avoid these in arid places, preferring acid soils because they are more moist; similarly, species which, in arid areas, grow in shade, where they suffer lower transpiration losses, grow in full daylight in humid areas, and so on.

To an ecologist, it is clear that the competitive ability of a species—and this is what is important in a plant community—*depends on the interaction of all the environmental factors*. If one factor changes, dependence on one or several others does not remain unaffected. This applies particularly when the competitors are dissimilar. *The ecological indicator value of individual plants should thus be treated with considerable caution*.

Ellenberg (1979), who has studied the competitive factor in detail, has expanded the method of indicator values to an almost dangerous level of elaboration. In his *Zeigerwerte der Gefäßpflanzen Mitteleuropas* he has attributed to practically every species in central Europe, including alpine and adventive species, an indicator value for the factors light, temperature, continentality, moisture, acidity of the soil and nitrogen. This he has extended to life form, structure and anatomy, as well as sociological position (character of Class, Order, Alliance or Suballiance). The factors have been arranged on a 10-point scale (1–9 and x = indifferent). These numbers have no quantitative value, however, for the position on the scale is determined by very vague qualitative assessment: for example, light 7 = "semi-shade plant, mostly in full sunlight but also in shade". Nevertheless, Ellenberg describes how the soil moisture rating (Feuchtigkeitszahl = F) of a beech wood can be calculated from those of the various herbaceous species found in the undergrowth. For example, 5 species with F3 = 15; 7 species with F4 = 28; 12 species with F5 = 60; together this makes 103, so that the mean F value of the 24 species is 103/24 = F 4.3. F3 is defined as a dryness indicator, "more frequent on dry than on damp soil, absent from very moist soil"; F4 is regarded as intermediate between 3 and 5; F5 is defined as a dampness indicator, "mostly found on moderately moist soil, absent from both wet soil and that which frequently dries out". How then are we to regard the conditions in a habitat with F = 4.3? Ellenberg expressly limits the applicability of his numbers to western parts of central Europe, but "including the central Alps". The climatic differences occurring from the lowlands of central Europe to the snowline of the Alps are so enormous that indicator ratings for individual species cannot remain constant. It is known, for example, that shade plants from the forests of central Eu-

rope occur on open meadows in full daylight in the mountains (vide Sects. 9.1 and 9.2).

The ecotype problem (vide Sect. 9.3) also precludes making precise statements about the behaviour of species. From an ecological viewpoint, the gravest doubt must be expressed about the use of numbers in this way. It gives an illusion of precision which does not exist in reality, and may be taken literally by uncritical readers.

The value of this method is particularly uncertain in central Europe, as this is an area where, as a result of constant interference by man, no plant community is in a state of ecological balance with its environment; examples of such interference are the use of herbicides, air pollution, uncontrolled mechanization in forestry and agriculture, and so on. The result is that a species is confronted with different competitors in different places, depending on the nature of the interference, and this immediately alters its indicator value.

Despite this, Ellenberg has added his indicator values to his *Vegetation of Central Europe from an Ecological Viewpoint* (Ellenberg 1982) without making any qualifications, although it is precisely "from an ecological viewpoint" that it is impossible to justify these sophisticated indicator values (see p 160, pH values).

7.6 Relationships Between the Synusiae of a Plant Community: Dominant, Dependent and Complementary Species

Let us consider the ecosystem of a deciduous forest, made up of two or more layers, each of which can be regarded as a synusia. The species in a synusia are usually in competition with one another, so that ultimately, when an equilibrium has been established, the competitively stronger species will be more numerous than the weaker. In natural conditions plant stands are usually mixed. Layers consisting of only one species are the result of floral poverty as, for example, in Europe, or the result of extreme conditions, as in the Rocky Mountains in North America where the natural *Picea engelmannii* forests occur only at the upper limits to the forest zone; on the lower slopes there is mixed forest.

Mixed stands are the rule, because environmental conditions fluctuate from year to year: for example, in dry years one species is at an advantage, in wet years another. Furthermore, it has been observed in the tropics that the seeds of tree species germinate less well in the shade of their own species than beneath others; this means that from generation to generation an "exchange of places" or "rotation" occurs, which contributes to the maintenance of mixed stands. A similar situation has been observed in the meadows of Europe (Lieth 1958). The distinction must always be borne in mind between dominant synusiae (layers) within a stand and dominant species within a synusia.

In Europe a common view is that those tree species which can tolerate the most shade and which have seedlings which can grow in shade, such as the beech, will be the competitively strongest and must eventually displace all other species which have a higher light requirement. In natural, virgin forests, which have not been interferred with by forestry, the situation is, however, far more complex.

In western Europe the forests with oaks in the upper storey, hornbeam in the lower, so-called Querceto-carpineta forest, are man-made stands, for oak seedlings normally die in the shade of the hornbeam. In the Caucasus, however, there are natural forests of this type, consisting of light-loving species of oak *Quercus iberica* and the shade-tree *Carpinus caucasica;* these are closely similar to the west European species (Walter 1974).

In the Caucasus, too, the oak and the hornbeam are in strong competition. Many young oaks die, but some manage to grow above the shorter hornbeam, at which point competition ceases. The oaks grow very old and at 180 years of age attain a trunk diameter of 50–100 cm; they constitute the greater part of the wood reserve of the forest. The younger hornbeam plants grow between these mighty oaks, while beneath them loose oak leaves accumulate on the forest floor; these afford an ideal seedbed for oak seedlings, but not for the hornbeam. As a result, many oak seedlings are found under the oaks, but they die after a few years as a result of lack of light. There are thus no young oaks at all, and the impression is gained that the oaks must die out. When an old oak falls, however, a large gap is formed in the canopy; this enables some of the 1–4-year-old oaks to increase rapidly in height and to grow above the hornbeam seedlings, eventually to form the upper tree storey. In this way, the many-storeyed mixed forest stand without young oaks is permanently maintained. It can be seen that only precise and long-term observations can provide a solution to such problems.

Mayer and Tichy (1979) investigated several oak-beech forests of central Europe: the nature reserve known as Lainzer Tiergarten, which is part of the Vienna Woods, the Unterhölzerwald near Donaueschingen in West Germany, the Bialowejesch forest in Poland, and the virgin forest reserve Krakova in Slovenia. Only in the latter, with a nearly natural population density of game animals and undisturbed development of young trees, is survival of the light-demanding oaks possible; in others they are replaced by shade trees, usually as a result of too high a density of game animals which destroy the young oaks.

The upper or tree storey is usually *dominant* to the herbaceous layer; the species of the tree layer are formative of the whole plant community; they determine the light conditions in which the dependent species of the herbaceous layer grow and give rise also to the relatively steady temperature and humidity conditions on the forest floor. The leaves which fall from the trees form a litter and humus layer above the mineral soil, and the roots of the herbaceous plants grow in these layers. It has already been emphasized that this herbaceous layer may consist of many synusiae (p 55). These herbaceous synusiae are also almost always mixed stands, because both light flecks on the forest floor and small unevenness in the terrain, with more moisture in the small hollows, give rise to a micro-mosaic of environmental conditions. There are no absolutely homogeneous areas within a forest, least of all within a natural forest.

As well as dominant and dependent species, there are also complementary species and synusiae in a stand. They complete the plant community by filling vacant niches in the stand.

The spring geophytes of deciduous forests are typical examples of such complementary species; these include *Galanthus, Leucojum* and *Scilla* in central Europe, as well as

Ficaria verna, Corydalis spp., *Anemone* spp., *Adoxa* and others. Their assimilate economy has already been discussed (p 55ff).

In spring, before the leaves appear on the trees, the litter layer warms up rapidly *because the heat capacity of the air-containing litter layer is small and, as a result, its "temperature conductivity" is very good.* Firbas (1928) found that in central Europe, at the end of April or beginning of May, before the leaves are out on the trees, the temperature in the leaf litter may rise to more than 25°–30°C, often even to 40°C; this encourages the development of the spring geophytes which root in this layer, and they are able to complete their normal development before the trees are in leaf. Their period of growth does not end as a result of shading by the trees, for in better light conditions their development is shortened. Some of these plants belong to mediterranean genera, which there make use of the brief period after the winter rains and before the onset of the summer drought for their development. They were thus preadapted for the niches they occupy in deciduous forests.

This synusia of spring geophytes is characteristic only of the deciduous forests of ZB VI; that is, the climatic zone with a cold season of the year, for a suitable niche is created by the delay in leaf development of the trees; this delay is, in turn, due to the fact that the tree roots lie in deeper layers that take longer to warm up in the spring. In the deciduous forests of the tropical zone, leaf development depends on the wetting of the deeper soil layers after the start of the rainy season. Since the first rains are usually quite heavy, wetting both of the usually poorly developed litter layer and of the deeper horizons in which the trees root occurs almost simultaneously.

In evergreen deciduous forests and in spruce forests there is no such niche created by a light period on the forest floor.

7.7 Some Critical Considerations of Plant Sociology with Remarks on Phytocenology in the USSR

It is clear from what has been said above that the herbaceous layer is very strongly influenced by the tree storey. The latter determines the microclimate on the forest floor, especially the light conditions, but also the quantities of nutrients available to the herbaceous layer. As has already been mentioned, the litter which falls from the trees has a marked effect on the herbaceous layer; furthermore, litter from different tree species differs greatly both in chemical composition and in the ease with which it is decomposed. It is from this litter that the humus layer, in which the herbaceous plants root, is formed.

This dependence of the herbaceous layer on the tree layer is true of both natural and managed forests. In the latter, however, the history of the stand has a considerable bearing on the character of the herbaceous layer; this involves the initial composition of the tree storey, the methods by which the trees are propagated and the frequency of thinning Plant sociologists, however, base their classification of forest communities entirely on the prevailing floristic composition of the herbaceous layer; deductions are even made from this about the most suitable trees to be planted in a particular stand. For trees, however, the decisive factors are macroclimate and the water content of the deeper soil layers below the humus horizon where the tree roots normally lie. The role of these ecological factors in the plant community is neglected by plant sociologists.

Braun-Blanquet, the founder of plant sociology, even saw a plant stand as a sort of "organism" and the plant communities as units which were as stable and readily defined as the plant species themselves. His hierarchical, floristic system, known as syntaxonomy, is based on this idea and is modelled on conventional taxonomy. Even today, plant sociologists abide by the notion of a "sociological" progression (Wilmanns 1978). It is, however, not the floristic rank of species which is important in plant stands, but their ecological characteristics; this applies especially to the dominant species which are the determinants in an ecosystem, but in no way to the very specialized species which mostly fill small, vacant niches; yet many of these are so-called "character species" of the herbaceous layer, used for determining the rank of a plant community.

Central Europe, the main area with which plant sociologists have been concerned is, in fact, one of the areas most intensively exploited by man. There are no natural plant communities here at all; even the smallest, unexploited areas are affected by air or water pollution, groundwater control and so on. Indeed, the most important ecological factor in central Europe is interference by man in its many forms. Any ecological subdivision of the vegetation must take this as its starting point, as has been done by Ellenberg (1982), while past interference must also be borne in mind. It would, in fact, even be preferable to avoid calling them "nearly natural" forest, for in the Middle Ages, as a result of the removal of trees for wood and the grazing of stock in the forests, almost all the forests of Europe were degraded to open stands with isolated groups of trees. Even in the uninhabited, forest-covered mountains the stands suffered greatly from the effects of charcoal burning, glass making and iron smelting. It was not until 1816 that H. von Cotta published his *Anweisung zum Waldbau* (Methods in Forestry), in which the "wooded field" served as a model for rational, long-term exploitation of a forest. As a result of such management, homogeneous stands arose and over large areas "high forest" (Hochwald) developed from the earlier "middle or low forest" (Mittelwald, Niederwald). It is these forest communities which have been the subject of study by plant sociologists over the past half century. Even the cathedral-like pure beech forests, in which all the trees are of the same age, are the result of deliberate aforestation by the "Schirmschlag" method (trees felled over a large area), and are by no means natural stands. The young stage of this kind of managed forest is very dense, so that there is no herbaceous layer at all. All managed forests lack the large numbers of standing or fallen trunks of dead trees which are important seedbeds in natural forests and essential for the propagation of certain tree species. This applies also to the otherwise almost natural "Plenterwald" (trees felled individually) of spruce and beech, as in the Black Forest, in which isolated trees are harvested singly, so that no large felled areas arise and trees of all ages are permanently present. Homogeneous "natural forest stands" are found only in areas where there has been a fire as, for example, the pine forests which have arisen after lightning fires in Yellowstone Park. They are closely similar in appearance to the managed pine forests of Europe, except that the trees are not arranged in such neat rows and show very varied rates of growth. Homogeneous stands, whether of forest plantations, grasslands or agricultural fields, are usually a sign of management (or in rare cases, the result of extreme environmental factors). In such managed places well differentiated plant communities develop and these have been described by plant sociologists. When the nature of the interference is altered, as has happened frequently during the past few decades, the plant communities also change. For this reason, detailed knowledge of the way in which forests have been altered in structure during the last century is essential. Yet plant sociologists provide no data on this whatsoever. For the plants of the herbaceous layer of the forest, the original composition of the stand, the methods of propagation used, the number of thinnings and many other factors are important. In fields the important factors are the sowing of the seed after ploughing, the methods used in management and so on. It is not sufficient simply to produce a list of the names and numbers of plants present at the time when the community is examined.

Common practices today are, in forestry, the establishment of mixed stands and the avoidance of felling over large areas, in agriculture, the use of machinery and herbicides. Grasslands have been reduced to a state of dull uniformity; unused wasteland is disappearing; the waters are becoming ever more polluted. Earlier plant sociological surveys are thus now of historical interest only. Plant scociologists devote far too little time to observing the changes continuously taking place in the plant cover of cultivated lands and too much time on the revision of old data from subjectively selected test areas, which anyway represent only a minute part of the whole area; 100 000 surveys, each covering $100\,m^2$ account for the vegetation of only $10\,km^2$.

As a practical exercise for students of botany, making an inventory on $100\,m^2$ test sites can be very useful. It forces the beginner to recognize plant species at all stages of development and not only when they are flowering. If such surveys are conducted, not simply at randomly selected points, but in a continuous series of sites in the direction of change of a particular ecological factor, the vegetation profile thus obtained shows that *individual species* disappear in turn to be replaced by others; in other words, the plant cover is a continuum. There are no "sociological" connections between the different species, nor are there any natural, clearly demarcated plant communities which can be arranged in a hierarchical floristic scheme; this is contrary to all ecological experience.

"Ecological plant sociology" is thus a contradiction in terms, as is clearly shown by the system set up by Wilmanns (1978). Her key is based on physiognomical and ecological characteristics; the main subdivision is based on "ecological formations" which, viewed ecologically, are very questionable. Orders and associations are determined floristically and the overall arrangement follows the sociological progression. This results in plants which belong within the ecological group of halophytes, the Salicornietum, being placed next to weed communities of fields and footpaths, while marine and salt-marsh grasses are put together with managed meadows and pastures. The alpine snow valleys are lumped together in the same "formation" as fenland, while the nitrophilic Curvuletum, which is ecologically similar to the flora of the adjacent snow valleys are grouped with the calciphilic Seslerietum; wooded fenland or "carr" is linked to the totally different heather-pinewood formation, and so on.

Phytocenology is a part of ecology but not of phytosociology.[9]

It should be the special task of phytocenology to provide a clear picture of the natural plant cover of a particular area and, ultimately, of the whole earth: this would include the physiognomy of the vegetation, the principles of its subdivisions and its dependance on environmental factors. At the same time, the geographical component must always be taken into account. Construction of such a composite picture is often difficult in areas where the human popula-

tion is dense and the natural vegetation destroyed. Plant sociology does justice to none of these demands, least of all to the geographical aspect. Its sole aim is to draw up a list of communities according to Classes, Orders, Alliances and Associations, although these degrees of rank are identified only by "character species" which are usually ecologically and productively unimportant. Such a list gives no idea of the plant cover; nor does a list of plants within an association, since it is not arranged according to layer, give any notion of the structure of the plant community. "Begleitpflanzen", that is, plants which are found in a particular layer but not regarded as "characteristic", are mentioned only in passing, even when they are dominant and formative.

Of all the various schools of phytocenology, which have often sharply attacked one another, that in the USSR alone has succeeded in attaining the above-mentioned goals, and has presented a clear survey of the vegetation of a vast area, representing one-sixth of the total land surface of the earth. While other schools started with the smallest units of the vegetation and got bogged down in trifles, the starting points in eastern Europe were the large vegetation zones—the tundra, boreal pine forest, deciduous forest, steppe, semi-desert and desert zones. These were subdivided down to the level of associations, which were characterized and named according to the dominant species of the individual layers; this immediately conveys a clear picture of the plant community and enables it to be placed in the phytocenological system. As a result, separate monographs giving a very clear picture of the plant cover of each large partial area have been produced. An investigation of the entire area, from the Baltic Sea to the Pacific and from the Polar Sea to Central Asia, was systematically organized from a central authority in the Geobotanical Department of the Academy of Science in Leningrad. Over 100 scientific workers were employed in the various departments: (1) extreme north, (2) forest zones, (3) arid zones, (4) experimental phytocenology, (5) biocenology (= ecosystem research), (6) phenology, (7) soil ecology, (8) cartography. Branch offices were established for Siberia in Novosibirsk and for the far east in Vladivostok. Similar centres have

---

9 The late Braun-Blanquet, although himself no ecologist, was very open-minded to ecological problems and it was possible to work well with him (Braun-Blanquet and Walter 1931).

been established in each of the Soviet Republics (e.g. in Kiev, Tbilisi, Askhabad, etc.).

During the vegetative growth season, a team of specialists worked closely together at field stations set up for a few years at a time, often in undeveloped areas. Vegetation and soil were always considered as an entity; usually the entire ecosystem with all its components was investigated and the area mapped. The results were published in a work of two volumes (Lavrenko and Soczava 1956) which includes a vegetation map in eight parts (scale 1:4 million) as well as detailed commentary. The coloured maps in this survey are extremely detailed with 109 main cartographic symbols, which are subdivided by letters into a number of variants, enabling the areas in which natural vegetation has been replaced by cultivated plants to be identified. There is, in addition, an excellent but still unpublished vegetation map of the Ukraine by Ju. D. Kleopov (vide Walter 1974, pp 134ff). In 1979 a vegetation map for Outer Mongolia with 98 symbols (scale 1:1.5 million) was published by A.A. Yunatov and B.Dashnyan et al. This area extends from 90° E to 120° E and from 52° N to 42° N and in the south includes the Gobi desert.

Finally, mention should be made of the new revised version of *Vegetation of the European Part of the USSR 1980*, 427 pp, Nauka-Leningrad (in Russian) with 9 maps (1:7.5 million). The distribution of the following communities is shown: (I) 21 tundra types; (II) 40 spruce forest types; (III) 30 pine forest types; (IV) 19 deciduous forest types; (V) 24 steppe types; (VI) 19 rock vegetation types; (VII) 23 desert vegetation types; (VIII) 5 moor vegetation types and (IX) 14 flood plane forest types; in each case there are detailed descriptions.

Because of the language barrier and the Iron Curtain, this work has, unfortunately, remained almost unknown in the West. Only a summary in German of extracts from the larger monographs gives an impression of the vegetation of the whole area of eastern Europe, north and central Asia (Walter 1974). There is, in addition, a more ecological summary *Overview of Eurasian Continental Deserts and Semi-Deserts* (Walter and Box 1983).

8 Succession—Climax—Zonal Vegetation

8.1 A Critical Consideration of Primary Succession and the Concept of Climax

In 1916, F. E. Clements published his great work *Plant Succession*. His dynamic approach became the orthodox view of plant geography in the USA and was largely adopted by Tansley and Braun-Blanquet as well. It was a grand theoretical conception, but it did not reflect real conditions in the natural environment and is a good example of the dangers inherent in theorizing in ecology without wide field experience. This may still arise today, particularly in the case of those authors who lack all experience of truly natural environments where there has been very little human interference.

Plant communities are not static but are in a state of dynamic ecological equilibrium. Single individuals die and are replaced by younger ones. In years when extreme conditions prevail, the species combination may undergo a marked quantitative and qualitative change, but in general, both the community and the ecotype maintain a reasonable constancy oscillating around a certain middle condition. Only when certain external factors change continuously in a particular direction, as, for example, a continuous rise in the ground water table, will a community of an entirely new species composition arise. Such a sequence of communities in one and the same area is known as a *succession*. If the change is initiated by factors which arise from the vegetation itself, such as accumulation of peat, it is known as autogenous succession; where a change in habitat conditions is the cause, it is known as allogenous succession.

According to Clements, every community represents a stage in a primary series. These primary successions should logically start with a substratum without vegetation (water, sand, rock or salt soil) and finally, after many intermediate stages, reach an end-stage or *climax community*, which will be characteristic of the particular area. This is a purely theoretical conception and in no way reflects reality.[10]

For millions of years, the greater part of the land surface has been covered with a layer of vegetation, which has changed in the course of time as different groups of plants evolved and climatic conditions altered. During the post-glacial period, the naked soil of the earlier glaciated Holarctic region was recolonized, but the pattern this followed was determined by the warming-up of the climate and by the particular species which immigrated to the area. It can thus not be regarded as a primary succession in Clements' sense of the term. True pioneer stages can still be seen today on the banks of water bodies, on sea coasts, on shifting sand dunes, on rock faces and on gravel heaps. The zonation of the vegetation around such areas is easily recognizable, but the mere fact that these zones lie adjacent to one another should not be interpreted as implying any sort of sequence in development. Pioneer stages, together with the adjacent zonation are, in fact, very stable, even in a landscape as apparently dynamic as the "bad lands" of North and South Dakota: these consist of almost vegetation-

10 The untenability of the climax concept as representing the final stage for all the series within a particular area—the theory of the "monoclimax"—was soon recognized. Since there are always several different final stages within an area, the term "polyclimax", or, alternatively, "paraclimax", was introduced. The term "climax swarm" is used for the various slopes of a mountainous area. The meaning of the climax has consequently become even less clear and should be dropped altogether.

less, heavily eroded, soft Tertiary strata. Old photographs of these areas show that over a period of 40 years there have been no significant changes in the shape of the numerous earth pyramids. Changes which have taken place over thousands of years cannot be regarded as a succession, leading from a primary stage to a climax community, because over so long a time span, the climate will not have remained constant. In the absence of changes in habitat conditions nearly all plant communities are in a dynamic equilibrium, unless the balance is disturbed by man. It is only possible to speak of successions when the development of a plant community can be directly followed over a relatively short period or can be established on the basis of old maps which might show, for example, drying up of water bodies, shift in dune areas, areas of mountain rock falls, or young volcanic rock, and so on. In a few cases, successions can be demonstrated on the basis of information from soil profiles as, for example, from peat profiles, which may extend over more than 10000 years. Yet even in this case the sequence is usually influenced by climatic changes or land subsidence.

The clear successional series which can be drawn up on the basis of zonation can, once stripped of any hypothetical interpretations, be retained as **ecological series**; that is, as a zonation which arises from change of an environmental factor, not in time, but in spatial terms; examples are increasing depth of the groundwater table around a water body, increasing thickness of the soil layer around areas of rock, decreasing salt content of the soil around salt pans, and a change in the mobility and increasing leaching of sand in dune areas. A dynamic approach should, indeed, be attempted, but it should not lose sight of reality and enter the realm of pure speculation. Such an approach is always appropriate when it comes to **secondary successions,** that is, those which arise when, on cultivated lands or anthropogenic communities, human intervention ceases (e.g. on abandoned fields, grassland which is no longer mown or grazed) or when habitat conditions are altered by some human interference (e.g. fenland drainage) or as a result of catastrophes (e.g. tornados, land slides, fire).

Here the succession of stages towards increasingly natural vegetation or in adaptation to the new habitat conditions can be directly followed. They are often of great practical and theoretical importance. Apart from this, the term "successional stages" should not be used.

Rejection of the theory of primary succession leads to rejection also of the "climax community" as a final stage. Instead, we use the term "zonal vegetation" (Walter 1954).

The best impression of the zonal vegetation of an area can be gained by flying at a low altitude over parts of the continents little influenced by man. If this is done along a line parallel to the climatic gradient, it can be seen how climatic change results in one particular zonal vegetation being replaced by another. This is usually shown on the vegetation maps of the individual continents. The zonal vegetation characteristic of a particular climate is, however, encountered only in areas where the particular regional climate has its full impact. Such biotopes may be called *euclimatopes* (Russian "plakor"). These are flat elevated areas with deep soils which are neither too porous to water, like sand, nor retain too much water, like clay soils, but are either clayey sand or sandy clay; rain thus penetrates the soil instead of running off the surface and most of the water is retained as capillary water. There must be no influence from groundwater. The zonal vegetation corresponds to the zonal soil type.

Since euclimatopes are particularly favourable for agriculture, they are mostly under cultivation in populated areas. The nature of the zonal vegetation can be gathered only from small remaining areas which have not been interfered with; in Islamic countries, for example, natural forest can be seen as the tree stands of the oldest graveyards or around the tomb of a saint, for felling of trees in such places has been forbidden since the beginning of the settlement. Similarly, the prairie vegetation in North America can be studied on the still unoccupied parts of graveyards, the steppe vegetation of eastern Europe on the prehistoric "kurgans" or burial mounds. Recourse may also be had, of course, to soil profiles, pollen analyses, historical data in archives—in fact, anything which may provide information about the original vegetation. With certain reservations, vegetation which has remained untouched on shallow soils can, if it happens to lie in the distribution centre of the zonal vegetation, provide evidence of the original vegetation; however, at the periphery of a zone it differs markedly from the vegetation on deeper soils.

In natural conditions there may, however, be widespread areas which have no zonal vegetation at all; this is so, for example, when the groundwater table is so high that the whole area is covered with bogs and fens, as in western Siberia and the Sudd area of the Nile; other examples are the areas of sand or alluvial soils of old river valley areas or volcanic rock not yet covered with soil, and saline soils of large undrained basins. These are pedobiomes with an "azonal vegetation" which is very strongly influenced by the special properties of the soil, but little affected by the climate.

Zonal vegetation is also found outside its climatic areas as "extrazonal vegetation", when habitat conditions in a particular localized part of the biotope are the same as those of the euclimatope of the zonal vegetation (see p 194). The occurrence of extrazonal vegetation is thus in keeping with what may be called the "law of the relative constancy of habitat and biotope change" (vide Sect. 9). Extrazonal vegetation can provide information about the nature of a zonal vegetation which has itself been destroyed.

The zonal vegetation should only be used for a large-scale subdivision of the vegetation of whole continents. Only in such cases does the effect of the climate make itself clearly apparent; small local differences due to soil, relief, and the aspect of the land then become unimportant.

8.2 The Structure of Zonal Virgin Forests

It should not be thought that the zonal vegetation comprises an entirely homogeneous plant cover. Quite apart from the secondary successions which arise after catastrophes such as fires, storm damage, flooding, pest epidemics and so on, a cyclic inner dynamic

process is also usually observed. This is particularly noticeable in virgin forests. Untouched virgin forests are today a rarity and of those that still exist, only very few have been observed over a long period. In Europe there is the 130 000 ha virgin forest area at Bialovjesh on the watershed of the Baltic Sea and Black Sea, 70 km north of Brest-Litovsk, on very heterogeneous but mainly wet soils. This was described in detail by Paczoski in 1928, but there have been no investigations of its dynamics (vide Walter 1974).

This type of dynamic process became apparent in the 300 ha virgin forest stand Rothwald, near Lunz-am-See, in Lower Austria. This forest lies 1000 m above sea level on dolomitic limestone in a mountain depression which is accessible only for a short period in summer; it has thus never been exploited for timber and is still only used for hunting (Zukrigl et al. 1963).

It is a mixed forest of beech, fir and spruce. The spruce seedlings develop mainly on rotting fallen tree trunks, which afford a suitably acidic seed bed. Accumulations of rotting litter from the spruce trees from localized areas of acidic soil, while elsewhere mull soils lie above the limestone. As a result, the undergrowth is a micromosaic of species from fir-beech forests and of others typical of a bilberry-spruce forest; the roots of the different tree species cross the boundaries of these mosaic units, so that distinction between different forest communities is no longer possible. Furthermore, the micromosaic is subject to constant change; beneath openings in the tree canopy snow collects, the open, snow-free period is very brief and species typical of acidic soils, such as *Vaccinium*, *Lycopodium* and *Blechnum*, establish themselves here; if, in the course of time, such a gap in the canopy closes, so that snow accumulation no longer occurs, these species are displaced by others.

The acidic humus soils (Moderboden), which arise from fallen tree trunks or from the litter of spruce trees, are also affected by the activities of soil organisms, such as earthworms, and transformed to mull soils; this, too, brings about changes in the herbaceous layer. Besides this microdynamic process, a macrodynamic change of the tree

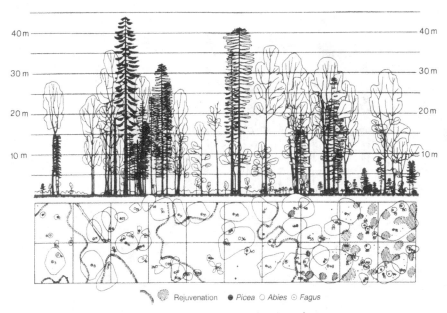

Rejuvenation   ● Picea ○ Abies ⊙ Fagus

Fig. 106. An early stage of the rejuvenation phase of a virgin forest

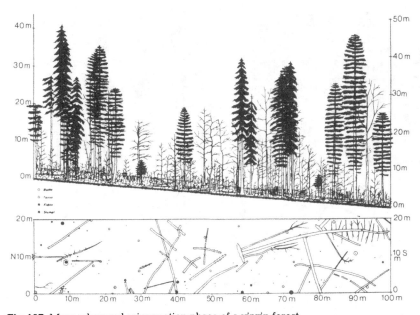

Fig. 107. More advanced rejuvenation phase of a virgin forest

Fig. 106–110. Structure of the different phases of a largely virgin forest in the Rothwald near Lunz am See (Lower Austria) at an altitude of 1000 m. (Zukrigl et al. 1963, from Walter 1976)

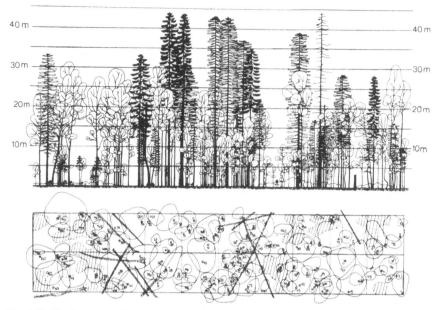

Fig. 108. Final stage of the rejuvenation phase of a virgin forest resembling a selection-type managed forest

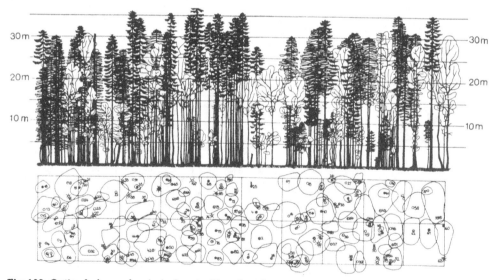

Fig. 109. Optimal phase of a virgin forest with a closed canopy

layer can also be observed: this consists of a cycle with several phases.

The most striking features of a virgin forest are, first, the large number of dead tree trunks lying on the ground in various stages of decay and, secondly, the very varied structure of the tree layer, with age differences of more than 300 years between individual trees. The rejuvenation of the forest is good, but the structure of the forest is variable. It is rare to find stands of equal density adjacent to one another; more usually very dense forest and light stands with trees of different ages ("Femelartig") are found

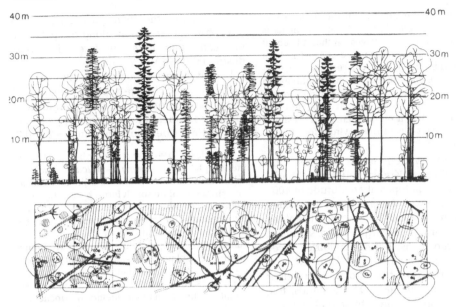

Fig. 110. Senile phase of a virgin forest showing signs of degeneration

growing together, depending on whether the stand is in a rejuvenation phase, an optimal phase, or a senile phase (Figs. 106–110).

The rejuvenation phase shows a distribution reminiscent of managed forests in which trees are felled over very small, irregularly spaced areas, so that young trees are found in groups—in the Rothwald at places where old spruce or fir trees have been blown over by the wind, for the trees do not root deeply on the shallow rock soil. In this virgin forest the beech belongs more to the undergrowth, but forms 58% of the total standing wood mass, while the firs and spruce each make up 21%. The stand shown in Fig. 107 is less dense, having both old and young windfalls; the beech dominates in the young growth. The next phase shows particularly large age differences ("Plenterphase"). The wood masses are: beech 39%, fir 34%, spruce 27% (Fig. 108).

Figure 109 shows the rarely encountered closed canopy stand, consisting mainly of conifers and almost no undergrowth; the standing wood masses are: beech 26%, spruce 36% and fir 38%. Figure 110 follows on this, showing the phase of decline which results from senescence; there is much dead wood, both standing and fallen and in places

large openings with developing young trees. The beech, with 67% now has the greatest share of the wood mass, while the firs make up 18% and the spruce 15%.

The following diagram shows the way in which this cycle may be envisaged.

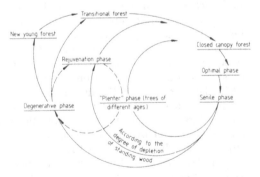

The excessive population of red deer constitutes a danger to this virgin forest, the young firs, in particular, suffering damage from being eaten. Mayer et al. (1980) drew attention to this in their investigation of various forests, including the 19 200-ha virgin forest reserve Corkova Uvada at an altitude of 850–1000 m near the Plitviz Lakes in Croatia. The 38-ha nucleus of this forest lies on a wind-protected easterly slope on karst, with dolomitic limestone. The composition of the trees in the stand is the same as in the Rothwald in Austria, but the population of game animals is controlled in winter by wolves, so that it is about five times smaller, and no tree species is here endangered by the game. As a result of the ever-increasing stream of tourists to the neighbouring lake area, the predators are withdrawing to less accessible, higher areas, so that here, too, the number of red deer is on the increase. This shows just how difficult it is to retain virgin forest in an unaltered condition.

These observations made in Europe are similar to those of Krauklis (1975) in Siberia. Here, on euclimatopes at an altitude of 400–450m, the snow cover remains on the ground for 6 months, while only 1½–2 months of the year are warm. The virgin forest stands of the dark taiga consist of *Abies sibirica, Picea obovata, Pinus sibirica* and *Larix sibirica*. The trees grow to a height of 25 m and their phytomass is 200t ha^{-1}. These stands, too, show cyclic phases: an activated phase with increase in phytomass (= rejuvenation and young tree phase), a normal (optimal) phase and a stagnating phase, with decline in the phytomass (old trees in a phase of senile decline).

The great heterogeneity of virgin forest, which is further increased by areas of fire and storm damage, as well as the large mass of standing and fallen dead wood, gives rise to so many different biotopes and small niches, that a large number of higher plants and also fungi are able to develop; this, in turn, makes possible a rich fauna of vertebrates, arthropods and others. Compared with this, a homogeneous managed forest is monotonous and barren.

Mueller-Dombois (1981) also found certain cyclic stages in *Metrosideros polymorpha* forests in Hawaii. Apart from this, little is known of the dynamics of tropical virgin forests. In such forests there are no cyclic phases incorporating large areas, but cyclic dynamic changes in small areas. Aubreville (1938) earlier pointed out this mosaic character of tropical forests.

With the frequent heavy storms in the tropics, lightning often strikes. This leads, however, not to forest fires, but to the formation of large gaps of up to 100m in the tree canopy; such spaces are easily recognizable from the air, for not only is the tree which is actually struck killed, but so, too, are others standing around it. A secondary succession

then occurs on the open space. Large areas of destruction arise only along the path of a hurricane and here the forest is totally destroyed (vide Vol. 2, ZB I).

Recent investigations have shown that cyclic dynamic change appears to be typical of most communities of living organisms. Examples of these on savanna, steppe and in the tundra are described in Volumes 2 and 3.

Pemadassa (1981) even describes a cycle for the simple *Anthrocnemum indicum* stands on saline soils. *Anthrocnemum* is a pioneer species on extremely saline soil. It forms horizontal shoots from which others arise and grow vertically, reminiscent of *Salicornia*; round about these, sand and organic material, blown by the wind, accumulate, so that little dune-like piles are formed around each plant. This sand is not very salty and retains rainwater, so that the grass *Cynodon dactylon* settles on the tiny dunes; this leads rapidly to the death of the ageing *Anthrocnemum* plants. Since *Cynodon* is unable to hold the sand together, the dunes are eroded by the wind, and on the naked sandy soil *Anthrocnemum* seedlings develop and form new dunes.

Cycles were to be seen also on the cultivated lands of man, when he was still at a stage closer to nature: for example, the rotation of crops in the three-field system, and in the Egert system with 2 years of grain cultivation and 4 years of grazing.

It is likely that such cyclic patterns are the norm for all life, including man. Technology, however, attempts to eliminate them altogether, and this will lead to an earth of lamentable uniformity.

8.3 The Vegetation of the Periglacial Zone in the Late Pleistocene

8.3.1 The Relationship Between Steppe and Tundra Species

The actual changes which have occurred in the plant cover in the course of the earth's history were not successions in Clements' sense, but were brought about by changes in climate; they can be reconstructed only in broad outline on the basis of fossil finds (Mägdefrau 1968).

The development of the vegetation in the northern hemisphere during the period following the last glaciation can, however, be reconstructed with rather more precision on the basis of pollen analysis and the many finds of plant and animal fossils.

In this regard, recent investigations in eastern Siberia have led to the conclusion that the arctic tundra vegetation of today first arose in the post-glacial period. The tundra climate is very humid (vide Vol. 3, ZB IX). Although the precipitation, mostly as snow, is not great, potential evaporation is even less, and the permafrost soil in areas with a mean annual temperature well below 0°C thaws out only at the surface during the short summer, and tends to become bog-like.

Only in eastern Siberia, around the cold pole of the northern hemisphere near Verkhoyansk-Omyakon, is there an extreme continental, cold climate; here the absolute minima are almost −70°C and the absolute maxima around +30°C. Another similar, but not quite so extreme area is in Alaska, near Fort Yukon (Fig. 111). There is much evidence that during the last glaciation of the Pleistocene (Würm glaciation of the Alps, Weichsel glaciation of NW Europe, Waldai glaciation of eastern Europe, Wisconsin glaciation of North America) the climate in the periglacial area around the great ice cap was even more continental.

It is very striking that in the extreme continental climate of zonobiom VII (central Asia), the forest altitude belt of the mountains is absent altogether; instead, the mountain steppe passes into the alpine belt without any sharp demarcation; this can be seen, for example, in Tjenschan (Walter 1975). The steppe extends to an altitude of almost 3000 m; here the steppe vegetation still has a period of vegetational growth of 4 months and the summers are hot. At this altitude some alpine elements are found amongst the steppe vegetation, while still higher, as a result of further shortening of the growth period, they displace the steppe plants altogether.

A very similar relationship is seen between the steppe and tundra elements in the extreme continental boreal-arctic zonobiomes VIII–IX of eastern Siberia, where the forest zone is often hardly developed.

It can be seen from the climatic diagram (Fig. 111) for Omyakon in the Yana basin (61° 67′ N) that the average daily temperature rises above 10°C for about 3 months in the summer, and in July almost reaches 20°C. In Verkhoyansk the sum of the temperatures on days when it was above 10°C is 1084°C, the hours of sunshine in July 310 h (Yurtsev 1981). At the same time, there is a dry period from May to August and in mid-summer even a 2-month period of drought. The climate is clearly semi-arid, as is shown by the occurrence of saline soils in undrained depressions. Here sodium indicators such as *Puccinellia tenuiflora* and *Hordeum brevisubulatum*, are found and with heavier salination, even *Atriplex litoralis, Salicornia europaea* and *Spergularia marina* (vide Walter 1974). The fact that forests of the light taiga, with *Larix dahurica* and *Pinus sylvestris*, are nevertheless dominant in this area is due to the permafrost soil. In the dry spring, the trees are supplied with water from the ice as it thaws to increasing depths in the soil, and later, from rains which are heaviest in late summer. Thus a light covering of forest is possible, even with an annual precipitation of only 200 mm. This is, however, a steppe forest on solod soils (cf. p 39) with an effervescent layer at only 20–40 cm depth; there is a steppe-type undergrowth consisting of *Artemisia laciniata, Lathyrus humilis, Vicia amoena, Pulsatilla multifida, Geranium jakutense, Pyrola incarnata* and the typical steppe shrub *Spiraea media*. Between are treeless areas with solonized soils (Russian "charany") and a grass steppe vegetation including *Koeleria gracilis, Poa botryoides, Leontopodium sibiricum, Artemisia commutata, Galium verum, Veronica incana,* etc.

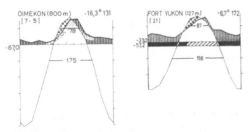

Fig. 111. Climatic diagrams for the coldest continental areas: Oimekon (Omyakon) in E. Siberia and Fort Yukon in inner Alaska. (Walter 1974)

Similar steppe forests are found even in the humid boreal area near Archangelsk on the White Sea on karst limestone in the central region of the Pinega basin (64°–65°N). Saburov (1972) has described larch and fir stands there with an undergrowth of *Pulsatilla patens*, *Thymus serpyllum*, *Astragalus danicus* and *Dianthus deltoides*, amongst others, or of *Atragene sibirica*, *Cypripedium calceolus* and *Anemone sylvestris*. All of these have the character of relics (vide Walter 1974).

Other plants which are regarded as periglacial relics are certain species found on chalk slopes of central Russian elevations which were not glaciated during the pleistocene. *Carex humilis*, a relic which is found scattered on dry limestone and loess slopes of central Europe, is regarded as the "character species" of periglacial steppe in this area. The following species are regarded as periglacial steppe relics on the lightly weathered chalk slopes on Oskol, near Kursk: *Daphne julia* (aff. *D. cneorum*), *Androsace villosa*, *Bupleurum ranunculoides*, *Potentilla tanaitica*, *Schiwereckia podolica*, *Scutellaria verna* (= *S. alpina* var. *lupulina*) and *Chrysanthemum arcticum* ssp. *alaunicum* (vide Walter 1974).

What is very remarkable is that, according to recent floristic investigations summarized by Yurtsev (1981), veritable islands of steppe are very widely distributed in the cold east Siberian region and relics of steppe vegetation are even found on Wrangel Island (71°N) (Fig. 112).

We have seen that in the extreme continental climate of the mountains of central Asia the steppe vegetation extends to a high altitude and, in the absence of a forest belt, mixes with the alpine vegetation until it is finally displaced by it. A similar situation is found in the extreme continental climate of eastern Siberia where, on favourable biotopes, steppe vegetation extends over 30 km along the valley of the middle Indigirka; that is, around the Arctic circle. Such vegetation is found even in the Kolyma and Anadyr areas, mainly in arid mountain basins. Still further north, steppe species are found distributed in mosaic fashion, mixed with tundra species, but in unfavourable habitats the latter form the only vegetation.

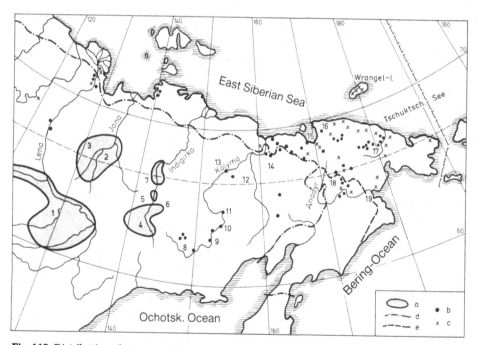

Fig. 112. Distribution of steppe and tundra-steppe communities in NE Asia: *a* areas of the northerly taiga in which steppe communities are a more or less regular occurrence; *b* areas in which tundra-steppe communities only are encountered; *c* areas with steppe and tundra-steppe communities; *d* northerly forest limit; *e* the north-east Asian boundary (after Yurtsev 1972). *Numbers* indicate the different regions investigated

The occurrence of these steppe islands does not depend on latitude, but on the local conditions of warmth and dryness of the habitats in the intermontane valleys and basins, the most important factor being high daytime temperatures in summer. In the Arctic, steep southerly slopes are favoured because in summer the sun's rays fall on such slopes almost perpendicularly at midday.

These steppe islands thus appear to be "extrazonal", like the steppe-heaths in humid central Europe. With an even greater degree of continentality, they would spread zonally over wider areas. There are, however, no such areas in the boreal arctic zone of today.

The wide distribution of solod soils in eastern Siberia indicate, however, that during the Pleistocene glaciations there was in this region an even more extreme continental climate with solonetz soils. We can thus assume that *at that time the steppes were zonally distributed, not only in eastern Siberia, but over the whole of the area which today lies south of the area previously covered by glaciers.*

Yurtsev (1981) gives a long list of these steppe species of Yakutia in eastern Siberia; a few of these, mainly the dominant species, are listed below (vide also Walter 1974).

Microthermic xerophytes: *Ephedra monosperma, Stipa krylovii, Koeleria cristata, Poa botryoides, Festuca kolymensis, F. lenensis, Agropyron (Elytrigia) cristatum, A. jacutensis, Psathyrostachys (Hordeum) juncea, Carex duriuscula, Arenaria meyeri, Potentilla nudicaulis, Chamaerhodos grandiflora* (Rosaceae), *Eritrichium sericeum* (Boraginaceae), *Artemisia frigida,* and others.

Endemic species: *Helictotrichon (Avena) krylovii, Astragalus* spp., *Oxytropis* spp., amongst others.

Microthermic grass steppe species: *Selaginella sibirica,* which fills in open spaces between the others, *Bromus pumpallianus, Pulsatilla flavescens, P. multifida, Draba* spp., *Potentilla arenosa, Linium perenne, Viola* spp., *Veronica incana, Scutellaria scordifolia, Galium verum, Leontopodium campestre, Aster alpinus* (typical of Siberian steppe), *Serratula marginata* and many *Artemisia* spp.

On stony ground species of *Dracocephalum, Thymus* and *Artemisia* are found, as well as *Gypsophila marginata* and *Orostachys spinosa* (Crassulaceae).

Very many of these species also occur on Wrangel Island, not, however, in steppe communities, but dispersed among typical tundra species, such as *Dryas, Kobresia myosuroides, Carex* spp. and others, particularly on limestone.

Yurtsev (1981) also shows on 20 area maps the distribution of species of the genera *Oxytropis, Carex, Smelovskia, Alyssum, Phlox, Chamaerhodos, Festuca, Agropyron, Plantago* and *Arabidopsis.* These species occur not only in continental Siberia. They, or vicarious species or subspecies, are found also in the continental area around Yukon in NW of North America, where it is almost as cold. This shows that there has been an intensive exchange of species between these areas and that this took place during the glacial period of the Pleistocene, when the sea level was lower and the Beringia land-bridge linked the two continents directly. It was along this route also that primitive man could migrate from Asia to America.

8.3.2 The Periglacial Steppe of the Late Glacial Period

According to Velichko (1973), the Sartang glaciation in Siberia (= Late Würm in Europe or the Wisconsin in North America) was characterized by marked continental climatic conditions as a result of the freezing of northern seas. This led to the development of a uniform, trans-continental arctic steppe, encompassing the lowlands of Yakutia and Chukotka. This has been clearly demonstrated by recent geological investigations of the Yedoma formations, which are made up of ice-loess deposits on high ground; they are especially rich in fossil remains of mammoths, bison, musk oxen, horses and other mammals. These Yedoma formations, which are today very wet, were formerly erroneously regarded as alluvial lake deposits. The view, first put forward by Shilo in 1964 (cited in Tomirdiaro 1977), that they are ice-loess formations, corresponding to the typical, periglacial, aeolian loess of Europe, is now considered to be confirmed (Tomirdiaro 1977). This was deposited on elements of the relief of a quite different age during the cold, dry Late Pleistocene; the deposits are characterized by the predominance of the silt fraction of 0.05–0.01 mm typical of loess; this is present even when the deposit has an overall depth of 30–40 m. There is no trace whatsoever of remains of fish or crustaceans, but fossils of steppe fauna are especially numerous. There is a marked vertical structure determined by the paths taken by the roots of herbaceous plants, but there are no woody remains at all. The columnar structure typical of loess is similarly very marked, as is also the salt content, which is four to five times greater than that of alluvial deposits. Pollen analysis of these formations showed an absolute predominance of pollen grains of grass and herbaceous plants, and the almost complete absence of tree pollen, even in layers in which fibrous ice lenses frequently occur. The presence of the latter might seem to contradict the notion of a loess deposit which dried out markedly in summer, but it has been established that this ice developed originally as hoar frost formed from air which penetrated dry cracks in the cold ground; this interpretation is confirmed by the fact that the ice deposits contain 8–10% air. In western Eurasia this ice disappeared after the thawing of the permafrost soil, the ground sunk and the "pods" of the steppe region were formed. In the colder north-eastern Siberia ice formed in this way persists today in the permafrost soil. This can all be seen as evidence that during the Late Pleistocene, over the whole of the circumpolar region from Alaska, across North America and Europe, to eastern Siberia, there was a very uniform loess-landscape; this carried a typical steppe vegetation and a correspondingly rich steppe fauna which included large herbivores. The climate at first became humid and warm only very much further south (Agachanjanz 1980).

Even species such as the Saiga antelope and the yak were able to cross the Bering land-bridge to reach Alaska. The occurrence of this antelope in Yakutia during the glacial period is evidence that the snow cover must have been meagre and the plant cover typical not of tundra, but of dry steppe. It has been possible, using ^{14}C-dating, to deter-

mine the age of herbaceous plant remains from the ice loess deposits. The remains from a depth of 15.5 m were 23 360 ± 720 years old, while those from a depth of 2 m were 11 500 ± 210 years old.

Similarly, fossil mammoths from the river Berelekh (a tributary of the Indigirka) were found to be only 12 000 years old. At the end of the last glacial period, a change in climate began; the frozen seas thawed, the Gulf Stream was established and the climate became more humid. Tundra replaced steppe and in Yakutia thermo-karst formations appeared; that is, there developed the Alasy landscapes of today, with undrained or semi-drained depressions with solonized soils and small remnant lakes (cf. Walter 1974, p 90). At the same time, the light taiga displaced the steppe, of which only small islands remained as relics in favourably dry biotopes. This sudden change in climate at the end of the Glacial Period was probably the main cause of extinction of the large mammalian fauna of the Pleistocene, in particular of the mammoths, as this fauna was adapted to snow-poor winters and dry summers and to feeding on steppe vegetation. Certainly the fauna could have survived further south, as it had done in previous interglacial periods, for example, in the Transbaikal-steppe where conditions today are similar to those of the Glacial steppe; but in the Postglacial Period, the game animals encountered the mounted huntsmen of nomadic tribes and were probably destroyed by them. This interpretation is also given by Kvasov (1977).

When an attempt is made to reconstruct the ecological conditions prevailing in the northern hemisphere during the last Glacial Period, the following picture emerges. Sea level had fallen markedly; as a result, North America was linked to eastern Siberia by a broad area of land known as Beringia (today the Bering Sea) and in Siberia the continent extended further northwards to include the present north-Siberian islands. The Arctic Ocean was completely separated from the Atlantic by a land barrier which included Greenland, Iceland, the Farre Islands and Scotland. The polar ice cap reached a thickness of 3 km. The winds which blew mainly in summer from the ice cap were berg winds, dry and warm, as is still the case today in the arctic steppe of northern Greenland, in the Trog valleys on Heklasund (80°N) and in Ingolf's Fjord (80° 30′ N) (cf. Walter 1968, pp 542–544). Here the high summer temperatures and dryness result in the absence of permafrost in the upper soil layers, and the plants develop deep tap roots which penetrate up to 1 m into the soil. Alkaline soils occur and also salt crusts with a flora of halophytes.

On the high Pamir plateau, too, at an altitude of 4000 m, the soils are, in general, not frozen because they are too dry. Permafrost is limited to valleys and hollows where the soils are wet as a result of water flowing into depressions (Sveshnikova 1962; Walter 1968, pp 809–824). Thus desert conditions prevail here.

We can imagine the climate of the zonal arctic steppe during glaciations to be as follows. The easterly winds blowing over the periglacial zone had the character of berg winds and were dry; they caused massive deposits of loess in the steppe. Frenzel (1964) was able to isolate pollen grains from such loess deposits in Lower Austria; these showed that the Glacial vegetation was not the same as the Arctic vegetation of today, but a herbaceous steppe vegetation.

Pollen analysis of food remains trapped between the teeth of woolly rhinoceros of the last Glacial Period from the region of Verkhoyansk (68° 30′ N) shows 69% of the pollen to be from Gramineae, 17% from *Artemisia* and the rest from other herbaceous plants. This rhinoceros had a broad upper lip, resembling that of the African white rhinoceros *Ceratotherium simum* which feeds on grasses and herbs. In this feature it differs from the other extant species which have a pointed upper lip; these feed upon shrubs and use the upper lip to break off twigs.

The presence of *Artemisia* and *Ephedra* in the pollen spectrum of the last Glacial Period suggests that the steppe islands of today, found in those parts of eastern Siberia which were never glaciated, are indeed relics of Glacial steppe. Finds of fossils of steppe rodents, such as *Citellus suslicus* and *Marmota* spp., confirm the steppe character of the zonal vegetation of that time (vide Walter and Straka 1970).

Temperature conditions may have been somewhat more favourable in summer than they are in eastern Siberia today (Fig. 111). More rain must have fallen in spring, when air humidity would have been high because the melting snows would have wetted the soil. At the end of the vegetational growth period of 3–4 months (or briefer at higher latitudes in the recent Arctic region with light nights) the steppe vegetation would probably have used up all the reserve water in the soil, thus excluding any possibility of tree growth. The soils would thus have been so dry and their heat capacity consequently so low, that they would have warmed in spring down to great depths; this would have been favourable to the development of steppe plants. Evapotranspiration must have exceeded annual precipitation; that is, the climate was probably semi-arid. The long winter must have been similar to that in eastern Siberia today: clear and very cold but with little snow.

There must certainly have been larger and smaller pedobiomes (that is, azonal soils) within this zonobiome, some of which were wet, permafrost soils; the latter would have warmed up very slowly in the spring, thus shortening the vegetational growth period in the particular locality. Tundra vegetation grew in such areas; dead remains fell into the lakes and gave rise to the rich fossil flora of the *Dryas* clay layers with leaves of dwarf birch and creeping *Salix* spp. The azonal vegetation of that time was largely the same as the tundra flora of today, but provides little information about the then prevailing climate.

Krause (1978) has made a detailed study of mammoth finds in the Arctic areas of today and has, rightly, emphasized that neither the existing arctic tundra nor the dark taiga would have provided sufficient food for the large Pleistocene herds, not only of mammoths, but also of woolly rhinoceros, bison, horses and other mammals. This difficulty can be resolved, however, if it is accepted that there was periglacial steppe in the Pleistocene. Even during the long winters there would then have been an adequate supply of food for these animals, for the aerial parts of the steppe plants, killed off by the first frosts, would have retained some nutritional value and, in the almost complete absence of snow, could have served as winter grazing.

Primary production of the steppe-complex relics in the cold-continental climate of today in NE-Yakutia (eastern Siberia), is $0.2\,t\,ha^{-1}$ dry substance where *Koeleria* predominates and $0.4\,t\,ha^{-1}$ in the herbaceous *Festuca-Carex*-steppes of lower mountain slopes (Skrjabin and Konorovski 1975; Yurtsev 1981). Galaktonova et al. (1975) have estimated the standing green phytomass of grass steppe with *Carex duriuscula* in the Sartang valley to be $1.22\,t\,ha^{-1}$ in mid-July and of the *Poa-Festuca* mountain steppe $0.75\,t\,ha^{-1}$ (Yurtsev 1981). Productivity, however, fluctuates markedly from year to year.

On southerly slopes the period of vegetational growth starts at the beginning of May. By mid-May the flowering plants on these slopes turn yellow, but in the second half of June the steppe is green as a result of the developing leaves. The *Carex* spp. flower at the beginning of June, the steppe grasses at the end of June, *Koeleria* and *Agropyron* only at the beginning of July. At the end of July, the grasses become dry, but species of the Lamiaceae and *Artemisia* spp. flower, while the other species fruit and seed; the seedings occur only in the following spring in the case of late-flowering species. At the beginning of August, when the heat of summer is over, the grasses may show a brief recovery.

Productivity of this steppe is thus similar to pasture production of the steppes of the euclimatopes of the cold, high plains of the Pamirs, for which Agakhanyanz (Walter and Box 1983) gives the following data.

1. In the montane steppe of the western Pamir, at altitudes of 3500–4200 m, productivity of air-dry hay is $410\,kg\,ha^{-1}$.
2. On small areas of montane steppe of the eastern Pamir, at altitudes of 4600–4800 m, productivity of air-dry hay is $300\,kg\,ha^{-1}$.

When, in the late Glacial, there was a sudden change to a more humid climate, tundra completely replaced the steppe; the herds of mammoth thus lost their source of food and began to die out. Large herds, probably overwhelmed by catastrophic floods, were covered with sand. Their corpses were thus buried in permafrost ground and have been preserved in a deep-frozen state. Elsewhere the bones were washed together after decay

of the soft parts of the body: the occurrence of mammoth graveyards may be explained in this way.

One particularly rich find of mammoth bones is situated on the banks of the Berelekh river in the catchment area of the Indigirka, at a latitude of 71°N and longitude of 145°E in NE Siberia. This is a veritable "mammoth graveyard". In an area of $5 \times 5\,m$ and to a depth of 80 cm, 954 mammoth bones were found (47–50 bones per m^3), while altogether 8431 bones were collected in 1970–1971 (Verestshagin 1977).

According to Lozhkin (1977), the age of the soft parts of these mammoths is $13\,700 \pm 400$ years, thus dating them at the end of the Sartang Glacial. Pollen analysis of the soils containing the bones showed that 90% was non-tree pollen; 60% was from grasses, up to 16% from Cyperaceae, up to 21% from Caryophyllaceae and up to 20% *Artimesia* pollen. Among the tree pollens, that from *Salix* predominated, indicating a transition to the conditions which now prevail. In the 10000–11000-year-old layers which lie immediately above this soil, the pollen spectrum shows that the vegetation at that time was similar to that of today, with light forest tundra and a humid climate.

In 1977 the complete and undamaged body of a young mammoth was found north of Magadan, at a depth of 1.8 m, on a 10 m high terrace of the Kirgilyakh stream (Shilo 1978). This animal, a male, had died at the age of 7–8 months. All soft tissues were markedly dehydrated. In the all but empty and contracted stomach a few grass and herbaceous plant remains were found, while the 3.2-m-long small intestine was empty; in the 1.4-m-long large intestine there was a black, earthy mass, 8–10% of which was made up of the stems and roots of herbaceous plants. The total weight of stomach and intestinal contents was 3.5 kg. The body showed no signs of physical damage, so that death was not the result of an accident. The young animal had clearly been starving for some time before its death (Shilo 1978; Verestshagin and Dubrovo 1978).

We have discussed this transition from the Glacial to the Postglacial in detail to show that the Arctic zonobiome IX of the tundra reached its present state of development only after the Pleistocene. This is true also for the ombrogenous high moors which are so typical of oceanic NW-Europe, their occurrence being associated with a damp, maritime climate, which depends upon the influence of the Gulf Stream.

The peat profile of the moors generally shows very clearly the succession from a lake vegetation in the Late Glacial, through a fen and lowland moor vegetation either to that typical of a wooded river bank or of ombrogenous high moorland; the latter has changed to heathland as a result of human intervention.

This provides indisputable evidence of a series of vegetational stages. Nevertheless, it cannot be regarded as an example of primary, autogenous succession, because it was determined to a very large extent by external factors, namely, the marked change in climate between the Late Glacial and the present time. The silting up of former lakes depended mainly on the sediment carried by inflowing rivers: while in some parts of a lake this occurred rapidly as a result of heavy sedimentation, in other places the banks were subjected to wave action, causing erosion of sedimentary deposits. Raising and sinking of the land also affected silting and moor formation. The Scandinavian platform, freed from the weight of the Glacial ice-cap, is still rising today, while the more southerly part on the North Sea coast is sinking; this causes a steady rise of the groundwater table, favouring the formation of moorland.

ZB IX is not the only zonobiome to have developed since the Pleistocene. As we show in Volume 3, this is true also of ZB IV with its Mediterranean-type climate.

8.4 Vegetational Series in the Postglacial

About 12000 years ago, tundra vegetation spread in central Europe, where the climate was now humid and the surface free of ice; as the process of warming-up continued, reforestation began, first with willows and pines during the Alleröd warm oscillation (10000–9000 B.C.). It became cooler again, however, and during the most recent sub-arctic period (9000–8250 B.C.), tundra predominated once more. During the subsequent postglacial warm period in central Europe, mean annual temperature was 2°–3°C above that of today. The development of the vegetation during this time can largely be reconstructed by pollen analyses. In both oceanic western Europe and continental eastern Europe this took a different course from the development in central Europe; here, on lower ground, the "willow" and "pine periods" were followed by a "hazel period"; then, during the warm period by mixed oak wood and later, in cooler conditions, by beech (vide summary by Walter and Straka 1970).

These vegetational stages cannot be regarded as a true succession, for they were determined by changes in climate and depended, furthermore, on the relative location of the places which had served as refuges for the plants during the ice age and from which the process of recolonization by tree species and their accompanying flora took place. The development of the vegetation was determined also by the rate of spread of the different species. It thus depends on processes which do not make it necessary to employ the postulates of Clements' climax theory.

True primary successions, which take place on uncolonized areas without fully developed soils, can certainly be observed in places where, for instance, a glacier recedes, or on volcanic ash or lava after an eruption or on the rocky slopes after a landslide. These usually develop so slowly that it takes many centuries before a plant cover is formed which is adapted to the climate and is in ecological balance with the environment.

Investigations of the recolonization of areas from which glaciers have receded have been frequently undertaken in the Alps. The rate and precise course of development of both vegetation and soil vary according to particular local conditions. It always commences, however, with a pioneer stage, which is reminiscent of the vegetation found on stony ground or on plains at low altitudes. After a time a stage is reached which resembles a grassy meadow, which gradually becomes more dense. The first woody plants establish themselves in lower and warmer positions, so that development leads to the formation of sub-alpine heaths and woodland and, at higher altitudes, to alpine grass communities. Older moraines

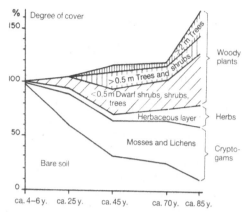

Fig. 113. Degree of coverage of different layers of the vegetation on increasingly old side moranes of the Great Aletsch Glacier. (Lüdi 1945, from Ellenberg 1982)

accordingly carry a denser vegetation, while on younger moraines the vegetation is more open. On the large Aletsch glacier, for example, 95% of the surface was still bare 4–6 years after the ice had receded, but after 85 years only 10% was still uncovered (Fig. 113). Lüdi (1945) demonstrated this clearly on test sites.

The Island of Krakatau still provides the best example of recolonization following a volcanic eruption. All life on this island of the Malay archipelago was destroyed in the eruption of 1883. The recolonizing flora had to reach it from the neighbouring islands which lie between 19 and 40 km away. After 1 year the island was still barren, but after 3 years 26 species of higher plants had established themselves. Ferns spreading by spores predominated, while the flowering plants belonged mainly to beach vegetation; that is, these were all hydrochoric species. The laval rocks became covered with bluegreen algae and mosses. By 1908 the island had a light plant cover, but the vegetation had not yet regained its pre-eruption condition. After 50 years, the island was covered with a forest reminiscent of secondary tropical forest. It will probably remain in this state for a long time. Initially, anemochoric species, borne by the wind, and others carried by ocean currents established themselves; these were followed by endozoochoric species distributed by birds and finally others introduced by man. Of the 271 species identified in 1934, 41% had been carried to the

island by wind, 28% by sea currents, 25% by birds and 6% by man.

Fridriksson (1975) made a very thorough investigation of the recolonization of the volcanic island of Surtsey, near Iceland, after the 1965/66 eruption there. The nearest small island is 5.5 km away. Very many viable seeds were carried to the island in the air or by water. Endozoochoric distribution by birds was less frequent. Only very few species succeeded, however, in establishing themselves. In 1968, 8 species of Cyanophyta and 100 other species of alga, including 74 diatoms, were found. The thermophilic alga *Mastigocladus laminosus* must have come from Iceland, 75 km away. The first lichens were observed in 1970. The most important pioneers are the mosses. As early as 1967, colonies of *Funaria hygrometrica* and *Bryum argenteum* were found and by 1970 there were as many as 16 species of moss. These developed well in the cool, moist climate and *Rhacomitrium canescens* and, to a lesser extent, *R. lanuginosum* were soon dominant. By 1972 there were 63 moss species on the island, 12 of which were of frequent occurrence. The mean biomass on the three most densely colonized areas was 0.518 g m^{-2} in 1971, and as much as 3.412 g m^{-2} in 1972. In 1971 the fern *Cystopteris fragilis* was found, as were also, on the beach, the first flowering plants, *Cakile edentula*, *Elymus arenarius*, *Honkenya peploides* and *Mertensia maritima*. By 1973 there were 13 species and 1273 individual plants. Among the arthropods, 112 insects and 24 arachnids were found in 1970. In 1967, 29 species of migratory birds were counted and 13 species which live on or near the shore.

Detailed investigations have also been made of the recolonization of rocky slopes after landslides in the Alps.

The most frequently encountered natural, secondary successions are, however, those which follow fires, and here they are of great importance in determining the structure of the vegetation (vide p 75).

Over large areas of wind damage, especially after tropical hurricanes, many plants are still alive; nevertheless, it takes a very long time for the plant cover to regain its former state, particularly in the case of forests. Constantly fluctuating secondary successions occur on the banks of rivers which repeatedly alter their course. In such places the vegetation never reaches a steady state. This applies also to sea coasts exposed to periodic heavy storms and to sandy beaches where dunes form and reform.

Successions determined by the activities of the plants themselves occur where peat is formed, and lead to a raising of the level of the ground. Such moor formation continues over a long period only in places where the climate is favourable, that is, continuously humid and cool, and the building up of new peat layers is compensated by a slow sinking of the continental plate as, for example, on the North Sea coast. Elsewhere the process of moor formation with a rise in ground level stops after a time as a result of increased drainage and consequent dryness.

Finally, we would again emphasize that the mere existence adjacent to one another of different vegetational belts, that is, a zonation with a slow change in a habitat factor, is not necessarily evidence of a succession. Far more commonly this is an ecological series which may itself be very stable.

We wish only to mention that there are many, readily recognizable secondary successions which result from such human activities as forest clearance, allowing fields to lie fallow, drainage schemes, soil improvement and so on.

9 The Law of Relative Constancy of Habitat—Ecotypes and Ecoclines

Many species are very widely distributed and grow under apparently very different climatic conditions. Waterplants have, to some extent, a very cosmopolitan distribution, but this is due to the fact that all over the world they find habitats where very similar conditions prevail. An example of this is the rigid hornwort, *Ceratophyllum demersum*, which grows not only in tropical waters but also in those of zones with a cold winter. In the latter, this plant forms dormant winter buds (turions) and overwinters at the bottom of the lake or pond. Annual weeds require only a short period for their development in biotopes where there is no competition; these may be fields in arid areas with a short rainy period, or in those areas in temperate zones which have short, hot summers or, frequently, uncultivated soils in warm climates.

Two other factors of great ecological significance are (a) biotope change in accordance with the law of relative constancy of habitat and (b) the formation of ecotypes and ecoclines.

9.1 The Law of Relative Constancy of Habitat and Biotope Change

Many species evade changes in the macroclimate by changing their biotope; that is, they exploit local microclimatic differences, so that their actual habitat conditions remain relatively constant. For example, a species adapted to a warm climate will, if its distribution extends into colder regions, avoid the euclimatope and grow only on warm southerly slopes, while species from colder regions grow only on the cool northerly slopes in warmer climatic regions.

This applies also to the vegetation as a whole—the zonal vegetation of one region grows well only extrazonally in some other zone, in biotopes where temperatures are similar to those of its native zone. In mountainous regions species or plant communities from higher latitudes will be found only at high altitudes or in cold ravines; in the Alps, arctic species are encountered in the alpine altitude belt; here the vegetational growth period is as brief as in the arctic region, although the climates are not identical (vide Sect. 3.3). The central European beech *Fagus sylvatica* is found as far south as Sicily; here it grows at the upper tree limit on Mount Etna where, in summer, the climate in the cloud belt is moist but not hot, and thus meets the normal requirements of the beech. The heather, *Calluna vulgaris*, which grows in open habitats in the oceanic climate of western Europe, is confined in eastern Europe to pine forests, where the soils remain moist during the summer. Species which in the eastern steppe are not limited to any specific soil type are found in western Europe on dry, southerly limestone or loess slopes which are warm in summer. Similarly, the beech, which in western Europe grows preferentially on limestone, is found in northern Greece on siliceous soils, which afford a better supply of water.

Some observations we made in Namibia drew our attention to the very general applicability of this law of biotope change and the relative constancy of habitat. In this arid area the rainfall decreases from 600 mm in the northeast to less than 100 mm in the southwest, but temperature conditions remain more or less uniform and the vegetation had been little influenced by man (Walter and Walter 1953). In arid areas water supply is least favourable on heavy, clayey soils (vide Sect. 6.10), somewhat better on light, sandy soils, better still on very cracked, rocky ground and particularly good on soils with a constant high groundwater table, thus enabling some tree species, by changing their biotope, to extend their distribution from the wettest area to the driest without incurring any disadvantage to their water supply.

On the western border of the Kalahari desert, where there is a relatively high rainfall, the thorntree *Acacia erioloba* (= *giraffae*) is found on sandy soils, whereas further south, where the rainfall is less, it is found on the dune ridges where there is an open vegetation cover; in even drier parts this tree is found on the terraces of the "riviers" (rivier = dry river bed), while in the rivier itself there are dense stands of *Acacia horrida*. The latter species is altogether absent in the almost rain-free, arid southwest, whereas *A. erioloba*, by growing in the groundwater-carrying riviers, has been able to penetrate right into the Namib desert. This species is not, however, adapted to rocky slopes or stony ground.

The kudu bush ("omuhama") *Terminalia prunioides*, with its brilliant red-winged fruits, requires a good supply of water and is thus mainly distributed in the northeastern part of the country. It is, however, found on the edge of the Namib desert, where the rainfall is only 100 mm; here its distribution is limited to narrow areas at the foot of granite slopes where water collects, or along small riviers, although in the forest found along larger riviers this species is unable to compete.

The "driedoring" *Rhigozum trichotomum*, a small shrub species typical of the dry parts of the country, behaves in the reverse manner. In the wet northeast it is found only on dry lime crusts ($CaCO_3$) while on the edge of the desert it is found on stony slopes and, finally, in erosion gullys into which some water drains after one of the rare falls of rain.

In arid areas the following series of biotopes, moving from the driest to the wettest, can be distinguished on the basis of soil type, relief and quantity of runoff:

1. slopes with impermeable soils and high rate of runoff;
2. terraces which are flooded from time to time and which have an upper layer of compacted soil;
3. clay soils on flat surfaces with a high rate of evaporation;
4. sandy, flat areas, where rain penetrates well;
5. sand dunes (non-shifting), with an open vegetation cover;
6. gravel slopes and stony areas from which there is little evaporation;
7. cracked, rocky surfaces, which store water well;
8. the base of slopes where the runoff water accumulates;
9. erosion gullys in which water flows at times and where there is reserve water in the soil;
10. dry valleys with a constant flow of shallow-lying groundwater.

If the distribution of individual species is traced from the wettest areas to the driest, changes in habitat are seen broadly to follow this sequence, although differences can, of course, be observed between different species, according to their particular water requirements and root systems.

The distribution of any two species seldom follows the same pattern. One may grow preferentially in rocky habitats, another in sandy places. Certain tree species of moist areas may suddenly be encountered in dry areas on a cleft outcrop of rock or at the foot of a slope in granite mountains; other species may be found exclusively in valleys where there is groundwater, but where they must also be able to withstand the force of periodic floodwaters. The general pattern is, however, unmistakable, and can be expressed as *the law of the relative constancy of habitat and biotope change as follows: when, within the area of distribution of a plant species, the climate changes in a particular direction, the species changes its biotope, thereby compensating for the change in climate; that is, the habitat conditions in the different biotopes occupied by the species remain more or less the same.* This law applies not only to hydrature conditions in arid areas, but generally to all those factors which are influenced by climate. The great disadvantage of distribution maps is that instead of emphasizing distribution within a certain euclimatope, they include all the extrazonal distribution areas as well, with the result that the main climatic

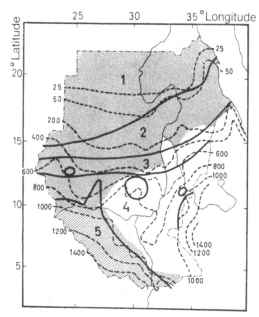

Fig. 114. Map of the Sudan showing annual rainfall isopleths *(broken line)* from 25 to 1400 mm, soil types *(stippled* sandy soil; *cross-hatched* ferrous sandstone; *open* clay soil) and the vegetation zones: *1* desert; *2* acacia semi-desert; *3* acacia-short-grass savanna; *4* acacia-long-grass savanna; *5* tropical dry forest. *Circles* mountains with forest within the acacia zone. (After Smith 1949, from Walter 1973)

range of a species usually appears far too large.

Smith (1949) provides some particularly good examples of the validity of this law from observations in the Sudan (Fig. 114). He distinguished between 14 different biotopes with increasingly favourable hydrature conditions. These make it possible for the same species to grow under very different climatic conditions. The range of precipitation within which individual species are found is as follows:

Acacia seyal	370– 800 mm
Khaya senegalensis	400–1050 mm
Prosopis africana	480–1200 mm
Tamarindus indica	500–1200 mm
Dalbergia melanoxylon	360–1200 mm

There is, of course, a change in biotope in the various climatic areas. On those with different annual precipitation, *Acacia tortilis,* for example, is found in areas with

50 mm, in desert erosion gullys
150 mm, near Khartoum, on sandy soil
300 mm, in the Kassala district, on clay soils
500 mm, on the dry slopes of the Butana hills.

Hyphaene thebaica occurs in areas with an annual rainfall of

100 mm, in erosion gullys and along river beds
600 mm, on red clay soils
900 mm, on heavy clays.

Sterculia setigera grows in areas with an annual rainfall of

300 mm, on rocky outcrops
1200 mm, on flat surfaces.

Albizzia aylmeri is found in areas with an annual rainfall of

300 mm, on sand dunes
800 mm, on clays.

Smith gives many examples of the more favourable conditions on sandy as compared with clay soils; the Sudan is very well suited for such a comparison, since the western part consists of broad sandy areas, while the eastern parts are characterized by clay soils. In general, it is found that on sandy soils plants require only two thirds of the precipitation necessary for their existence on clay soils. Thus the major occurrence of *Acacia senegal* on sandy soils is in areas where the rainfall is 400 mm, whereas on clay soils it is limited to areas where the rainfall is 600 mm.

The water requirements of different species can be assessed only by comparing their occurrence on the same soil type in flat areas where there is neither accumulation of water nor runoff. The *Acacia* species of the Sudan can be listed in order of occurrence with increasing rainfall as follows: *A. flava, A. orfota, A. tortilis, A. raddiana, A. mellifera, A. fistula, A. senegal, A. seyal, A. drepanolobium, A. campylacantha, A. sieberiana, A. albida, A. hebecladoides, A. abyssinica.*

Of course for any particular annual rainfall, several species may occur, but they will then be distributed over different biotopes, those which require less water will be in the drier biotopes, those with higher water demands in the moister ones. Since clay content has an adverse effect on soil water content, there is a clear relationship be-

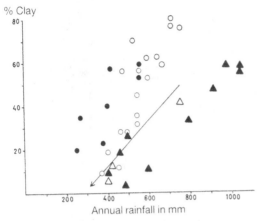

Fig. 115. Relationship between the distribution in the Sudan of various woody species, clay content of the soil and the annual rainfall in mm. ○ *Acacia seyal*; ● *Acacia mellifera*; △ *Prosopis africana*; ▲ *Khaya senegalensis*. Long arrow separates the woody species of the acacia zone from those of the tropical dry forest. (Smith 1949, after Walter 1973)

tween the distribution of any species and both the rainfall and the clay content of the soil (Fig. 115).

The greater the clay content of the soil, the greater the annual rainfall required for the occurrence of any particular species. The conditions on clay soil may become more favourable for woody species if sand, blown by the wind, is caught by the plants and collects around them as dunes which slowly increase in size, until finally, the shrubs form the crown of a fairly large sand dune.

The crests of non-shifting dunes form a particularly favourable habitat. This can be seen, for example, in the southern Kalahari, where the dunes are 5–51 m high and up to 3 km long; they run parallel to one another, NW to SE, about 230 m apart, and cover an area of 117 000 km². The sand on the crest of a dune is coarser than that in the valleys, so that grasses no longer occur as competitors and the soil is also more open. For this reason, trees can thrive here and species such as *Albizzia anthelminthica*, *Terminalia sericea*, *Acacia erioloba*, *Boscia albitrunca* and *Acacia haematoxylon* occur. Of these, *Albizzia* and *Terminalia* have their main area of distribution in the more rainy northern parts, since their water requirements are

greater than those of the other species. The indigenous *Acacia haematoxylon* is competitively weak and its distribution is limited to this favourable habitat on the dune crest (Leistner 1961, 1964).

9.2 Ecotypes and Ecoclines

The basic unit of plant systematics or taxonomy is the *species*. The ecologist, too, makes use of this unit to describe the plants he studies in the field and he attempts to explain the distribution of this unit in terms of causal environmental relationships. It is known, however, that a species is not genetically entirely uniform, but is composed of separate plant populations and is thus a mixture of many different biotypes. A "species" is an abstract concept, determined by taxonomists who, on the basis of extensive material, group together those individuals which are identical in the morphological characteristics which they deem most important: it can, on the basis of these characteristics, be clearly differentiated from any other group of individuals. A species may often extend over a very large area within which there are clear climatic differences from one part to another. Within such an area one and the same species may be found in habitats with different soils or different microclimatic conditions.

The pine *Pinus sylvestris* is, for example, found in Europe from Lappland to Spain and across the Urals into Siberia. It is found in the boreal forests together with spruce, above all in areas where there has been fire, but also mixed with oaks in the forest steppe area. It grows both on dry outcrops of rock and on wet bogs.

If specimens of this pine of different origins are cultivated in the same environmental conditions, they are found to differ markedly in vigour, cold- and drought-resistance and so on. They can usually also be distinguished morphologically by such characters as the shape of the needles, form of the crown and shape of the trunk. Such differences are, however, of no significance for taxonomists in distinguishing one species from another, for they are characteristics which are subject to variation. Thus characteristics of the pine which are established in ecological investigations may not apply to all individuals which belong to the species *Pinus sylvestris,* but, strictly speaking, only to specimens of the particular race or from the same provenance.

Kerner (1891) was the first to draw attention to this. He cultivated plants of various species which occurred both in low-lying places and in the mountains and had developed clearly distinguishable lowland and mountain forms, and observed their development over 6 years at three different altitudes: at 180 m near Vienna, at 569 m near Innsbruck and at 2195 m in the Tyrolean Alps. It was found that when grown in the alpine nursery, the lowland forms showed certain changes such as smaller size, narrow leaves, fewer flowers, darker green, but were still clearly distinguishable from the alpine forms growing adjacent to them. A similar situation was found with the alpine forms grown at a low altitude alongside lowland forms. It was concluded that the lowland and alpine forms represented *genetically different races* of the same species, although their particular characteristics could be altered within certain limits.

Bonnier (1920) came to a different conclusion, namely, that the lowland and mountain forms are merely environmentally induced *modifications* (Fig. 116). Since his results have, however, not been confirmed by more recent investigations, Bonnier's opinion must be regarded as having been based on insufficiently precise observations.

Turesson (1926) examined this phenomenon in great detail. He, too, found that individuals of the same species from different climatic areas or very different habitats retained their differences in appearance when cultivated under the same conditions. He suggested that such genetically determined races, occurring in nature under very different ecological conditions, be called *ecotypes,* and distinguished between:

a) climatic ecotypes, such as lowland and mountain races;

b) edaphic ecotypes, such as races which grow on sandy or saline soils;

c) biotic ecotypes, determined by the different forms of cultivation. One example is *Phleum pratense* which has a flat form

Fig. 116. *Taraxacum officinale:* P lowland form; M alpine form (both ⅓ natural size). M' as M but ½ natural size. (Bonnier 1920, from Walter 1973)

in pastures, a more erect form in meadows; another is the red clover *Trifolium pratense* which has a form in southern Europe which can be cropped twice, while in northern Europe it has a frost-hardy form which can be cropped once only.

It must, however, be pointed out that non-heritable modifications may also be found in ecologically different habitats and that it is not possible to distinguish these from true ecotypes on the basis of appearance alone. Thus, for example, of three *Succisa pratensis* plants which showed dwarf forms on an infertile coastal meadow, two retained their dwarf form after being transplanted to a garden, while the third assumed the normal form; in the latter case, the dwarf form was clearly a non-heritable modification. We know, too, that the land forms of water plants are only such modifications. These are especially well-developed in *Polygonum amphibium* which has three quite different, non-heritable forms, one which grows in

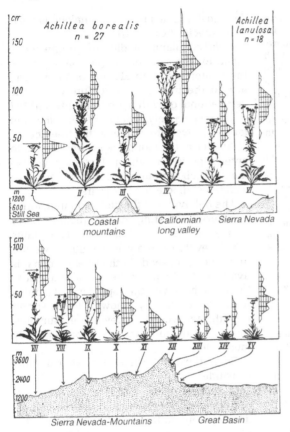

Fig. 117. Profile from the Pacific Ocean to the Great Basin showing the natural habitats of various forms of *Achillea millefolium* plants cultivated in the Stanford nursery; an average-sized plant and distribution curve for height is shown in each case. All plants were grown in identical conditions but, to a large extent, retained their original form. (Clausen et al. 1948, from Walter 1960)

water, another on moist soil and a third on dry dune sand.

Further investigations have, moreover, shown that even the concept of the ecotype as a particular genetic-ecological unit can, in most cases, only be upheld with reservations. Particularly illuminating in this regard have been the very careful investigations initiated by Hall in 1922 in California, and continued over some 30 years, latterly, from 1932 onwards, by Clausen et al. (1948). We take as an example only the results obtained from cultivating *Achillea millefolium* s.l. This plant is distributed from Stanford, on the Californian coast, to an altitude of over 3000 m in the Sierra Nevada. Seeds were collected from all of its habitats and sown in a garden near Stanford, 22.5 km from the coast and at an altitude of 30 m (vide Walter 1960b, p 91 ff). The results, as growth form and height of the plants, are shown diagrammatically in Fig. 117. Additional histograms show the variation in height: the

arrows indicate mean height reached by each group of plants.

Various main types can be distinguished and are indicated in Fig. 117 by Roman numerals.

I. The coastal type developed strongly without a rest period, displayed a squat growth form and was succulent. These characteristics are somewhat more marked in the natural habitat under the influence of salty sea winds; that is, they are further strengthened by a modificative response.

II, III, V. The foothill type is exposed to summer drought in its natural habitat. The plants are more hairy and more grey in colour. They grow mainly during the winter rain period and are dormant during the summer. They retained these characteristics in Stanford, even when watered in summer.

IV. The valley type grows in moist habitats in the Californian Long Valley and in this habitat reaches 2 m, an unusual height for this species. In the garden in Stanford it was also exceptionally tall. These plants lack a dormant period not only in their natural habitat, but also with continuous watering in the garden.

VI–XV. The mountain type, however, behaves differently. It is not uniform, yet cannot be subdivided into distinct sub-types. Rather, it shows continuous change with increasing altitude, both morphologically and ecophysiologically. The rainfall is higher in the mountains, so that the plants of the lower mountain slopes are larger than those of the foothills, and summer dormancy is less marked; in the mild winter of Stanford growth did not cease. When grown from seed in the garden in Stanford, plants from increasing altitudes showed ever more stunted growth, until true rosette plants were obtained from seeds collected in the highest mountain areas. At the same time, the period of winter dormancy became ever more marked, even in the experimental garden. If one descends the eastern slope of the Sierra Nevada, one reaches a continental climate. The winters are very cold, the summers very hot, and the rainfall low. *Achillea* grows here in moist meadows. In the garden in Stanford these plants had a short period of winter dormancy and flowered most luxuriantly.

For the sake of comparison, plants from an altitude of 365m from the northern Aleutian island of Kiska were grown. These plants, which come from a cold, moist, maritime climate, were morphologically similar to the high mountain type, but their winter dormancy period was relatively short.

Thus we can see that morphological characteristics do not necessarily go hand in hand with ecophysiological characteristics. What is particularly striking is the smooth transition from one genetically fixed type to another in the mountain forms, and this was all the more gradual, the closer the places from which seed samples were taken. In this case it is impossible to speak of well-defined ecotypes, but only of a continuous ecotype series, which is described as an *ecocline*. This might suggest that ecotypes frequently do not occur in reality, but simply appear to exist as a result of too low a rate of sampling or because samples have not been taken from intermediate zones. For example, *Campanula rotundifolia* is linked by intermediate forms to *C. scheuchzeri,* which is described as an ecotype. Similarly, there are transitional forms between *Prunella vulgaris* and *P. grandiflora* on lime-rich soils and *P. laciniata* in south-eastern Europe. There are also hybrid forms between these species, as there are between all ecotypes. They could be grouped together as an ecospecies.

Amongst animals there are analogous polytypic species (Rassenkreise).

Achillea plants of different origin were, again for the sake of comparison, grown in the mountains at an altitude of 1400m and also at the tree line, at 3050m, where the vegetational growth periods are 147 and 167 days respectively. None of the lowland forms survived the winter at the tree line.

At 1400m the foothill types were forced to enter a period of winter dormancy, but retained their summer dormancy period as well; most died after 1–2 years.

The behaviour of the high mountain forms was very interesting. At the treeline winter dormancy lasts for 9 months. The plants there develop rapidly in spring, flowering and producing seeds in the brief time that is available to them. At the low altitude near Stanford, however, they spent only 3 months in winter dormancy, showed weak spring growth, recovered somewhat in summer, but their overall growth remained only moderate.

From these investigations we can thus conclude:

a) *Achillea* plants of different origin are in each case adapted, in the periodicity of their development, to the climatic conditions prevailing in their normal habitats. The adaptation is not a modificative response, but genetically determined, for it is retained even under quite different climatic conditions. The simplest way to explain how such hereditary adaptations arise is to assume that within a population in which new mutant forms constantly arise, those genotypes are selected which are best adapted to the prevailing climatic conditions. This selection involves both morphological and ecophysiological characteristics, although by no means all morphological features are necessarily of selective value. They may be due to pleiotropic genes; that is, they are determined by the same hereditory factor as a selectively important ecological characteristic, or linked to ecologically adaptive features.

b) It is not possible to speak of optimal climatic conditions or favourable growth conditions which will apply to all members of the species, irrespective of their origin. For the high mountain race, for example, the hard high mountain climate is far more favourable than the mild climate of Stanford, for in their natural habitat there is a harmonious balance between assimilative and degradative processes, and this is disturbed in different climatic conditions. Cold-resistance is, as an adaptation to the climate in the natural habitat, far greater in high mountain than in lowland forms. Foothill forms, on the other hand, are far more resistant to drought, as a result of their summer dormancy period.

c) Where the climatic or habitat conditions show gradual and continuous change, this applies also to the hereditary morpho-logical-ecological adaptations. In this case the term ecotype is usually used for single representatives taken from a more or less continuous series or ecocline.

d) In addition to the genetically determined adaptations, there are also always functional adaptations which are, however, due to modificative responses. It is not possible to predict the nature or extent of these. Moreover, they cannot be distinguished on the basis of appearance from genetically determined adaptations, and can be established only by changing the external conditions experimentally. Experiments in which different ecotypes were crossed have shown that polyhybrids arise; indeed, most morpho-logical and ecophysiological characteristics are determined by polygenes. Segregation in the F2 generation is very complicated, so that a genetically diverse population arises; this is then subjected to selection by the environment. It is the ecophysiological characteristics which are critical for competition, but by no means all the morphological features. The latter are often of no significance; when, however, they are determined by the same gene as the ecophysiological characteristics, then it is also possible to recognize an ecotype from its morphological features. Subdivision of species into subspecies or variants can be ecologically meaningful, but frequently it is not. As the investigations in California have shown, polyploidy in this case is often of no ecological significance.

e) There are species, each with a limited or small area of distribution, which together form a super-species, which will show a similar distribution pattern to that of a single species which has a series of ecotypes, which collectively form an ecocline. The dif-

ference lies in the fact that the ecologically limited species (ecospecies) cannot cross; they are genetically isolated and have sharply differentiated areas of distribution. They are described as vicarious species.

Lawrence (1945) made a study of *Deschampsia caespitosa*, which is distributed over the whole of the northern hemisphere. It is divided into many subspecies; in *D. caespitosa genuina* it is possible to distinguish five ecotypes, and these were studied in the three Californian nurseries, at Stanford, at 1400 m and at the treeline. Of these ecotypes, one came from the Californian coast, a second from the subalpine belt of the Sierra Nevada, a third from the alpine belt, a fourth from the northerly temperate zone of Europe and the fifth from the subarctic zone of Europe (Lappland). One very interesting phenomenon is that all the European races become viviparous in California, without any change in the genome. In this respect they resemble *Deschampsia alpina* from Spitzbergen or *D. rhenana (D. caespitosa* ssp. *litoralis)* from Lake Constance in central Europe, where it is often submerged for long periods. Both of these subspecies are similarly viviparous. It can thus be seen that in certain genetic forms viviparity is released by the environment (cf. *Poa alpina-vivipara* and *Polygonum viviparum* in the Alps). A further finding was that ecotypes from the Sierra Nevada were killed by rust in Stanford. This did not occur with the others. Thus not only genetically determined resistance, but also environmental factors are of importance in determining susceptibility of this plant to infection by rust: the subalpine and alpine ecotypes are rust-resistant in the mountains, but not at lower altitudes.

Summing up, it can be said that species with a wide distribution are not good indicators of environmental conditions; ecotypes are, but cannot always be distinguished on the basis of morphological features (cf. indicator value on p 171).

These results of ecotype research have been confirmed in recent times for a large number of species (vide the summary of Davis and Heywood 1964 and Heslop-Harrison 1964).

We wish to return once more to the pine *Pinus sylvestris*. Langlet (1936) showed that the Lappland pine, the central Swedish and those from upper Italy and Spain differed in their cold-resistance and growth rate and that these characteristics are paralleled by differences in the dry weight and sugar content of the seedlings. With increasing growth period in their original habitat from 105 days in the north to 189 in the south, the dry substance content of the seedlings falls from 40% in Lappland to 31% in southern Sweden, 29% in upper Italy and 27% in the Spanish mountains. In Sweden the dry substance content was determined at 580 different places. It was found to decrease continuously from north to south, just as do ecologically important characteristics. Thus here, too, there are no ecotypes, but ecoclines.

The same probably applies to spruce, which is differentiated taxonomically into the European *Picea abies* and the Siberian *Picea obovata*, although they are linked by intermediate forms *(P.fennica, P.medioxima, P.uralensis)* in the intervening area. In the beech, too, the intermediate Balkan form *Fagus moesiaca* (Moulopoulos 1965) and the Crimean form *Fagus taurica* (Czeczott 1933) are intermediate between the European *Fagus sylvatica* and the eastern *F.orientalis*. Even *"Fagus sylvatica"* has five ecological races in as small an area as Belgium (Galoux 1966). All these examples may involve ecoclines.

The recognition of ecotype series is of great practical significance. In forestry only seeds of a provenance corresponding to the area to be planted are used. In Sweden the forests have been divided into zones, each extending 200 km from north to south. In reforestation only seeds from the same zone are used. In agriculture, too, the provenance of the seeds is borne in mind. For the ecologist, however, the existence of ecotypes creates the difficulty that the results ob-

tained from plants of a particular species are not valid for all individuals of the species. One must thus be very cautious about making generalizations. This applies to all characteristics which are regarded as adaptations to the environment, and thus in particular to adaptations which are of particular ecological interest.

There is today a strong tendency to undertake experiments using climate chambers or phytotrons to study ecophysiological characteristics and to establish the degree of adaptation to particular environmental factors. It is dangerous to make inferences on the basis of such experiments. Even in the best climate chambers and phytotrons it is not possible to imitate ecological conditions precisely. Experiments conducted by Clausen et al. (1948) with *Achillea millefolium* races in a phytotron in Pasadena have shown that the reactions of the individual races to different temperature conditions and day length, although different, were not at all what might be expected. High mountain ecotypes from an area still south of 40°N, grown under hot daytime temperatures and warm nights, could be induced to flower only with a daylength of 24 h, while for the Lappland ecotype, in the same temperature conditions, an 8 h day sufficed. Thus even physiological experiments conducted in carefully controlled conditions, with variation of a single factor, can be evaluated ecologically only with great caution. Adaptations involve an interaction with a great many constantly changing factors, which makes analysis very difficult.

Chouard (1979) similarly mentions that *Artemisia herba-alba* behaves as a short-day plant in the phytotron, but in the field as a long-day plant, flowering only in autumn, since it requires a *large number of successive long days* to induce flowering.

10 Example of a Vegetation Mosaic in the Tropics: Venezuela

10.1 General

We can illustrate some of the general principles we have been discussing by reference to Venezuela. At the same time, this chapter leads directly to Volume 2, in which tropical vegetation is dealt with in detail.

Venezuela, which was explored by Alexander von Humboldt 180 years ago, reaches almost to the equator along its southern border with Brazil, while in the north, on the Caribbean coast, it extends just beyond 10° latitude. It thus lies within the northern hemisphere in the equatorial zonobiome I, and has a diurnal climate (see p 21). It must be borne in mind, however, that the temperature equator of South America does not coincide with the geographical equator, but lies at 7.5°N. During the winter months of the northern hemisphere, the northeast trade wind brings sunny weather to Venezuela, while the summer months are very rainy, because the inter-tropical convergence in July lies only a little to the south of the lower Orinoco river. The trade wind is diverted from the northern coast of Venezuela in an easterly direction, so that it blows parallel to the coast, with the result that rain falls only on more northerly projections of the coast and where the wind blows against a mountain. Since the northern part of Venezuela is very mountainous (Fig. 118), extreme differences in climate occur within a very small area. The eastern and to some extent also the northern slopes and the lowland area to the east of the mountains have a high rainfall as a result of the trade wind, while the slopes in the wind shadow, the inner mountain valleys and all that part of the coast which runs in an east–west direction are extremely dry.

Figure 119 depicts typical climatic diagrams for Venezuela: 1–10 on a northerly cross-section, 11–15 on a westerly diagonal transect, 16–20 on a north-south transect. A glance suffices to show how very diverse the climate is. The only two diagrams which are typical of the equatorial zone are 11 (Rio de Oro) and 20 (San Carlos de Rio Negro) on the edge of the Amazon basin (Promedios Climatologicos 1965; Walter and Medina 1971).

Although it lies in the otherwise very humid equatorial zone, there are very great differences in the rainfall of different parts of Venezuela: only 150 mm on the coral island of La Orchila (16), which barely projects above the surface of the sea but over 4500 mm on the Rio de Oro (11) just beneath the western branch of the Andes on the Colombian border (Fig. 119).

We know of no other comparable diversity of climatic conditions and of tropical vegetation types in a relatively small area. This prompted us to include a discussion of Venezuela in this volume, for it enables a number of important problems for the tropics as a whole to be raised; these will be considered in detail in Volume 2. Here we have been able, with the kind permission of V. Vareschi, to reproduce many of the very clear vegetation profiles from his book (Vareschi 1980b).

10.2 The Relief of Venezuela

From the eastern Colombian chain of the Andes, two mountain ridges branch off towards Venezuela: the Sierra de Perija on the western border of Venezuela with Colombia, and the Andes of Táchira and Mérida, which form the southeastern border of the Maracaibo basin, breaking up in the north into separate, diverging, low finger-shaped chains (Fig. 118); of these, the Sierra San Luis almost reach the north coast. Be-

Fig. 118. Schematic representation of the mountains of Venezuela. The numbers correspond to the climatic diagrams shown in Fig. 119. (Walter 1973)

tween them lie the dry Carora and Barquisimeto basins (13 and 14) (Vila 1960; Marrero 1964).

The northeastern branch, the Sierra de Aroa, after a gap near San Felipe, continues as coastal mountains, although these do not in fact belong to the Andes system. Among these are the Cordillera de la Costa, running close to the coastline and reaching, on Naiguatá, height of 2763 m. This range then disappears below the surface of the sea, to reemerge on the isthmus between the peninsulas of de Araya and de Paria and as Mount Aripo on Trinidad. Parallel to these, but somewhat further south, are the various mountain ranges of the Cordillera del Interior, which in the east form the mountains around Barcelona, Cumanacoa and Caripe, and reach a height of 2600 m. South of these mountains and stretching to

the Orinoco is a large plane, the llanos. The 2140-km-long Orinoco skirts three sides of the Guiana shield, containing the high, flat-topped mountains, tepuis, up to 2810 m. The southwestern border of Venezuela is formed by the Rio Negro, a tributary of the Amazon river, but linked to one fork of the Orinoco by the Casiquiare (Fig. 118).

10.3 The Climate of Venezuela

During the winter half-year of the northern hemisphere, that is, from November to April, the climate of the northern half of Venezuela is affected by the northeasterly trade wind; passing over the continent, this wind changes direction and blows from the east, while in the Maracaibo basin it is more a north wind.

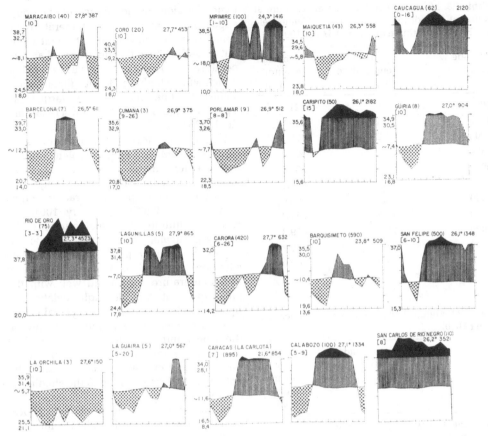

Fig. 119. Climatic diagrams for weather stations in Venezuela (for their situation, see numbers in Fig. 118): stations *1–10 (upper two rows)*, northern transect; stations *11–15 (middle row)*, western oblique transect; stations *16–20 (bottom row)*, north to south transect. (Walter 1973)

The trade wind brings no rain with it. Although the water content of the air masses which move across the sea is relatively high, the fact that they move from higher, cooler latitudes to lower, warmer parts and warm up further over the continent means that precipitation will only take place if the air masses are forced upwards by the mountains. As the air masses rise, they cool, clouds are formed, with resulting rain.

It is thus possible to make a general distinction between two seasons based on rainfall (Fig. 119, 18 Caracas, 19 Calabozo): a 5-month dry season (November–March), known in Venezuela, by analogy to the mediterranean summer dry period, by the Spanish word verano (= summer), and a 7-month rainy season (April–October), known as invierno (= winter), corresponding to the winter rain period of the mediterranean area. By contrast, temperature remains almost constant throughout the year. The difference between the monthly averages for the coldest and warmest months is 1.5–2.5°C, while the average daily fluctuation in temperature for Venezuela as a whole is put at 8.5°C. Thus we have here a typical diurnal climate and this applies also to the mountains where the only difference is that the mean annual temperature falls with increasing altitude. Table 23 shows the relationship between altitude, mean temperature and barometric pressure; it must be borne in mind that marked local deviations arise as a result of the position (valley, peak), the aspect (eastern and western slopes are the warmest) and the number of hours of sunshine or degree of clouding.

Table 23. Mean temperature and air pressure at different altitudes in the mountains of Venezuela (cf. Fig. 120). Altitude gradient 0.57 per 100 m

Altitude (in m)	Mean temperature (°C)	Barometer reading (mm Hg)
5000	−1.5	421.4
4500	1.4	448.0
4000	4.2	476.5
3500	7.0	506.2
3000	9.9	537.5
2500	12.8	570.4
2000	15.6	605.0
1500	18.5	641.2
1000	21.3	679.3
500	24.2	719.2
0	27.0	760.0

Table 24. Altitude belts of orobiomes in Venezuela. (Vila 1960)

Altitude belt	Altitude (m)	Mean annual temperature (°C)
Permanent snow	>4850	<0
Tierra helada (alpine)	3200–4850	9– 0
Tierra fria (subalpine)	2200–3200	14– 9
Tierra templada (a, montane)	1500–2200	18–14
(b, submontane)	800–1500	22–18
Tierra caliente (lowland plain)	0– 800	28–22

The various altitude belts distinguished in Venezuela are shown in Table 24.

A diurnal climate provides a very simple method for assessing local deviations in mean annual temperature. Because there are, in general, only daily fluctuations in temperature, these do not penetrate deep into the soil. Where the soil surface is shaded, soil temperature, even at 30 cm depth, is therefore almost constant throughout the year, and is the same as the mean annual temperature measured at a meteoro-

logical station. If the surface of the soil is exposed to sunlight, soil temperature becomes constant at a depth of about 60 cm, but at 2–3°C above the mean annual air temperature. We used this method in 1934 in East Africa when climbing Mount Kilimanjaro. In Venezuela it has been used on a larger scale by Boussingault (1849), Jahn (1934) and Röhl (1952).

Figure 120 shows the temperatures recorded by Jahn and Röhl, complemented by measurements made at higher elevations by Walter and Medina (1971). The straight line gives the mean temperature gradient which decreases by 0.57°C for every 100 m increase in altitude. The actual gradients will, in any particular case, be higher or lower, depending on the aspect of the slope and its situation. The individual measurements of soil temperature correspond well with the mean gradients and although deviations of up to 4.5°C from these average values do occur, the extent of the variation depends on the aspect and is usually less than 2°C. The fact that the deviations were mostly in the positive direction may indicate that Jahn did not pay sufficient attention to the shading of the soil.

In measuring rainfall, the effect of local factors, particularly of the relief, is far more marked. Wherever mountains run at right angles to the direction of the trade wind, both windward slopes and the adjacent lowlands receive plentiful rains from the rising air masses. This applies to the foreland and eastern slopes of the mountains to the west of the Gulf of Paria, to the "Barlovento" area on the Tuy river below Caucagua, to the eastern slopes of the Sierra de Aroa near San Felipe and to the lower Toxuyo river up to the eastern slopes of the Sierra de San Luis. The whole of the southeastern slope of the Andes chain which descends into the llanos of Orinoco has a particularly high rainfall (Fig. 119, diagrams 3, 5, 9, 15).

Further west, in the Maracaibo basin, the wind blows mainly from the north. The result is that even at the northwestern foot of the Andes, the rainfall is high. The maximum is, however, reached southwest of Lago de Maracaibo, against the mountains on the Colombian border, where the average over 3 years was 4523 mm (Fig. 119, diagram 11).

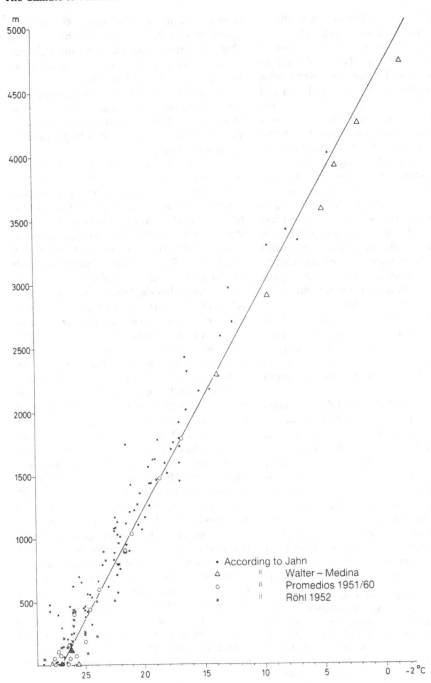

Fig. 120. Relationship between altitude and constant soil temperature or mean daily temperature data from Promedios (see text). (Walter 1973)

In the whole of the llanos plain, the 5-month dry season is sharply separated from the 7-month rainy season, and the annual rainfall is about 1300 mm. The most southerly weather station of Venezuela, San Carlos de Rio Negro, lies in the Amazon basin and almost on the equator (1° 54' N). Here the mean monthly rainfall over the year as a whole is 200 mm, and from April to August as much as 300 mm. The mean annual rainfall is 3521 mm (Fig. 119, diagram 20).

The driest areas are the low offshore coral islands (La Orchila, for example, has an annual rainfall of 150 mm) and the Paraguana peninsula over which the trade wind blows; indeed, the whole of the coast which lies parallel to the direction of the trade wind is very dry[11], even where the terrain is steep as, for example, near La Guaira. This applies also to the mountain slopes and basin areas (Carora, Barquisimeto), which lie in the wind shadow (Fig. 119, diagrams 2, 4, 6, 13–14). Particularly complicated climatic conditions are found in the inner valleys of the Andean mountains, with the rivers Mocotin, Chama, Motatan and the upper San Domingo.

There is thus in N Venezuela, a climate mosaic which it is not possible to depict accurately on any climatic map. All that can be done is to describe the climate in those places where weather factors are actually measured, that is, to produce climatic diagrams for the individual meteorological stations (vide the climatic diagram map of Walter and Median 1971). In Venezuela there are 700 stations for measuring rainfall, but only relatively few where long-term temperature measurements are made. Since, however, mean monthly temperatures hardly change and since the mean annual temperature can be judged on the basis of altitude if no values for constant soil temperature are available, data on rainfall suffice to form an idea of the climate.

We will have to content ourselves with few examples here, and concentrate on those stations which make clear the great variety in the climate of this tropical-equatorial area (Fig. 119).

What is important for the climate as a whole is not only the quantity of rainfall, but also whether it is light or heavy. It is characteristic of the tropics that the rain usually falls as short, sharp showers or as violent storms, after which the sun very soon reappears. Most of the rainwater then runs off the surface. Evaporation is interrupted only briefly and continues again as soon as the sun shines. This explains why, even with a relatively high annual rainfall, the climate may be arid, for, over the year as a whole, the potential evaporation is often higher than the rainfall. There is, for example, no drainage from Lake Valencia near Maracay, in the long valley between the Cordillera de la Costa and the Cordillera del Interior, although Maracay has a mean annual rainfall of 914 mm; the potential evaporation is estimated at 1280 mm, with the result that Lake Valencia has fallen by 10 m over the last 200 years. Table 25 shows the annual water balance for the most important meteorological stations for the period 1951–1960. This shows how varied is the climate of Venezuela. It fluctuates from extremely humid areas with a very positive water balance of up to +3000 mm, to extremely arid areas with a negative water balance of more than −2000 mm. The higher mountains are mostly humid: for example, Caracas, at an altitude of 1035 m, is very nearly in water balance (Table 25), while nearby, the Pico Avila rises to over 2000 m, and has a very humid climate, as can be seen from the moist and misty forests with bamboo stands.

11 Wilhelmy (1954) discusses the origin of this "trade wind desert", which includes also the offshore islands (Curacao 559 mm, rainy season October–November) and north Colombia (Guajira peninsula 366–373 mm, Santa Marta 386 mm). The climate in this area has become increasingly dry since the last Pluvial period in the Pleistocene; the level of the sea has risen by 100 m, so that many bays were formed from flooded river valleys. More recently, a raising of the land by 4–5 m has become noticeable. These fluctuations in sea level have had an important effect on the formation of the Venezuelan coast and on the distribution of mangroves on the north coast of South America.

10.4 Vegetation Types in Venezuela (Zonobiomes and Subzonobiomes)

Apart from the high mountains, the predominant type of soil in Venezuela is red

Table 25. Water balance for the years 1951–1960 for some meteorological stations in Venezuela. Altitude of each station shown in brackets. Potential evaporation according to the Promedios Climatologicos 1951–60 (1965) measured with a "type-A" tank. (Rainfall data after Promedios Climatologicos 1965)

Station	Annual rainfall (mm)	Potential evaporation (mm)	Balance
Barcelona (7 m)	611	1453	− 842
Barquisimeto (590 m)	509	1610	−1101
Caracas-la Carlota (895 m)	854	867	− 13
Caracas-Cagigal (1035 m)	874	785	+ 89
Carrizal (160 m)	1111	1813	− 702
Ciudad-Bolivar (50 m)	972	1633	− 661
Colonia-Tovar (1790 m)	1271	518	+ 753
Coro (20 m)	453	2153	−1700
Guiria (8 m)	904	1122	− 218
La Orchila (3 m)	150	2258	−2108
Maiquetia (43 m)	558	1014	− 456
Maracaibo (40 m)	387	1582	−1195
Maracay (442 m)	914	1280	− 366
Maturin (70 m)	1320	1296	+ 24
Mene Grande (18 m)	1310	1254	+ 56
Mérida (1479 m)	1633	805	+ 828
Moron (3 m)	1064	851	+ 207
Puerto-Ayacucho (99 m)	2041	1073	+ 968
San Antonio (404 m)	655	1714	−1059
San Carlos de Rio Negro (110 m)	3521	520	+3001
San Fernando (73 m)	1533	1420	+ 113
Santa Elena (910 m)	1796	655	+1141
Tumeremo (180 m)	1236	837	+ 399

tropical clay; it is acidic and poor in nutrients. Except in the coastal regions, no salt accumulation takes place, even in the most arid areas; this is because the rainfall is so heavy during the rainy season that all soluble salts are washed into the rivers. A light salinization is indeed found around the undrained Lake Valencia. The lake itself contains 350 mg l^{-1} of sulphates and, according to Bonazzi, the salt content has doubled in the past 100 years (Gessner 1955, 1956). There is also a small, undrained soda lake, Lagunillas, in the dry inner-Andian long valley, below Mérida, at an altitude of 1079 m; here the rainfall is only 463 mm.

In Venezuela a whole range of climatic vegetational types are found, from the clearly arid to the extremely humid. These are shown in Fig. 121. For the lowlands, starting with the most arid, these are:

1. the cactus semi-desert, known as cardonales,
2. the thornwoods or espinar,
3. the deciduous forest, the selva decidua,
4. the semi-evergreen forest,
5. the evergreen forest, which in the most humid areas is an evergreen tropical rainforest.

The vegetation-belt series associated with increasing altitude is less clear. Although, as we know, the temperature falls continuously with increasing altitude, the rainfall varies with inclination of the slope and the geographical situation of any mountain. We will

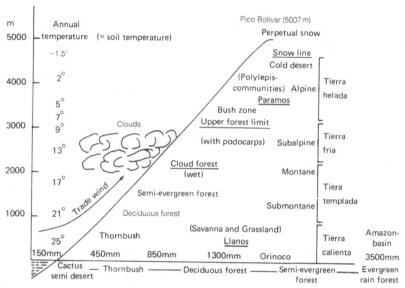

Fig. 121. Schematic representation of the altitude belts in Venezuela. The scale shows altitude and mean annual temperature in °C. Along the abscissa the change in vegetation from north to south with increasing rainfall (in mm) is shown. (Walter 1973)

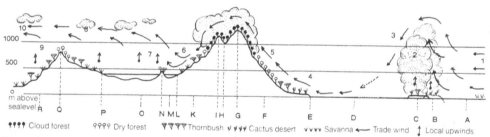

Fig. 122. Diagrammatic transect through the cloud-forest area of Rancho Grande and its surroundings in a SE–NW direction; vertical scale increased 13 times. A–R Geographical features; 1–10 air movements. A Atlantic Ocean; B,C,D Caribbean coastal sea (reduced in size); E dry coastal zone; F northern slope of the Cordillera; G ridge by Rancho Grande; H Portachuelo Pass; I Periquito peak; K southern slope; L north bank of Lake Valencia; M,O Lake Valencia; N Chambero Island; P southern bank; Q Cordillera del Interior (Cero Maria); R southern slope down to the llanos. 1 Trade wind from NE; 2 cloud tower above the bare Antille Islands; 3 Island Föhn; 4 hot upward air movement; 5 air cooling as it rises; 5–6 typical Föhn cap; 6 wind redirected by local upward air movement (7); 8 air cooling again; 9 local upwind; 10 trade wind towards the llanos. (Vareschi 1980b)

limit ourselves to the series on the windward side of a mountain against which the trade wind blows, and where the rainfall increases rapidly up to a certain height, while the foot of the mountain extends into the arid area. The following altitude belts are then encountered, starting from the bottom (see also Fig. 122):

1. semi-desert with cactus or thornbush,
2. deciduous forest,
3. semi-evergreen forests,

4. very moist, misty forests, the selva nublada, at a height where cloud-formation occurs; they are cool as a result of both high altitude and little sun,
5. high montane or subalpine forests up to the treeline,
6. transitional shrub zone, known as "chivi-rital",
7. alpine belt of páramos,
8. subnival belt which takes the form of a vegetationless cold desert,
9. firn areas.

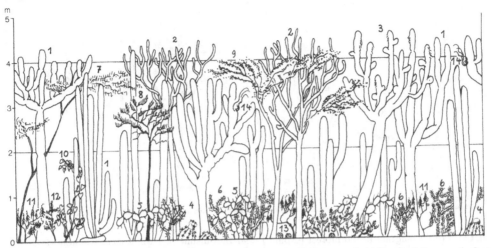

Fig. 123. Profile diagram of a strip, about 10 m across, through the succulent forest of Carora. *1–6* and *13* Succulents. For detailed explanation, see text. (Vareschi 1980b)

Disregarding for the moment the mountains, it is helpful to examine the vegetation of the lowlands on the basis of a profile running from the driest areas in the north to the most rainy parts in the south. This shows there to be a series of three subzonobiomes (sZB) which all belong to zonobiome I with a diurnal climate, but which differ markedly from one another with regard to rainfall. The driest can be designated I(rIII), that is, with a rainfall (r) almost the same as that in the desert of zonobiome III; the second, with a 5-month drought period, as I(II), that is, it has a rainfall distribution as in the tropical summer rain zone (II). It is only the third, with no or almost no dry period, which is typical of ZB I. Furthermore, these subzonobiomes are linked by transitional areas or ecotones. Very similar conditions, found in the dry area of north Colombia near Santa Martha, have been described by Schnetter (1968, 1970, 1971), who also made microclimatic and some ecological measurements.

The vegetation of the cactus semi-desert of the sZB I(rIII) is often named the *cardonales* after the dominant column-like cactus (cardones). This dry area is found along the coast, mainly in the lee of the offshore islands, because heat from the islands causes an upward movement of the air masses blown by the trade wind, this leads to the formation of cloud towers above the islands and to dry winds descending on the coast (Fig. 122).

Almost desert conditions also prevail, however, in the wind shadow of mountains as, for instance, in the Carora and Barquisimeto basins or in inner-Andean valleys (Fig. 119, diagrams 1, 2, 4, 7, 8, 13, 14, 16, 17). The rainfall may actually exceed 400 mm, but it falls as infrequent, sharp downpours and thus has little lasting effect. It is striking that the rainfall curve has two maxima, one small and one larger; this is particularly advantageous to succulents which are able to replenish their water reserves twice a year.

Figure 123 gives an impression of the vegetation in a very dense stand of succulents, with 1 *Cereus lemairei*, 2 *Cephalocereus moritzianus*, 3 *Cereus griseus*, 4 *Melocactus caesius*, 5 *Opuntia wentiana*, 6 *O. caribea*, 13 *Mammilaria simplex*, and amongst the shrubs 7 Mimosaceae spp., 8 Capparidaceae spp., 9 *Cercidium praecox* with green assimilating bark, 10 *Ipomoea carnea* as peripheral shrub, 11 Euphorbiaceae (*Croton, Jatropha urens* with stinging hairs), 12 herbs (*Sida, Bastardia*) and 14 of the xerophytic epiphytes *Tillandsia flexuosa* or *T. recurva*. Typical are also the many crustose and fructose lichens, indicative of high air humidity or fog. Usually the stands are more open. In the driest biotopes the only plant is *Melanocactus*. The opuntias often cover large areas, for their thorny axil joints become attached to the hair of goats and are dispersed by them.

The thornwood or *espinar* forms a transition to the deciduous forests. Dominant here are those shrubs which are found only sparsely scattered in the cardonales—*Acacia, Mimosa* and the Caesalpiniaceae spp. (Fig.

Fig. 124. Profile diagram of thornforest (espinar) near Barquisimeto. *1* Mimosaceae genera; *2 Cereus jamacaru; 3 Plumeria alba* in flower; *4 Peireskia guamacho* with leafless tips to its branches; *5 Pitcairnia madifolia; 6 Croton rhamnifolius; 7 Bastardia viscosa; 8 Pfaffia iresinoides; 9* Gramineae and Cyperaceae; *10 Arrabidiaea* sp.; *11 Evolvulus filipes.* (Vareschi 1980b)

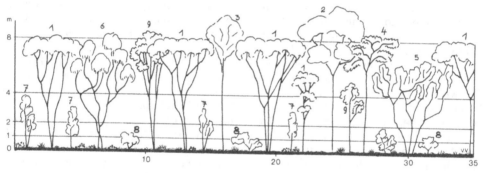

Fig. 125. Profile diagram of the dry trade wind forest of Quiriquire, in the llanos of Venezuela. *1 Luehea candida* (Tiliaceae); *2 Spondias mombin* (Anacardiaceae); *3 Tabebuia spectabilis* (Bignoniaceae); *4 Astronium graveolens* (Anacardiaceae); *5 Pithecolobium caraboboensis* (Mimosaceae); *6 Libidibia* sp. (Mimosaceae); *7 Lonchocarpus* sp. (Papilionaceae); *8 Randia* sp. (Rubiaceae); *9 Platymiscium* sp. Papilionaceae). (Vareschi 1980b)

124); the column cacti are represented only by *Cereus jamacaru.* Of especial interest is the leafed cactus *Peireskia guamacho,* which is believed to be close to the ancestral form of all the Cactaceae. Here one also finds the beautifully fragrant Apocynacean *Plumeria alba* with its succulent branches; it is cultivated in gardens throughout the tropics and is known as frangipanni. *Guaiacum officinale* also occurs, and Bromeliaceae (*Bromelia, Pitcairnia,* and so on) often grow on the ground.

The very xeromorphic *Commelina repens,* as well as the poikilohydric *Selaginella sollowii* also occur. Very striking is *Bursera simaruba* with its succulent stem and a flaking, brown bark. This tree grows to a height of more than 4 m, as do several other tree species, forming a transition to the 5–20 m high deciduous forests—*selva decidua*—of sZB I(II) (Fig. 125). Typical for these is the climatic diagram for Caracas (Fig. 119, No. 18) with more than 800 mm rainfall, which falls mainly in the months May to November. Many tree species flower in a leafless condition at the end of the drought period and before the first rain falls. The releasing factor may be the particularly high temperature at this time of the year. Especially impressive then are the yellow blossoms of

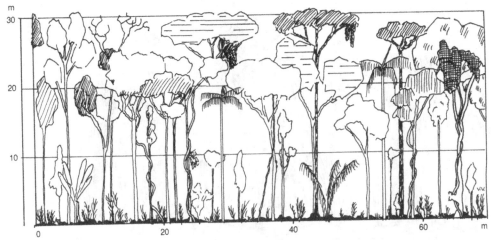

Fig. 126. Profile diagram of the wet trade wind forest of Rio Caura. (Vareschi 1980b)

Cochlospermum vitifolium and the pink flowers of *Triplaris caracana* and of the *Tabebuia* spp. The euphorbiacean *Hura crepitans*, the fruits of which explode on drying, flinging out the seeds, is particularly noticeable because of the spiny projections on the trunk. When it grows in moist biotopes, it remains evergreen. Indeed, leaf fall seems to be facultative in many species and the duration of the leafless period can change according to rainfall in different years.

The crowns of species such as *Ceiba pentandra, Bombax, Bombacopsis* and *Erythrina glauca* project above the canopy of these forests. The mimosacean *Pithecellobium saman* has a very wide, umbrella-like crown; in forest clearance it is frequently left standing as a shade tree; its strong horizontal branches are then densely covered with epiphytes (orchids, bromeliads, ferns). Other important species are *Cassia moschata, Copaifera officinalis* and *Pterocarpus podocarpus*, which flower abundantly. Vareschi puts the number of species in a stand, excluding the 14 lianes, at about 62.

A distinction is made between dry deciduous forests like those of Caracas and the moist deciduous forests in which the drought period lasts only 2 months (with an annual rainfall of 1500 mm) or 3 months (with an annual rainfall of over 2000 mm). The latter are richer in species (80 species) and may reach a height of 30 m (Fig. 126). They are also more dense and formed of two tree storeys.

These forests contain trees which are of economic importance, such as *Anacardium excelsum, Bombacopsis sapium, Cedrela* spp., *Pithecellobium* spp. and the mahogani tree *Swietenia macrophyllum.* The lovely pink-flower of *Tabebuia pentaphylla* is the national flower of Venezuela. The deciduous forests typical of the lowlands of northern Venezuela extend southwards just beyond the lower Orinoco and its tributary, the Apuri. Extensive areas of forest have been cleared and the land cultivated or the burnt-off areas may be used for grazing. The cactus semi-desert and the thornbush are far too dry for agriculture and are only used for grazing goats. In the region of the deciduous forests coffee can be grown under the shade of *Erythrina* trees; in drier parts sisal-agaves can be grown; in the moister areas, intensive grazing is possible after sowing *Panicum maximum;* sugar cane usually requires additional watering, but then gives a very high yield.

The climate in the region of the dry deciduous forests is very pleasant as a result of the long drought, especially in the mountains, and is therefore the part most heavily settled by whites. The black population is found in the lowest hot, moist areas, where cocoa and hevea can be grown. The Andean Indian half-castes live at high altitudes in the Andes at about 3000 m.

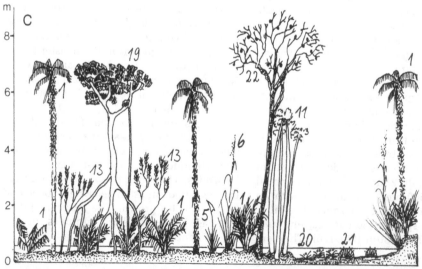

Fig. 127

Ecologically, the deciduous forests of Venezuela have hardly been investigated. In the centre of this deciduous forest area, north of the lower Orinoco, and in the region of its many tributaries from the Venezuelan Andes in the large basin of the Llanos del Orinoco, the forests are replaced by a savanna, which stretches far to the west towards Colombia. This savanna is not determined by climate, but by edaphic factors; in the central part, which has been investigated in detail, it is due to the formation of a hard laterite layer, known as the arrecife. We will discuss this in Volume 2 under tropical pedobiomes.

With increasing rainfall and a shortened period of drought, there is a smooth transition from the moist deciduous forests to the *semi-evergreen tropical forests*. These are most readily recognized during the drought period as the dark evergreen shrub layer is then to be seen beneath the bare tree layer. Such forests with the evergreen rutacean *Pilocarpus* resemble a coffee plantation under deciduous shade trees.

The more moist the climate, the denser the evergreen shrub and lower tree layer. Finally, the whole forest becomes evergreen and the large deciduous Bombacaceae project above the irregular canopy. The semi-evergreen forests form a transitional type of vegetation to the *evergreen tropical rain forests* which are found in southern Venezuela and also in the northern parts of the country in those low-lying areas where the path of trade wind is interrupted by mountains, so that rain falls during the dry months.

We have already mentioned that of all the meteorological stations, only that on the Rio de Oro (11) and that in Carlos de Rio Negro (20) show the typical climatic diagram of tropical rain forest. The forests of the Rio de Oro on the Colombian border are, however, almost impenetrable and have thus been little investigated. San Carlos de Rio Negro, as the name suggests, lies in the region of the black water rivers, coloured dark brown by humus colloids and extremely poor in nutrients. This is an area of tropical ombrogenous high moor forest on bleached sandy soils with a pH of 3.8 to 4.5, on which no typical evergreen rain forest can grow. There are, rather, low forests with close-standing tree trunks, often very much thickened and carrying a thin foliage. There is likewise a very poor fauna: no crocodiles, piranhas or mosquitoes, and even birds are rare. These forests are at present being studied ecologically. Here we do not have the climatically determined vegetation of a zonobiome, but rather a pedobiome (peinohelobiome) on wet soil which is very poor in nutrients and corresponding to central European raised bogs. We will return to this in Volume 2. Here we will limit ourselves to reproduction of the forest profile of Vareschi (Fig. 127).

We are unable to report the existence in Venezuela at low altitudes of any typical, virgin evergreen rainforests with tree species which show no seasonal rhythm with respect to leaf fall, flowering and fruiting. In the extensive forests of Guiana the effect of the drought period is always noticeable, although it differs in intensity from year to year. This is illustrated for the rain forest of El Dorado in Fig. 128. Even a very short drought period, which may be absent altogether in certain years, brings about a certain annual periodicity in the living processes in the forest. These are thus better known as "seasonal rainforests". Rainforests such as that near El Dorado do not reach a great height; they have a smooth outer canopy and are rich in species, with a total number of about 100 species, most of them trees. Almost all their leaves have drip tips and the number of living forms is large:

Fig. 127. Three profile diagrams of the black-water forest of Cano Monomi, Territorio Amazonas (100 m): **A, B, C** are examples with increasing height of the water table. *1: Leopoldinia pulchra* (Palmae); *2: L. piassava; 3: Mauritia armata; 4: Desmoncus horridus; 5: Spathanthus bicolor* (Rapateaceae); *6: Rhynchospora* cf. *corymbosa* (Cyperaceae); *7: Maranta arundinacea* (Marantaceae); *8: Anthurium* sp. (Araceae); *9: Ravenala guianensis* (Musaceae); *10: Pachira aquatica* (Bombacaceae); *11: Malouetia glandulifera* (Apocynaceae); *12: Thurnia polycephala* (Thurniaceae); *13: Macairea stylosa* (Melastomaceae); *14: Cephalostemon affinis* (Rapateaceae); *15: Montrichardia arborescens* (Araceae); *16: Ambelania laxa* (Apocynaceae); *17: Tococa ciliata* (Melastomaceae); *18: Tachigalia cavipes* (Caesalpiniaceae); *19: Clusia amazonica* (Guttiferae); *20: Dichronema acaulis* (Cyperaceae); *21: Eriocaulaceae* genera; *22: Eschweileria rigida* (Lecythidiaceae). (Vareschi 1980b)

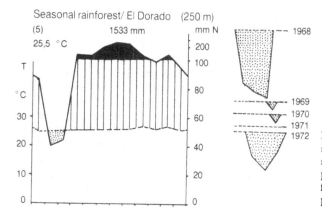

Fig. 128. Climatic diagram of the seasonal rainforest of El Dorado. *Right* schematic representation of the dry periods from 1968 to 1972; the marked fluctuations in the extent of the dry periods is clear. (Vareschi 1980b)

Fig. 129. Seasonal rainforest of El Dorado in Guiana, Venezuela (130 m altitude), showing the most important plant families and life forms: *1, 6* Lauraceae; *2, 5, 17* Leguminosae, Papilionaceae; *3* Meliaceae; *4* Bignoniaceae; *7* not identified; *8 Olfersia; 9 Cyathea; 10 Ravenala guyanensis; 11* young plants; *12, 13* Araceae; *14* shrubs (Piperaceae, Melastomaceae; Rubiaceae); *15, 19* Rubiaceae; *16 Geonema* (Palmae); *17* mosses and liverworts on the lower part of tree trunks; *18* herbs (Rubiaceae, ferns); *19 Desmoncus* (Palms); *20* lianes (Bignoniaceae, Sapindaceae). (Vareschi 1980b)

lianes and epiphytes, and strangler species (e.g. *Clusia*) also occur (Fig. 129).

10.5 The Orobiome of Venezuela

The diurnal character of the climate does not change with increasing altitude in the mountains, but the mean annual temperature decreases continuously (Fig. 120). The rainfall pattern changes with altitude in a quite different way. Since the trade wind blows all the year round from the same direction, conditions on the windward side are completely different from those in the wind shadow, that is, on the lee slope.

Figure 121 shows conditions on the wind-ward side of a mountain; here the air masses rise, cooling in the process, so that above a particular altitude condensation occurs and rain falls, frequently even over the foot of the mountain.

Figure 122 shows the complicated con-ditions found in the Cordillera de la Costa around Rancho Grande. On the northern slopes the rainfall increases continuously with increasing altitude, and the duration of the drought period decreases. There is con-sequently an altitude-belt series which is closely similar to the zonation of the vegeta-tion from north to south. The lower slopes with large column-shaped cacti and thorn-bush are followed by deciduous forest, and this in turn is replaced by semi-evergreen and then evergreen forest. This applies to the windward slope, but the lee side, in the wind shadow, does not show this series; here it is extremely dry. Some differences from the vegetational zones of the plains are found also on the windward slopes; clouds often remain at a particular altitude even in the drought period; this results in the forma-tion of fog and dew. This belt of *fog forest* is typical for all tropical mountain forests af-fected by trade winds. In Venezuela the fog forests are known as *selva nublada,* or cloud forests, although Vareschi makes a distinc-tion between a "cloud forest" and a "fog forest" or "selva de niblas". This distinction meets with difficulties, however, for clouds hang in the air, but when they reach down to the ground, it is known as fog. On the east-ern slopes of the Mexican highlands, Lauer observed with two altitudes at which cloud formation and condensation occurred with rising air masses. At the lower, where air temperature was still relatively high, the clouds hung over the slope and rain fell from them, so that rainfall was maximal at this altitude. Higher, at the upper forest limit, air temperature was low, so that the rising air masses were cooled further. Here further condensation occurred, with resultant fog and drizzle, but since the colder air contains far less water vapour at saturation point, the precipitation is much less than at the lower altitude. In this case the lower zone could be regarded as cloud forest and the upper as mist forest.

According to Vareschi (1980b and Fig. 122G), the windward side of the whole of the upper part of the Cordillera de la Costa near Rancho Grande is covered in cloud; this starts at about 1000 m above sea level, below the Passes Porta Chuela (1136 m) and extends up to the peak of Mount Paraiso (1814 m). This Vareschi calls a "cloud forest".

The only data available on weather condi-tions over a long period are for March to August 1946 (Beebe and Crane 1958). These are shown in Table 26. Only the month of March represents a period of drought, but even then with much fog during the day. The most foggy month is April, the least foggy August, which also has the highest rainfall (511 mm).

The senior author visited Rancho Grande in 1969, in the middle of the drought period. The nights were cloudless and so too were the early hours of the morning, but from 08.00 h fog started to form; at noon the fog cleared a little, but during the afternoon, it became denser and turned to fine drizzle; this lasted until 18.00–19.00 h, then the sky became cloudless once more. During the

Table 26. Weather conditions near Rancho Grande, March–August 1946. (Beebe and Crane 1958)

Months	March	April	May	June	July	August
Mean daily temperature (°C)	18.6	19.8	19.4	18.9	18.5	18.3
Mean humidity (in %)	88.3	94.8	94.0	92.6	92.7	92.3
Rainfall per month (mm)	16	363	330	132	179	511
Hours of fog per month between 06.00 and 22.00 h	194	274	194	134	120	84
Days without fog	6	1	3	11	14	12
Days without sun	2	12	5	3	4	4
Hours of sunshine per month	102	47	82	96	109	103

drought period, the fog lasts for 7 to 9h of the day. The term fog forest should thus be used in this case. If is probably different in August during the rainy season when there are frequent showers, but, on average, not even 3h of fog each day. In this case the appropriate description would be a cloud forest. It is thus very difficult to distinguish between cloud forest and fog forest. Only when precise data on weather conditions over a whole year and for different mountains are available will it be possible to come to a more precise conclusion.

Huber (1976) has made a detailed study of the forest of the Rancho Grande and found that, depending on aspect and altitude, these fog forests differed markedly in their composition and structure. Some have tree ferns, others palms, and so on. It is thus not surprising that the forest La Carbonera on the northwest slopes of the central Venezuelan Andes at 2300m, cited by Vareschi as an example of a fog forest, is markedly different from the forest by Rancho Grande, which lies 1000m lower in the Cordillera de la Costa; this is seen alone in the presence of *Podocarpus*. The fog belt may lie even higher amongst the windy peaks, where the trees are smaller, bent and densely covered with mosses or lichens, so that they look like elfin woods.

The height of the fog belt depends above all on the humidity of the air rising up the slope. Compared with Venezuela, the air on Mount Kilimanjaro in East Africa is very dry and the fog belt coincides with the upper tree line at more than 3000m; above this fog belt the rainfall decreases rapidly (vide Vol. 2). Because of the low temperatures there are no phanerogamic epiphytes at this height and the trees are mostly hung with lichens.

Fog forests are thus not a particular type of forest, but are simply characterized by reference to a fog factor. Not only is humidity high in these forests, but the light intensity is also low; in dense fog there is twilight. The question thus arises as to whether the plants of these forests, particularly in the interior of the forest, have the character of shade plants. Huber (1976, 1978) investigated this in the fog forest of Rancho Grande, which lies at 10° 16′ N between 1000 and 1800m on the windward slope. Mean

annual temperature is 16°–20°C, the annual rainfall 1600–2000mm. He determined the light compensation point of 54 common herbaceous and woody species in the interior of the forest, using the method of Alvik (see p 114).

On clear days, light intensity on the forest floor is, as a result of oblique light falling on the slope, greater in the fog forest (1–3% of daylight) than on the floor of the lowland rain forest (0.5–1%). In the mist forest on clear days illumination was 20–50 μ-einstein m^{-2} s^{-1}, but in thick fog only 3–5 μ-einstein m^{-2} s^{-1} (1 μ-einstein m^{-2} s^{-1} = about 60lx). The light compensation point at 20.8–24.4°C for 80% of all species from the depths of the forest was found to be greater than that for typical shade plants; it lay between 1.5 and 5 μ-einstein m^{-2} s^{-1} or about 90 to 260lx. These are thus not extreme shade plants, but are adapted to markedly fluctuating light intensities. There is, however, a very noticeable difference in size between the leaves of young plants of tree species on the forest floor in deep shade and the later leaves of the fully grown trees which receive far more light. In the colorimetric measurements, it was very striking how sluggishly all palm leaves reacted; these seem to have a very low rate of gaseous exchange, probably as a result of peculiarities in leaf anatomy. While the plants of the fog forest must receive sufficient light to maintain a positive balance in their assimilation economy, it was not possible to draw up an exact daily balance for the assimilation economy on misty days.

We return to a discussion of the altitude-belt series of individual tropical mountains in the discussion of zonobiome I in Volume 2. Here discussion will be limited to the unique flat-topped mountains in the southern part of Venezuela in the provence of Guiana. These are remains of the Guiana shield, which was separated from the giant Brazilian shield by the sinking of the Amazon basin and, as peripheral area, underwent heavy erosion. A hard quarzite sandstone layer, more than 2000m thick, was deposited on granite which was about 1600 million years old. Today the only remaining evidence of this is the separate table mountains of 500–1000m. With their completely vertical sides, they resemble large towers, and are called tepuis by the Indians (tepui = house of the gods).

At the foot of the tepuis is a seasonal rain forest with mean annual temperature of 27°C. On the top of the tepuis, however, the mean temperature is only 10°C and stormy trade winds rage there constantly. The steep sides are almost entirely free of vegetation, so that there are no altitude belts. The flat table top is also almost impossible to cross, because of the ravines and cracks, often hundreds of metres deep, which form grotesque sandstone sculptures. There have been very few successful ascents, and only recently, by the use of helicopters, has it been possible to study the vegetation (Steyermark 1967). The highest tepui, Roraima, on the Venezuelan-Brazilian border reaches 2810 m above sea level.

Vareschi (1980a) gives some information about the flora on the top of the tepuis, 5–10% of which consists of indigenous species. On weathering, the quarzite-sandstone forms a white sandy soil, very poor in nutrients and acidic. As a result, only species typical of oligotrophic soils are found, insectivorous plants and others which are related to epiphytes. There are no trees at all. *Bonnetia* is a dwarf tree with large, scented flowers; *Barbacaenia tubiflora* is reminiscent of tufted tress, but its "trunk" is formed of adventitious roots. It belongs to the Velloziaceae, most of which are poikilohydric. *Chimantea cineraria*, a member of the Compositae, has a resinous dwarf trunk which can be set alight even in rain. *Cyrilla racemiflora* has a trellis-like growth with dark pink clusters of flowers. *Tepuia* is the only member of the Ericaceae in these mountains; *Stegolepis*, with its strangely bent sword-like leaves, belongs to the Rapateaceae. Those families which include large numbers of epiphytes are especially well represented. Amongst the Orchideae there are 195 species on the Auyan Tepui alone, while of the Bromeliaceae, 25 species of *Cottendorfia* spp. are found exclusively on the tepuis. Families with species of oligotrophic habitats, such as Rapateaceae, Eriocaulaceae, Cyperaceae and so on are represented by very many species. Amongst the insectivorous plants, 7 of the 16 Venezuelan species of *Drosera* have been found on the tepuis; *Utricularia* spp. are found in ponds, *Genflisea* spp. (Iridaceae) with their spiralled fork traps are found in streams. The genus *Heliamphora* (Sarraceniaceae) is indigenous; its "pitchers", in which nitrogenous substances from decaying insects are absorbed, can contain up to 200 ml of fluid. With almost every expedition new species are found.

Steyermark and Dunsterville (1980) present a detailed floristic analysis of the tepuis of Venezuela (Fig. 118, see Guiana), listing 41 genera of ferns and 422 of flowering plants; 103 of the latter genera include species which are abundant at lower altitudes of 50–500 m. This is particularly marked on the Guaiquinima tepui; its high plateau reaches 1680 m, but falls away to 700 m in parts and a valley which leads off from this height affords ready access to plants from the plains. Of the 449 species found at 700 to 750 m on this tepui, 21.6% are lowland species. Furthermore, most of the species indigenous to the high plateaux belong to genera which have their main distribution on the low plain.

It can thus be assumed that the higher plateaux were colonized from lower regions during an earlier warm period and that evolution of new species took place subsequently to this.

In the west, from the lower to the central area of the Orinoco, the sandstone layer is completely eroded, but from the granite base single 200–300 m high inselbergs arise —the Lajas. Huber (1980) reports that, on the slopes of these granite mountains, the weathered rock is freed as scales, so that extremely dry habitats are formed.

Weathered sandstone and scree, however, collect, especially at the base of the bare slopes and in rocky clefts, and here more water accumulates; biotopes are formed which, although poor in nutrients, are moist, so that trees can grow there; an example is *Pseudobombax croizatii* with bright green bark that, during the 3–4-month-long leafless period, can assimilate CO_2. Von Humboldt and Bonpland recorded the occurrence of: *Bursera orinocensis*, the melastomatacean *Acanthella sprucei* and *Graffenrieda rotundifolia*, two *Tabebuia* spp., *Jacaranda obtusifolia* (Bignoniaceae), *Certoria dendrina* (Papilionaceae) and *Tocoyena orinocensis* (Rubiaceae).

On dry granite rock *Barbacenia tubiflora* (probably a poikilohydric species) sometimes forms colonies covering several hundred square metres. A near relative is the *Schiekia orinocensis* (Haemodoraceae). Several cacti *(Cereus hexagonus, Melocactus)* and many Bromeliaceae occur here, even though the rainfall reaches 2000 m during the rainy season. Of the terrestrial Bromeliaceae, various *Pitcairnia* species have inflorescences up to 2 m high. In small pools during the rainy season, short-lived species form a floral carpet; these include *Utricularia limbriata, Xyris stenostachya, Syngonanthus alleni* (Eriocaulaceae), the indigenous *Borreria pygmaea* (Rubiaceae) and an indigenous Bulbostylis (Cyperaceae) as well as the more widely distributed *B. leucostachya*. The poor nutrient content of the water is reflected in the low pH values of 3.5 to 4.2. Lianes are also present in small islands of forest and bush. We do not yet know all the species on these inselbergs which are in part relict habitats of the vegetation of previous dry periods.

In northeast Venezuela there is an unusual pedobiome with a unique vegetation. This is the asphalt lake of Guanaco, which has an area of more than 4.4 km² (Lasser and Vareschi 1959). The asphalt mass (pH 6.2) extruded at the centre of the lake is a sterile substratum for plants and has consequently no vegetation. The first settlers are microorganisms, ants and termites; these are soon joined by lichens *(Cladonia furfuracea)* and mosses *(Campylopus arctocarpus, Sphagnum acutifolium)*. They prepare the ground for the first stunted specimens of *Cyperus polystachyus; Andropogon bicornis* and *Rhynchanthera acuminata* (Orchidaceae) follow. At the same time, the pH falls to 5.4, the water capacity of the soil rises to 20% and nutrient content increases. Colonization in the neighbourhood of sulphur springs (pH = 2.5) passes through a *Cladonia* stage to *Eleocharis caribea* stands. The termite heaps are settled by *Clusia* spp. which are tree stranglers. *Anthurium rugosum* and various epiphytes (Bromeliaceae, Orchideae, ferns) while around them a Rhynochosporetum develops,

Fig. 130. Asphalt Lake of Guanaco: transect through the lower part of the *Pterocarpetum officinalis*. *Black* asphalt; *dotted* soil and mud. *A: Pterocarpus officinalis; B: Clusia* sp.; *C: Rhizophora mangle; D: Mauritia setigera* palms; *E: Montrichardia arborescens* (Araceae); *F:* seedlings of *A; G: Asplenium serratum* (epiphytic fern on *Rhizophora* stilt roots; *H: Anthurium rugosum* (on *Clusia* stilt roots); *J: Campyloneuron phylliditis* (epiphytic fern on *Clusia*); *K: Phyllodendron laciniatum,* also shown at *Q* as a liane, climbing up the trunk of *Pterocarpus; L: Vittarea lineata* and *M: Nephrolepis biserrata; O: Cyclopeltis semicordata; P: Microgramma persicariaefolia* (the latter four are epiphytic ferns). (Lasser and Vareschi, from Walter 1973)

with *Jussiaea nervosa, Xyris communis* and others, or else *Andropogon bicornis* becomes dominant.

The vegetation of the flooded parts of the asphalt lake in the region of the Orinoco delta changes with increasing accumulation of soil. The first plants to arrive are *Nymphaea ampla, Limnanthemum humboldtianum* and *Utricularia*; then follow *Eleocharis ellipticus* or *Lycopodium meridionale* with *Drosera* spp.–all indicators of nutrient-poor soil. As growth conditions become more favourable, *Blechnum serrulatum* and *Heliconia psittacorum* appear, until finally *Montrichardia arborescens*, with its large rhizomes and, typical for the delta as a whole, establishes itself. To these are added trees with buttress or stilt roots, *Pterocarpus officinalis, Clusia* spp. and the palm *Mauritia setigera*.

A 10–20 m raised bog forest has developed around the asphalt lake. Where the water has become slightly brackish, *Rhizophora mangle* has also established itself.

On the trees there are many epiphytes: ferns *(Asplenium serratum, Nephrolepis biserrata, Cyclopeltis semicordata, Microgramma persicariaefolia, Vittaria lineata),* Araceae *(Anthurium rugosum, Philodendron laciniatum, Peperomia* spp.) and mosses (Fig. 130).

We will return in Volume 2 to the alpine belt of the Andes in Venezuela—the páramos.

Thus we conclude our general observations with this example of Venezuela, an ecologically particularly interesting and varied tropical area.

In Volumes 2 and 3, the special ecological conditions of the individual zonobiomes, with their oro- and pedobiomes, stretching from the tropics to the arctic, will be dealt with, to lead finally to consider the dangers to the geobiosphere today which stem from the activities of man. Man, as part of the geobiosphere, is thereby seriously endangering his own existence. It is the duty of every globally minded ecologist to sound this warning.

References

Abaturov BD (1980) The effect of the small ground squirrel on plant productivity in semi-deserts (in Russian). In: Sokolov VE (ed) pp 111–127

Abd-El-Rahman AA, Batanouny KH (1965) The water output of the desert vegetation in the different microhabitats of Wadi Hoff. J Ecol 53: 139–145

Agachanjanz OE (1980) Die geographischen Ursachen für die Lückenhaftigkeit der Flora in den Gebirgen Mittelasiens. Petermanns Geogr Mitt (Leipzig) 1 1980, pp 47–52

Ahlgren HL, Aamodt OS (1939) Harmful root interactions as a possible explanation for effects noted between various species of grasses and legumes. J Am Soc Agron 31: 982

Albert R (1982) Halophyten. In: Kinzel 1982, pp 33–204

Albert R, Kinzel H (1973) Unterscheidung von Physiotypen bei Halophyten des Neusiedlerseegebietes (Österreich). Z Pflanzenphysiol 70: 138–157

Altmann PL, Dittmer DS (1966) Environmental biology. Fed Am Soc Exper Biol, Bethesda-Maryland

Andersson G (1944) Gas exchange and frost hardening studies in winter cereals. Lund

Aragno M (1981) Responses of microorganisms to temperature. In: Lange et al (eds) Physiol Plant Ecol, pp 339–369

Aubreville A (1938) La forêt équatoriale et les formations forestières tropicales africaines. Scientia (Como) 63: 157

Barbour MG (1970) Is any angiosperm an obligate halophyte. Am Mid Nat 84: 105–120

Batanouny KH (1963) Water economy of desert plants in Wadi Hoff. Ph D thesis, Cairo University

Batanouny KH, Batanouny MH (1968, 1969) Formation of phytogenic hillocks: I and II. Acta Bot Acad Sci Hung 14: 243–252; 15: 1–18

Batanouny KH, Zaki MA (1969) Root development of two common species in different habitats in the mediterranean subregion in Egypt. Acta Bot Acad Sci Hung 15: 217–226

Baumgartner A, Reichel E (1975) Die Weltwasserbilanz. München Wien

Bazilevich NI, Titljanova AA (1978) Comparative studies of ecosystem formation. Grassland system analysis and man. In: Breymeyer AJ, Van Dyne GM (eds) International biological programm. Cambridge Univ Press, pp 713–758

Beadle NCW (1981) The vegetation of Australia. Fischer, Stuttgart. In: Walter H, Breckle SW (eds) Bd 5 der Vegetationsmonographien der einzelnen Großräume

Beebe W, Crane J (1958) Ecologia de Rancho Grande, una selva nublada subtropical en el Norte de Venezuela. Bol Soc Venez Cienc Natur, No. 73: 218–257

Bell KL, Hiat HD, Niles WE (1979) Seasonal changes in biomass allocation in eight winter annuals of the Mojave Desert. J Ecol 67: 781–787

Bergh JP van den (1968) An analysis of yields of grasses in mixed and pure stands. Versl Landbouwk, Onderz (Wageningen), Rep 714, 71 p

Beug H-J (1977) Waldgrenzen und Waldbestand in Europa. Göttinger Universitätsreden 61: 23 pp

Bewley JD, Krochko JE (1982) Desiccation-tolerance. In: Lange et al (eds) pp 325–378

Björkman E (1956) Über die Natur der Mykorrhizabildung unter besonderer Berücksichtigung der Waldbäume. Forstwiss Centralbl (Hamb) 75

Blackman GE, Wilson GE (1951) An analysis of differential effects of light intensity on the net-assimilation rate, leaf-area ratio and relative growth rate in different species. Ann Bot, NS 15: 373–408

Blagowestschenski WA (1935) Über den Verlauf der Photosynthese im Hochgebirge des Pamirs. Planta (Berl) 24: 276–287

Bonnier G (1920) Nouvelles observations sur les cultures experimentales à diverses altitudes. Rev Gèn Bot 32: 305–326

Börner H (1960) Excretion of organic compounds from higher plants and its role in the soil-sickness problem. Bot Rev 26: 393–424

Bornkamm R (1961a) Zur quantitativen Bestimmung von Konkurrenzkraft und Wettbewerbsspannung. Ber Dtsch Bot Ges 74: 75–83

Bornkamm R (1961b) Zur Lichtkonkurrenz von Ackerunkräutern. Flora 151: 126–143

Bornkamm R (1963) Erscheinungen der Konkurrenz zwischen Höheren Pflanzen und ihre begriffliche Fassung. Ber Geobot Forschungsinstitut Rübel (Zürich) 34: 83–197

Boussingault JB (1849) Viajes cientificos a los Andes equatoriales. Paris

Boyce SG (1954) The salt spray community. Ecol Monogr 24: 21–67

Boysen-Jensen P (1932) Die Stoffproduktion der Pflanzen. Jena

Boysen-Jensen P (1949) Causal plant ecology. Dan Vidensk Selskab, Biol Medd 21, Nr. 3

Braun-Blanquet J (1928) Pflanzensoziologie. Berlin

Braun-Blanquet J, Walter H (1931) Zur Ökologie der Mediterranpflanzen. Jahrb wiss Bot 74: 697–748

Breckle S-W (1976) Zur Ökologie und zu den Mineralstoffverhältnissen absalzender und nicht absalzender Xerohalophyten. Diss Bot 35

Breckle S-W (1981) The time-scale of salt-accumulation in a desertic hydrotope in north-central Iran. In: Frey W, Uerpmann H-P (eds) Beihefte Tübinger Atlas des Vorderen Orients, Reihe A, Nr. 8. Beiträge zur Umweltgeschichte des Vorderen Orients, pp 51–60

Brock TD (1978) Thermophilic microorganisms and life at high temperatures. Springer, Berlin Heidelberg New York

Budyko MI (1980) Global ecology. Progress Publishers, Moscow

Bünning E (1947) In den Wäldern Nordsumatras. Bonn

Burcik E (1950) Über die Beziehungen zwischen Hydratur und Wachstum bei Bakterien und Hefen. Arch Mikrobiol 15: 203–235

Burschell P, Schmaltz J (1965a) Untersuchungen über die Bedeutung von Unkraut- und Wurzelkonkurrenz für junge Buchen. Forstwiss Centralbl (Hamb) 84: 201–264

Burschell P, Schmaltz J (1965b) Die Bedeutung des Lichtes für die Entwicklung junger Buchen. Allg Forst- Jagdztg 136: 193–210

Cannon JR, Corbett NH, Haydock KP, Trackey JG, Webb LJ (1962) An investigation of the effect of the Dehydroangustione present in the leaf litter of Backhousia angustifolia on the germination of Araucaria cunninghamii — an experimental approach to a problem in rainforest ecology. Aust J Bot 10: 119–128

Carey SW (ed) (1968) Continental drift. A symposion. Geol Dept Univ Tasmania, Hobart

Cartellieri E (1940) Über Transpiration und Kohlensäureassimilation an einem hochalpinen Standort. Sitzber Akad Wiss Wien, Math Nat Kl Abt I, 149: 95–143

Chernov Yu I (1975) The natural zonation of animals on land (in Russian). Moskow

Chouard P (1979) Flowering and morphogenesis in two desert species. In: Goodall et al, vol 1, pp 640–641

Clausen J, Keck DD, Hiesey WM (1948) Experimental studies on the nature of species. III. Environmental response of climatic races of Achillea. Carnegie Inst Wash, Publ No. 581

Clements FE (1916) Plant succession. Carnegie Inst Washington

Clements FE, Weaver JE, Hanson HC (1929) Plant competition. Carnegie Inst Washington, Publ No. 398

Corliss JB, Ballard RD (1977) Oases of life in the cold abyss. Nat Geogr Mag 152, No. 4, pp 440–453

Coutinho LM (1964) Untersuchungen über die Lage des Lichtkompensationspunktes einiger Pflanzen zu verschiedenen Tageszeiten mit besonderer Berücksichtigung des «de Saussure-Effektes» bei Sukkulenten. Beitr zur Phytologie (Walter-Festschr). In: Arb Landwirtsch Hochsch, Stuttgart, vol 30, pp 101–108

Coutinho LM (1965) Algunas informações sôbre a capacidade ritmica diaria da fixação e acumulação de CO_2 no escuro em epifitas e erbáceas terrestre da mata pluvial. Bolet No. 294, Fac Fil Ciênc e Letras da Univ São Paulo, Botan 21: 397–408

Coutinho LM (1969) Novas observações sôbre a ocorrencia do «Efeito de de-Saussure» e suas relações com a suculencia, a temperatura folhear e os movimentos estomaticos. Bol 331, Fac Fil Ciênc et Letr da Univ São Paulo, Botan 24: 77–102

Czeczott H (1933) A study on the variability of leaves of beeches: F.orientalis Lipsky, F.silvatica L., and intermediate forms, I–II. Ann Soc Dendr Pologne 5 u 6

Dauvillier A (1956) Sur le cycle du sal. C R Acad Paris 242: 47–49

Davis PH, Heywood VH (1964) Principles of angiosperm taxonomy. Edinburgh London

Deneke H (1931) Über den Einfluß bewegter Luft auf die Kohlensäureassimilation. Jahrb wiss Bot 74: 1–32

Donald CM (1946) Competition between pasture species, with reference to the hypothesis of harmful root interactions. J Comm Sc Ind Res (Australia) 19: 32–37

Drautz R (1935) Über die Wirkung äußerer und innerer Faktoren bei der Kohlensäureassimilation. Jahrb wiss Bot 82: 171–232

Duvigneaud P et al (1971) La chênai mélangée calcicole de Virelles-Blaimont. Productivité des écosystemes forestiers. Actes Coll Bruxelles 1969, UNESCO

Duvigneaud P (ed) (1974) Productivité des écosystemes forestiers. Actes Coll Bruxelles 1974, UNESCO: La synthèse écologique. Paris

Duvigneaud P, Denayer-de-Smet S (1968) Essai de classification chimique (éléments mineraux) des plantes du bassin de l'Ebre. Bull Soc R Bot Belg 101: 279–291

Dylis NV, Utkin AN (1968) Experimentelle Methode zur Untersuchung der Natur von Breitlaub-Fichten-Wäldern (in Russian). Probleme der Botanik (Leningrad) X: 141–153

Edelmann G (1982) Fünfzig Spitzenleistungen der Wetter-Maschine. Geo-Special No. 2 (1982), pp 142–143

Ehleringer J, Björkman O (1977) Quantum yields for CO_2-uptake in C_3 and C_4 plants. Plant Physiol (Bethesda) 59: 86–90

Ehrendorfer F (1970) Mediterran-mitteleuropäische Florenbeziehungen im Lichte zytotaxonomischer Befunde. Feddes Repert 81: 3–32

Elias P (1978) Shoot biomass of the ruderal plant communities at human settlements. Acta Bot Slovaka, Ser A/3: 127–130

Ellenberg H (1939) Über Zusammensetzung, Standort und Stoffproduktion bodenfeuchter Eichen-Buchen-Mischwaldgesellschaften Nordwestdeutschlands. Mitt florist-soziol Arb-gem Niedersachsen 5: 3–135

Ellenberg H (1950) Kausale Pflanzensoziologie auf physiologischer Grundlage. Ber Dtsch Bot Ges 63: 24–31

Ellenberg H (1954) Über einige Fortschritte der kausalen Vegetationskunde. Vegetatio 5-6: 199–211

Ellenberg H (1956) Aufgaben und Methoden der Vegetationskunde. Stuttgart

Ellenberg H (1968) Wege der Geobotanik zum Verständnis der Pflanzendecke. Naturwissenschaften 55: 452–470

Ellenberg H (ed) (1971) Integrated experimental ecology. Methods of ecosystem research in the German Solling project. Ecological studies 2. Springer, Berlin Heidelberg New York

Ellenberg H (1975) Vegetationsstufen in perhumiden bis perariden Bereichen der tropischen Anden. Phytocoenologia 2: 368–387

Ellenberg H (1977) Stickstoff als Standortsfaktor, insbesondere für mitteleuropäische Pflanzengesellschaften. Oecol Plant 12: 1–22, 79–85

Ellenberg H (1979) Zeigerwerte der Gefäßpflanzen Mitteleuropas, 2nd edn. Lehrst für Geobot, Univ Göttingen

Ellenberg H (1982) Vegetation Mitteleuropas mit den Alpen in ökologischer Sicht, 3rd edn. Stuttgart

Evenari M, Shanan L, Tadmor N (1971) The Negev, the challenge of a desert. Cambridge, Ma

Evenari M, Yaalon DH, Gutterman Y (1974) Note on soils with vesicular structure in deserts. Z Geomorph N F 18: 162–172

Fedorov VD, Gilmanov TG (1980) Ecology (in Russian). Moscow University Press

Firbas F (1928) Über die Bedeutung des thermischen Verhaltens der Laubstreu für die Frühjahrsvegetation des sommergrünen Laubwaldes. Berl Bot Zbl 44 (II): 179–198

Flohn H (1973) Klimazonen und atmosphärische Zirkulation. In: Walter H (ed) Die Vegetation der Erde, vol 1, 3rd edn. Jena Stuttgart, p 71–81

Flowers S, Evans FR (1966) The flora and fauna of Great Salt Lake region, Utah. In: Boyko H (ed) Salinity and aridity. The Hague, pp 367–393

Freijsen AHJ (1971) Growth-physiology, salt-tolerance and mineral nutrition of Centaurium littorale (Turner) Gilmore; adaptations to its oligotrophic and brackish habitat. Acta Bot Neerl 20: 577–588

Frenzel B (1964) Über die offene Vegetation der letzten Eiszeit am Ostrande der Alpen. Verh Zool Bot Ges Wien 103/104: 110–143

Frey-Wissling A (1935) Die Stoffausscheidung der Höheren Pflanzen. Berlin

Fridriksson S (1975) Surtsey, evolution of life on a volcanic island. Butterworths, London

Friedmann EI, Galun M (1974) Desert algae, lichens and fungi. In: Brown G (ed) Desert biology, vol II. New York, pp 165–212

Gabriel A (1957) Zur Oberflächengestaltung der Pfannen in den Trockenräumen Zentralpersiens. Mitt Geogr Ges Wien, pp 146–160

Gaff DF (1971) Dessication tolerant flowering plants in Southern Africa. Science (Wash DC) 174: 1033–1034

Gaff DF (1980) Protoplasmatic tolerance of extrem water stress. In: Turner NC, Kramer PJ (eds) Adaptation of plants to water and high temperature stress. Wiley, New York, pp 207–229

Galaktionova TF, Solovjeva AP, Stepanova FT (1975) Seasonal changes in the steppe phytocenoses of the Sartang valley (in Russian). In: Bot Mat of Yakutsia, pp 61–67

Galoux A (1966) La variabilité génécologique du hêtre commun (Fagus silvatica L.) en Belgique. Trav Stat Rech Eaux et Forêts, Sér A, No. 11

Gams H (1927) Von den Follatères zur Dent de Morcles. Beitr Geobot Landesaufn Schweiz 15

Gedroiz KK (1929) Der absorbierende Bodenkomplex. Dresden Leipzig

Geiger M (1928) Studium zum Gaswechsel einer extremen Schattenpflanze (Aspidistra) und zur Methodik der Gaswechselversuche. Jahrb wiss Bot 67: 635–701

Geiger R (1928) Das Klima der bodennahen Luftschicht, 3rd edn. Braunschweig

Gessner F (1955) Die limnologischen Verhältnisse in den Seen und Flüssen von Venezuela. Verh Intern Verein f theor u angew Limnologie 12: 284–295

Gessner F (1956) Der Verlauf der Venezuela-Expedition 1952. Ergeb deutsch limnol Venez Exped (Berlin) 1952, 1: 1–22

Goodall DW, Perry RA (1979) Arid land ecosystems, vol I, vol II (1981)

Goryschina TK (1969) Spring ephemeroids of the oak forests of the forest steppe (ecological,

225

physiological and phytocenological investigations) (in Russian). Leningrad

Goryschina TK (1972a) Changes in the rate of development of Scilla sibirica in relation to temperature in early spring as a cause of the stability of the spring phenology (in Russian). Ekologia (Akad Nauk SSSR) No. 5:93–95

Goryschina TK (1972b) Comparative geographical review of the seasonal changes in photosynthesis in deciduous forest herbs (in Russian). Bot Z (Leningr) 57:446–456

Goryschina TK (1974) Factors affecting biological productivities in an oak wood of the forest steppe (in Russian). Nach Isvest Leningr. Univ 367, Ser Biol Nauk 53:216pp

Graw D (1978) The influence of soil pH on the efficiency of vesicular-arbuscular mycorrhiza. New Phytology 82:687–695

Graw D, Rehm S (1977) Vesikulär-arbuskuläre Mycorrhiza in den Fruchtträgern von Arachis hypogaea L. Z Acker- Pflanzenb 144:75–78

Graw D, Moawad M, Rehm S (1979) Untersuchungen zur Wirts- und Wirkungsspezifität der VA-Mykorrhiza. Z Acker- Pflanzenb 148:85–98

Grodzinski AM (1965) Allelopathy in the life of plants and their communities (in Russian). Kiev

Grodzinski AM (1968) Dynamic changes in the colline and vegetational changes in the Ukraine steppe (in Russian). Problems in Botany (Leningrad) X:286–290

Grümmer G (1955) Die gegenseitige Beeinflussung Höherer Pflanzen, Allelopathie. Jena

Haber W (1958) Ökologische Untersuchungen der Bodenatmung. Flora 146:109–157

Haber W (1959) Vergleichende Untersuchungen der Bodenbakterienzahlen und der Bodenatmung in verschiedenen Pflanzenbeständen. Flora 147:1–34

Hamilton WJ, Seely M (1976) Fog basking by the Namib Desert Onymaeris unguicularis beetle. Nature (Lond) 262:284–285

Harder R (1923) Bemerkungen über die Variationsbreite des Kompensationspunktes beim Gaswechsel der Pflanzen. Ber Dtsch Bot Ges 41:194–198

Harmer PM, Benne EJ, Laughlin WM, Key C (1953) Factors affecting crop response to sodium applied as common salt on Michigan muck soil. Soil Sci 76:1–17

Harper JL (1967) A Darwinian approach to plant ecology. J Ecol 55:247–250

Harris JA (1932) The physico-chemical properties of plant saps in relation to phytogeography. Minneapolis

Hartesveldt RJ, Harvey HT (1967) The fire ecology of Sequoia regeneration. Tall Timbers Fire Ecol Confer Proceed No. 7:65–78

Harvey GW (1980) Photosynthetic performance of isolated leaf cells from sun and shade plants.

Carnegie Inst (Stanford Cal) Annual Report (Plant Biol) for 1979–1980, pp 160–164

Hedenström HE von, Breckle S-W (1974) Obligate halophytes? A test with tissue culture methods. Z Pflanzenphysiol 74:183–185

Heintzeler I (1939) Das Wachstum der Schimmelpilze in Abhängigkeit von den Hydraturverhältnissen unter verschiedenen Außenbedingungen. Arch Mikrobiol 10:92–132

Hemming CF (1965) Vegetation areas in Somaliland. J Ecol 53:57–67

Henning I, Henning D (1976) Die klimatologische Trockengrenze. Meteor Rdsch 29:142–151

Henning I, Henning D (1977) Klimatologische Wasserbilanz von Nordamerika. Arch Met Geophs Bioklimatol, Ser B, 25:51–66

Henning I, Henning D (1977a) Ibid., pp 117–125

Heslop-Harrison J (1964) Forty years of genecology. Adv Ecol Res 2:159–247

Hillel D, Tadmor N (1962) Water regime and vegetation in the Central Negev Highlands of Israel. Ecology 43:33–41

Hodge CAH (1962) Vegetation stripes in Somaliland. J Ecol 50:465–474

Hoffmann C (1949) Über die Durchlässigkeit dünner Sandschichten für Licht. Planta (Berl) 37:48–56

Houerou H-N le (1959) Ecologie, phytosociologie et productivité de l'olivier en Tunisie méridionale. Bull Serv Carte phytogéogr, Sér B, Tome IV, Fasc 1.7.72

Huber B (1952) Der Einfluß der Vegetation auf die Schwankungen des CO_2-Gehalts der Atmosphäre. Arch Meteorol, Geophys Bioklimatol Ser B 4:154–163

Huber O (1976) Pflanzenökologische Untersuchungen im Gebirgsnebelwald von Rancho Grande (Venezolanische Küstenkordillere). Diss Innsbruck

Huber O (1978) Light compensation point of vascular plants of a tropical cloud forest and an ecological interpretation. Photosynthetica (Prague) 12:382–390

Huber O (1980) Die Felsvegetation am oberen Orinoco in Südvenezuela. In: Reisigl H (ed) «Blumenparadiese». Pinguin, Innsbruck, pp 200–203

Huber B, Pommer J (1954) Zur Frage eines jahreszeitlichen Ganges im CO_2-Gehalt der Atmosphäre. Angew Bot 28:53–62

Illies J (1971) Einführung in die Tiergeographie. UTB

Irmscher E (1922) Pflanzenverbreitung und Entwicklung der Kontinente I. Mitt Inst Allg Bot Hamb 5:17–235

Irmscher E (1929) Ibid Teil II, 8:169–374

Iwaki H (1959) Ecological studies in interspecific competition in plant community I. JPN J Bot 17:120–138

226

Jagnow G (1957) Streptomyceten. Arch Mikrobiol 26:175–191

Jahn A (1934) Las temperatures médias y extremas de las zonas altudinales de Venezuela. Biol Soc Venez Ciencias Naturales, Nr. 14

Jerusalimov EN (1980) Kompensierende Vorgänge bei Forstschäden durch blattfressende Insekten (in Russian). In: Sokolov VE (ed) pp 74–87

Jordan CF, Herera R (1981) Les forêts ombrophiles tropicals: les éléments nutritifs sont-ils réellement un facteur critique? Nat Ressour (UNESCO) 17:8–15

Kaben H (1959) Die Schwankungen der CO$_2$-Assimilation von Lamium galeobdolon (L.). In: Crantz (ed) Im Laufe eines Jahres. Biol Zentralbl 78:213–231

Karpov VG (1961) The influence of tree root competition on assimilation in the herbaceous layer in spruce forests (in Russian). Dokl Akad Nauk 140:1205–1208

Karpov VG (1962) Several questions on the phytocenology of spruce forests from an experimental viewpoint (in Russian). Akad Nauk Isvest Labor Lesh Inst No. 6:35–61. Moscow

Karpov VG (1962) Assessment of ^{32}P uptake in an investigation of root competition of trees and undergrowth in the southern taiga forests (in Russian). Dokl Akad Nauk 146:717–720

Karpov VG (1962) Some results from an experimental investigation of the composition and structure of the lower storeys of the bilberry-spruce forests (in Russian). Problems in Botany (Leningrad) 6:258–276. Leningrad Moscow

Karpov VG (1969) Experimental phytocenology of the dark coniferous taiga (in Russian). Leningrad

Karpov VG (1969) Various lines of approach and some results of experimental investigation of phytocenoses of the dark coniferous taiga. Symposium: Experimental plant sociology 1965. The Hague, pp 181–190

Katz JR (1918) Die Gesetze der Quellung. Koloidch Beih 9:1–182

Kearney TH, Briggs LJ, Shantz HL, McLane JW, Piemeisel RL (1914) Indicator significance of vegetation in Tooele Valley, Utah. J Agric Res 1:365–417

Keller B (1926) Die Vegetation auf Salzböden der russischen Halbwüsten und Wüsten. Z Bot 18:113–137

Kerner von Marilaun (1891) Pflanzenleben, vol 2, pp 489–507

Killian CH, Feher D (1935) Recherches sur les phénomènes microbiologiques des sols sahariens. Ann Inst Pasteur 55:573

Killick DJB (1963) An account of the plant ecology of the Cathedral Peak area of Natal Drakensberg. Bot Surv S Afr Nem No. 34, Pretoria

Kinzel H (1982) Pflanzenökologie und Mineralstoffwechsel. Ulmer, Stuttgart

Kluge M, Ting IP (1978) Crassulacean acid metabolism. Ecol Studies 30

Kluge M, Lange OL, Eichmann M v, Schmid R (1973) Diurnaler Säurerhythmus bei Tillandsia usneoides. Planta (Berlin) 112:357–372

Knapp R (1967) Experimentelle Soziologie und gegenseitige Beeinflussung der Pflanzen, 2nd edn. Stuttgart

Kobak NI (1964) Probleme der Ökologie und Physiologie der Waldpflanzen (in Russian) 2:61–98

Koch C (1961) Some aspects of abundant life in the vegetationless sand of the Namib Desert. Sci Pap Namib Desert Res Stat 1:8–92

Kohler MA, Nordenson PJ, Baken DR (1959) Evaporation maps for the United States. Weather Bureau, Techn Paper No. 37, Wash DC

Komarek EV (1966) The meteorological basis for fire ecology. Tall Timber Fire Ecol Proceed No. 5. Tallahassee, pp 85–125

Komarek EV (1967) The nature of lightning fire ecology. Tall Timber Fire Ecol Proceed No. 5. Tallahassee, pp 85–125

Komarek EV (1971) Lightning and fire. Ecology in Africa. Ann Proc Tall Timber Fire Ecology, Conf No. 11. Tallahassee, pp 473–516

Komarek EV (1973) Ancient fires. Dass No. 12. Tallahassee, pp 219–240

Köppen W, Wegener A (1924) Die Klimate der geologischen Vorzeit. Berlin

Kornas J (1977) Life-forms and seasonal patterns in the pteridophytes in Zambia. Acta Sol Bot Pol 46:669–690

Korte F (1980) Ökologische Chemie. Thieme, Stuttgart

Kostytschew S, Kudriavzewa M, Moissejewa W, Smirnowa M (1926) Der tägliche Verlauf der Photosynthese bei Landpflanzen. Planta (Berl) 1:679–699

Kovalev (ed) (1976) Structure, function and evolution of the Baraba biogeocenosis (in Russian). Nanka, Novosibirsk, 495 pp

Kovda WA (1939) Solonetz soils (in Russian). Soils of the USSR, vol 1, pp 299–348

Krauklis AA (1975) Local geographical structure of the taiga in the Angara area (in Russian). Rep Geog Inst Siberia and Far East (Irkutsk) 41:3–16

Krause H (1978) Das Mammut in Eis und Schnee? Selbstverl Stuttgart

Kreeb K (1964) Ökologische Grundlagen der Bewässerungskulturen in den Subtropen. Stuttgart

Kreeb K (1974) Pflanzen an Salzstandorten. Naturwissenschaften 61:337–343

Krinsley DB (1970) A geomorphological and palaeoclimatological study of the playas of Iran. University of Maryland, PhD

Kühnelt W (1975) Beiträge zur Kenntnis der Nahrungsketten in der Namib (Südwestafrika). Verh Ges Ökologie, Wien, pp 197–210

Kuroiwa S (1960) Intraspecific competition in artificial sunflower community. Bot Mag Tokyo 73:300–309

Kuznetsov GV (1980) The role of ungulates in forest ecosystems (in Russian). In: Sokolov VE (ed) pp 88–110

Kvasov DD (1977) Increase of climate moisture during the Pleistocene-Holocene transition as a cause of mammoth extinction (in Russian). Proceed Zool Inst, Acad Science USSR, vol 73:71–77

Lamberti A (1969) Contribuicão ao conhecimento da ecologie das plantas do manguezal de Itanhaém. Facult Filos Ciênc e Letr, Univ São Paulo, Bul No. 317, Botan No. 23

Lange OL (1965) Der CO₂-Gaswechsel von Flechten bei tiefen Temperaturen. Planta (Berl) 64:1–19

Lange OL (1980) Water relations and photosynthetic response of a desert moss. Flora 169:329–335

Lange OL, Zuber M (1977) Freria indica, a stem succullent CAM plant with deciduous C₃-leaves. Oecologia (Berl) 31:67–72

Lange OL, Zuber M (1980) Temperaturabhängigkeiten des CO₂-Gaswechsels stammsukkulenter Asclepiadaceen mit Säurestoffwechsel. Flora 179:529–553

Lange OL, Koch W, Schulze ED (1969) CO₂-Gaswechsel und Wasserhaushalt von Pflanzen in der Negev-Wüste am Ende der Trockenzeit. Ber Dtsch Bot Ges 82:39–61

Lange OL, Schulze E-D, Koch W (1970) Untersuchungen an Flechten der Negev-Wüste. II. CO₂-Gaswechsel und Wasserhaushalt von Ramalina maciformis (DEL). In: Bory (ed) Am natürlichen Standort während der sommerlichen Trockenperiode. Flora 159:38–62

Lange OL, Nobel PS, Osmond CB, Ziegler (eds) (1981) Physiological plant ecology I (In: Encyclopedia of plant physiology, new series); Daselbst: II 1982. Springer, Berlin Heidelberg New York

Langlet O (1936) Studien über die physiologische Variabilität der Kiefer und deren Zusammenhang mit dem Klima. Medd Skogsförsokanst (Stockholm) 29:219–470

Larcher W (1980) Untersuchungen zur Frostresistenz von Palmen. Anz Österr Akad Wiss Math Naturw Kl 117:37–48

Larcher W (1982) Frost-drought and its ecological significance. In: Lange et al (eds) pp 379–400

Lasser T, Vareschi V (1959) La vegetacion del Lago de Asfalto de Guanaco. Acta Biol Venez (Caracas) 2:407–452

Lastschinski NN (1968) The effect of tree root competition on the structure of the herb, dwarf-shrub and moss layers in the dark coniferous forests of the lower Angara area (in Russian). Problems in Botany (Leningrad) X:154–161

Lattin G de (1967) Grundriß der Zoogeographie. VEB Gustav Fischer Verlag, Jena

Lauer W (1975) Vom Wesen der Tropen. Akad Wiss Mainz, Abh Math-Naturw Kl, Jg 1975, Nr. 3

Lavrenko EM, Soczava VB (1956) Descriptio Vegetationes URSS ad Tabulam m 1:4000000 annotiones (in Russian)

Lawrence WE (1945) Some ecotypic relations of Deschampsia caespitosa. Am J Bot 32:298–314

Leistner OA (1961) Zur Verbreitung und Ökologie der Bäume der Kalaharidünen. J SWA-Wiss Ges 15:35–40

Leistner OA (1964) The plant ecology of Southern Kalahari. Dissert Stellenbosch 1964

Lieth H (1958) Konkurrenz und Zuwanderung von Wiesenpflanzen. Z Acker- Pflanzenb 106:205–223

Lieth H (1960) Über den Lichtkompensationspunkt der Landpflanzen I und II. Planta (Berl) 54:530–576

Lieth H, Vogt M (1959) Der Lichtkompensationspunkt einiger Waldschattenpflanzen im Sommer und Frühjahr. IX. Int Bot Congr Montreal

Lindeman GV (1980) Wood-eating insects in forest communities (in Russian). In: Sokolov VE (ed) pp 63–73

Liphschitz N, Waisel Y (1982) Adaptations of plants to saline environments: salt excretion and glandular structure. In: Sen DN (ed) Contributions to the ecology of halophytes. T:VS 2, Junk, Den Haag, pp 197–214

Livingston BE (1907) Relative transpiration in cacti. Plant World 10:110

Lozhkin AV (1977) Condition of habitat of Berelek mammoth population (in Russian). Proceedings of the Zoological Institute of the USSR, vol 72, pp 67–68

Lüdi W (1945) Besiedlung und Vegetationsentwicklung auf den jungen Seitenmoränen des großen Aletschgletschers. Ber Geobot Forsch Inst. Ruebel, Zürich 1944, pp 35–112

Lundegårdh H (1921) Ecological studies in the assimilation of certain forest plants and shore plants. Sven Bot Tidskr 15:46ff

Lundegårdh H (1924) Der Kreislauf der Kohlensäure in der Natur. Jena

Lundegårdh H (1925) Klima und Boden in ihrer Wirkung auf die Pflanzenwelt, 5th edn 1957. Jena

Mägdefrau K (1968) Paläobiologie der Pflanzen, 4th edn. Stuttgart

Maiwald K (1930) Nährstoffaufnahme und Stofferzeugung beim Buchweizen (Fagopyrum esculentum). Z Pflanzenernähr Düng Bodenkd 17:12–42

Marrero L (1964) Venezuela y sus recursos. Madrid

Martin P (1957) Die Abgabe von organischen Verbindungen, insbesondere von Scopoletin aus den Keimwurzeln des Hafers. Z Bot 45:475–506

Mayer A, Plantifol L (1924) Équilibre des constituants cellulaires et intensité des oxydations de la cellule. Inhibition et oxydation. CR Sc Acad Sci 178:1385

Mayer H, Tichy K (1979) Das Eichen-Naturschutzgebiet Johannser Kogel im Lainzer Tiergarten, Wienerwald. Zentralbl Gesamte Forstwes 96:193–226

Mayer H, Neumann M, Sommer H-G (1980) Bestandsaufbau und Verjüngungsdynamik unter dem Einfluß natürlicher Wilddichten im kroatischen Urwaldsreservat Corkova Uvala, Plitvicer Seen. Schweiz Zeitschr Forstwes 131:45–70

Medina E (1974) Dark CO_2 fixation, habitat preference and evolution within the Bromeliaceae. Evolution 28:677–686

Medvecka-Kornas A (1967) Ecosystem studies in a beech forest and meadow in Ojców National Park (polnisch). Kraków

Meyer FH (1974) Physiology of mycorrhiza. Annu Rev Plant Physiol 25:567–586

Migahid AM, Abdelwahab AM, Batanouny KH (1972) VII. Water relations of Leptadenia pyrotechnica (Forsk). Dence growing in the Egyptian desert. Oecologia (Berl) 10:79–91

Moawad M (1979) Nutzung der vesikulär-arbuskulären Mykorrhiza im tropischen Pflanzenbau. Angew Bot 53:99–109

Monod Th (1954) Modes «contractés» et «diffus» de la végétation saharienne. Biology of deserts, London, pp 35–44

Monsi M (1960) Dry-matter production in plants. 1. Schemata of dry-matter reproduction. Bot Mag (Tokyo) 73:81–90

Moulopoulos CH (1965) The beech species and their distribution in Greece. Thessaloniki

Mount AB (1969) Eucalypt ecology as related to fire. Tall Timbers Fire Ecol Conf No. 9:75–108

Müller D (1928) Die Kohlensäureassimilation bei arktischen Pflanzen und die Abhängigkeit der Assimilation von der Temperatur. Planta (Berl) 6:22–39

Müller D (1932) Analyse der verminderten Stoffproduktion bei Stickstoffmangel. Planta (Berl) 16:1–9

Müller D, Larsen P (1935) Analyse der Stoffproduktion bei Stickstoff- und Kalkmangel. Planta (Berl) 23:501–517

Mueller-Dombois D (1981) Spatial variation and succession in tropical island rain forests.—A progress report. Dep of Botany, Univ Hawaii, Honolulu

Muller CH, Moral R del (1966) Soil toxicity induced by terpenes from Salvia leucophylla. Bull Torrey Bot Club 93:130–137

Nordhausen M (1912) Über Sonnen- und Schattenblätter. Ber Dtsch Bot Ges 30:483–503

Novichkova-Ivanova LN (1980) Soil algae of the Sahara-Gobi desert area (in Russian). Nauka, Moscow

Nuernbergk EL (1960, 1961) Endogener Rhythmus und CO_2-Stoffwechsel bei Pflanzen mit diurnalem Säurerhythmus. Planta (Berl) 54:530–576 und 56:28–70

Oertli JJ, Kohl HC (1961) Some considerations about the tolerance of various plant species to excessive supplies of boron. Soil Sci 92:243–247

Oppenheimer H, Halevy AH (1962) Anabiosis in Ceterach officinarum Lam et D. Bull Res Counc Isr D 11:127–142

Orlenko EG (1955) On the interaction of oaks in dense stands (in Russian). Dokl Akad Nauk USSR 102:841–844

Orshan G (1953) Note in the application of Raunkiaer's system of life forms in arid regions. Palest J Bot, Jerus, Ser 6, pp 120–122

Orshan G (1954) Surface reduction and its significance as a hydroecological factor. J Ecol 42:442–444

Osmond CB, Björkman O, Anderson DJ (1980) Physiological processes in plant ecology — towards a synthesis with Atriplex. Ecol Studies 36. Springer, Berlin Heidelberg New York

Paczoski J (1928) La flore de la forêt de Bialowieža. V-me Excurs. Phytogeogr Intern, 87 ps Varsovie

Pavletic Z, Lieth H (1958) Der Lichtkompensationspunkt einiger immergrüner Pflanzen im Winter und im Frühjahr. Ber Dtsch Bot Ges 69:309–314

Pemadassa MA (1981) Cyclic change and pattern in an Arthrocnemum community in Sri Lanka. J Ecol 69:565–574

Penck A (1910) Versuch einer Klimaklassifikation auf physio-geographischer Grundlage. Sitz Ber Kgl Preuß Akad Wiss 12:236–246

Phillips J (1965) Fire — as master and servant: Its influence in the bioclimatic regions of Trans-Saharan Africa. Tall Timbers Fire Ecol Confer, Proceed No. 4:7–109

Pisek A, Rehner G (1958) Temperaturminima der Nettoassimilation von mediterranen und nordisch-alpinen Immergrünen. Ber Dtsch Bot Ges 71:188–193

Pisek A, Winkler E (1956) Wassersättigung, Spaltenbewegung und Photosynthese. Protoplasma 46:597–611

Pisek A, Winkler E (1958) Assimilationsvermögen und Respiration der Fichte (Picea excelsa Link) in verschiedenen Höhenlagen und der Zirbe

(Pinus cembra L.) an der alpinen Waldgrenze. Planta (Berl) 51:518–543

Prusinkiewicz Z (1959) Investigations on the limit of the biological utility of soil water (polnisch). Poznanske Towarz Przyjaciol Nauk 5 (H 6)

Rachkowskaya EI (1977) Desert forms of the extremely arid Transaltai-Gobi (in Russian). Problemy Ekologii, Geobotaniki i Botanicheskoy Geografii i Floristiki Leningrad, pp 99–124

Rademacher B (1959) Gegenseitige Beeinflussung Höherer Pflanzen. Hdb Pflanzenphys II, pp 655–706

Ranwell DS (1972) Ecology of salt marshes and sand dunes. Champmann and Hall, London

Rauh W (1973) Über die Zonierung und Differenzierung der Vegetation Madagaskars. Akad d Wiss Mainz, Klasse Trop u Subtrop Pflw 1: 1–146

Remmert H (1978) Ökologie. Springer, Berlin Heidelberg New York

Repp G (1939) Ökologische Untersuchungen im Halophytengebiet am Neusiedler See. Jahrb wiss Bot 88:554–632

Ried A (1960) Thallusbau und Assimilathaushalt von Laub- und Krustenflechten. Biol Zentralbl 79:129–151

Risser P, Cottam G (1968) Carbohydrate cycle in the bulbs of some spring ephemerals. Bull Torrey Bot Club 95(4):359–369

Röhl E (1952) Climatografia de Venezuela. Bol Acad Cienc Fis, Mat Nat (Caracas) No. 46

Romell L-G (1922) Die Bodenventilation als ökologischer Faktor. Medd Stat Skogförsöksanst 19: 125–139

Romell L-G (1932) Mull and duff as biotic equilibria. Soil Sci 34:161–188

Romose V (1940) Ökologische Untersuchungen über Homalothecium sericeum, seine Wachstumsperiode und seine Stoffproduktion. Dan Bot Ark (Copenhagen) 10, Nr. 4

Rouschal E (1939) Eine physiologische Studie an Ceterach officinarum Willd. Flora (Berl) 132: 305–318

Rozema J (1975) An eco-physiological investigation into the salt tolerance of Glaux maritima L. Acta Bot Neerl 24:409–416

Rozema J (1976) An ecophysiological study on the response to salt of four halophytic and glykophytic Juncus species. Flora (Berl) 165:197–209

Rozema J (1978) On the ecology of some halophytes from a beach plain in the Netherlands. Diss Amsterdam

Rubey WW (1951) Geological history of sea water. Bull Geol Soc Am 62:1111–1148

Ruhland W (1915) Untersuchungen über die Hautdrüsen der Plumbaginaceen. Ein Beitrag zur Biologie der Halophyten. Jahrb Wiss Bot 55:409–498

Rundel PW (1981) Fire as an ecological factor. In: Lange et al (ed) Physiological plant ecology, pp 501–538

Rundel PW, Lange OL (1980) Water relations and photosynthetic response of desert moss. Flora 169:329–338

Saburov DN (1972) The forests of the Pinega (in Russian). Akad Nauk Leningrad

Schaarschmidt F (1974) Mosellophyton hefteri n.g.n.sp. (Psilophyta), ein sukkulenter Halophyt aus dem Unterdevon von Alken an der Mosel. Paleontol ZH 48:188–204

Scharlau K (1958) Zum Problem der Pluvialzeiten in Nordost-Iran. Z Geomorph 2:258–277

Schirmer U, Breckle S-W (1982) The role of bladders for salt removal in some Chenopodiaceae (mainly Atriplex species). In: Sen DN (ed) Contributions to the ecology of halophytes. T:VS 2, Junk, Den Haag

Schnell R (1963) Sur une structure anomale de Casuarina equisetifolia L. et al question des halomorphoses. Bull IFAN, Sér A, No. 2:301–306

Schnetter R (1968) Die Vegetation des Cerro San Fernando im Trockengebiet bei Santa Marta, Kolumbien. Ber Dtsch Bot Ges 81:289–302

Schnetter R (1970) Untersuchungen zum Standortsklima im Trockengebiet von Santa Marta, Kolumbien. Mitt Inst Colombo-Aleman Invest Cient 4:45–75

Schnetter R (1971) Untersuchugnen zum Wärme- und Wasserhaushalt ausgewählter Pflanzenarten des Trockengebiets von Santa Marta (Kolumbien). Beitr Biol Pflanzen 47:155–213

Scholander PF (1966) Sap concentration in halophytes and some other plants. Plant Physiol (Bethesda) 41:529–532

Scholander PF (1968) How mangroves desalinate seawater. Physiol Plant 21:251–261

Schönborn A von (1965) Die Atmung der Samen. Allg Forstzeitschr 1/12:16–20

Schröder J (1938) Über natürliche und künstliche Änderung des Interzellularvolumens bei Laubblättern. Beitr Biol Pflanz 25:75–124

Schulze E-D (1982) Plant life forms and their carbon, water and nutrient relations. In: Lange et al (eds) pp 615–676

Seely MK (1978) The Namib Desert: an unusual ecosystem. J arid Envir 1:117–128

Seely MK, Hamilton WJ (1977) The catchment sand trenches constructed by Tenebrionid beetles Lepidochora from Namib Desert. Science (Wash DC) 193, No. 4252

Seely MK, Louw GN (1980) First approximation of the effects of rainfall in the ecology and energetics of Namib Desert dune ecosystem. J of Arid Environment 3:25–54

Shilo MA (1978) The mammoth find on the Kirgiljach stream in the Mogadan area (in Russian). Priroda, pp 18–20

Shreve EB (1915) An investigation in the causes of autonomic movements in succulent plants. Plant World 18:297a–321

Sieverding E (1979) Einfluß der Bodenfeuchte auf die Effektivität der VA-Mykorrhiza. Angew Bot 53:91–98

Simonis W (1941) Über den Einfluß des Bodenwassergehalts auf die CO_2-Assimilation und den Ertrag der Einjährigen. Ber Dtsch Bot Ges 59:52–62

Simonis W (1947) CO_2-Assimilation und Stoffproduktion trocken gezogener Pflanzen. Planta (Berl) 35:188–224

Skrjabin SZ, Konorovski AK (1975) Cited from Yurtsev 1981 (in Russian)

Slavikova J (1958) Einfluß der Buche (Fagus silvatica) als Edifikator auf die Entwicklung der Krautschicht in Buchenphytozönosen. Preslia (Prague) 30:19–42

Slavikova J (1965) Die maximale Wurzelsaugkraft als ökologischer Faktor. Preslia (Prague) 37:419–428

Slavikova J (1966) Wechselbeziehungen der Wurzelsaugkraft bei einigen Komponenten der Eschenphytozönosen. Preslia (Prague) 38:15–22

Smith J (1949) Distribution of tree species in the Sudan in relation to rainfall and soil texture. Khartoum

Smith MK, McComb JA (1981) Effect of NaCl on the growth of whole plants and their corresponding callus cultures. Aust J Plant Physiol 8:267–275

Snow D (1949) Keimung der Pilzsporen nach 2 Jahren bei 62–66% hy. Ann Appl Biol 36:1

Sokolov VE (1980) Phytophages in plant communities (in Russian). Nauka, Moscow

Specht RL (1957, 1958) Dark island heath (ninety-mile-plain), South Australia, I–VI. Aust J Bot 5:52–114, 137–172; 6:59–88

Spoehr HA (1913) Photochemische Vorgänge bei der diurnalen Entsäuerung der Sukkulenten. Biochem Z 57:95ff

Stålfelt MG (1924) Untersuchungen zur Ökologie der Kohlensäureassimilation der Nadelbäume. Medd Stat Skogsforsökanst, H 21:181–258

Stålfelt MG (1935) Die Spaltöffnungswerte als Assimilationsfaktor. Planta (Berl) 23:715–759

Stålfelt MG (1938) Der Gasaustausch der Moose. Planta (Berl) 27:30–60

Steiner M (1934) Zur Ökologie der Salzmarschen der nordöstlichen Vereinigten Staaten von Nordamerika. Jahrb wiss Bot 81:94–202

Stelzer R, Läuchli A (1978) Salt- and flooding tolerance of Puccinellia peisonis. III. Distribution and localization of ions in the plant. Z Pflanzenphysiol 88:437–448

Steyermark JA (1967) Flora del Auyantepui. Acta Bot Venez 2

Steyermark JA, Dunsterville GCK (1980) The lowland floral element on the summit of Cerro Guaiquima and others cerros of the Guaiana highland of Venezuela. J Biogeog 7:285–303

Stocker O (1928) Das Halophytenproblem. Ergeb Biol 3:265–353

Stocker O (1931) Über die Assimilationsbedingungen im tropischen Regenwald. Ber Dtsch Bot Ges 49:267–273

Stocker O (1937) Über die Beziehung zwischen Wasser- und Assimilationshaushalt. Ber Dtsch Bot Ges 55:370–376

Stocker O (1954) Der Wasser- und Assimilationshaushalt südalgerischer Wüstenpflanzen. Ber Dtsch Bot Ges 67:288–298

Studt W (1926) Die heutige und frühere Verbreitung der Coniferen und die Geschichte ihrer Arealgestaltung. Mitt Inst Allg Bot Hamb 6:168–307

Sukatchev VN, Dylis NV (1964) Basic forest typology and biogeocenology (in Russian). Leningrad

Sullivan CY, Levitt J (1959) Drought tolerance and avoidance in two species of oaks. Physiol Plant 12:299–305

Sveshnikova WM (1962) On some special ecophysiological features of the dominant Stipa species in steppe communities (in Russian). Problemy Botaniki 6:388–398

Tadmor NH, Shanan L, Evenari M (1960, 1961) The ancient desert agriculture of the Negev. VI and VII. Isr J Agric Res 10:193–206, 11:9–31

Talsma T, Lelij A van der (1972) Water balance estimates of evaporation from ponded ricefields in semiarid regions. Agric Water Manage 1:89–97

Tamm CO (1950) Growth and plant nutrient concentration in Hylocomium proliferum (L.). In: Lindb. (ed) In relation to tree canopy. Oikos 2:60–64

Teakle JH (1937) The salt (sodium chloride) content of rain water. J Dep Agric West Aust 14:115–133

Thenius E (1977) Meere und Länder im Wechsel der Zeiten. Verständl Wiss, vol 114

Thornthwaite CW (1948) An approach towards a rational classification of climate. Geogr Rev 38:55–94

Tischler W (1976) Einführung in die Ökologie. Stuttgart

Titljanowa AA (1977) The carbon cycle in herbaceous biocenoses (in Russian). Nauka, Novosibirsk

Tomirdiaro SV (1977) Changes of the physical and geographical conditions on the plains of the north-eastern Asia during the Pleistocene-Holocene transition as the main cause of the extinction of the mammoth complex theriofauna (in Russian). Proc Ecol Inst Acad USSR, vol 73, pp 64–71

Tranquillini W (1955) Die Bedeutung des Lichtes und der Temperatur auf die Kohlensäureassimilation von Pinus cembra-Jungwuchs an einem hochalpinen Standort. Planta (Berl) 46: 154–178

Troll C (1973) The upper timberlines in different climate zones. Arc Alp Res 5: A3–A18

Tumanov JJ (1979) Physiologie der Abhärtung und Frostresistenz der Pflanzen (in Russian). Nauka, Moscow

Turesson G (1926) Die Bedeutung der Rassenökologie für die Systematik und Geographie der Pflanzen. Beih Rep spec nov regni veg 41 und Arbeiten in Hereditas 3, 4, 6, 7, 14 und 15: 1922–1931

Vareschi V (1980a) Tafelberge: Floreninseln Venezuelas. In: Reisigl H (ed) Blumenparadiese. Pinguin, Innsbruck, pp 198–199

Vareschi V (1980b) Vegetationsökologie der Tropen. Stuttgart

Velichko AA (1973) Natural processes in Pleistocene (in Russian). IX. Kongr INQUA, Novaya Zelandia Nauka, Moscow 256

Verestshagin NK (1977) Berelekh «cemetary» of mammothes. Proceedings of the zoological Institute, Academy of Science of the USSR, vol 72, pp 5–50

Verestshagin NK, Dubrovo IA (1978) Palaeontological description of the find (in Russian). Priroda No. 1, p 21

Vila P (1960) Geografia de Venezuela, tomo I. Ministero de Educacion, Caracas, 1965, dto tomo II

Vogel S (1955) «Niedere Fensterpflanzen» in der südafrikanischen Wüste. Beitr Biol Pflanz 31: 45–135

Vogel R (1967) Fire adaptations of some Southern California plants. Tall Timber Fire Ecol Confer, Proceed No. 7, Tallahassee, pp 79–109

Volk OH, Geyger E (1970) Schaumböden als Ursache der Vegetationslosigkeit in ariden Gebieten. Z Geomorph N F 14: 79–95

Wagner WE (1961) Past fire incidents in Sierra Nevada forests. J For 59(10): 739–748

Walter H (1923) Protoplasma- und Membranquellung bei Plasmolyse. Jahrb wiss Bot 62: 145–213

Walter H (1924) Plasmaquellung und Wachstum. Z Bot 26: 353–417

Walter H (1929) Die osmotischen Werte und die Kälteschäden unserer wintergrünen Pflanzen während der Winterperiode. Ber Dtsch Bot Ges 47: 338–348

Walter H (1929) Die Winterschäden an unseren immergrünen Pflanzen während der Kälteperiode Januar–März 1929 und ihre Ursachen. Naturwissenschaften 17: 854–859

Walter H (1931) Die Hydratur der Pflanze und ihre physiologisch-ökologische Bedeutung. G Fischer, Jena

Walter H (1936) Nährstoffgehalt des Bodens und natürliche Waldbestände. Silvia 24: 201–205, 209–213

Walter H (1936a) Zur Frage nach dem Endzustand der Entwicklung von Waldgesellschaften. Naturforscher 13: 151–155

Walter H (1936b) Die ökologischen Verhältnisse in der Namib-Nebelwüste (Südwestafrika). Jahrb wiss Bot 84: 58–222

Walter H (1939) Grasland, Savanne und Busch der ariden Teile Afrikas in ihrer ökologischen Bedingtheit. Jahrb wiss Bot 87: 750–860

Walter H (1943) Die Vegetation Osteuropas, 2nd edn. Parey, Berlin

Walter H (1952) Eine einfache Methode zur ökologischen Erfassung des CO_2-Faktors am Standort. Ber Dtsch Bot Ges 65: 175–182

Walter H (1954) Klimax und zonale Vegetation. Angew Pflanzensoziologie (Festschrift Aichinger) I: 144–150

Walter H (1955) Die Klimadiagramme als Mittel zur Beurteilung der Klimaverhältnisse für ökologische, vegetationskundliche und landwirtschaftliche Zwecke. Ber Dtsch Bot Ges 68: 331–344

Walter H (1960a) The adaptation of plants to saline soils. Salinity problems in the arid zones. Unesco Symposium in Teheran

Walter H (1960b) Standortslehre, analytisch-ökologische Geobotanik (Einführung in die Phytologie, vol 3), 2nd edn. Ulmer, Stuttgart

Walter H (1963) Climatic diagrams as a means to comprehend the various climatic types for ecological and agricultural purposes. Symposium: The water relations of plants (Oxford), pp 3–9

Walter H (1963) The water supply of desert plants. Symposium: The water relations of plants (Oxford), pp 199–205

Walter H (1964) Productivity of vegetation in arid countries. IUCN Bull ns No. 4. Part III, pp 221–229

Walter H (1967a) Das Pampaproblem in vergleichend ökologischer Betrachtung und seine Lösung. Erdkunde 21: 181–203

Walter H (1967b) The pampa problem and its solution. Publ ITC-UNESCO Centre. Delft. The Netherlands

Walter H (1967c) Die physiologischen Voraussetzungen für den Übergang der autotrophen Pflanzen vom Leben im Wasser zum Landleben. Z Pflanzenphysiol 56: 170–185

Walter H (1968) Die Vegetation der Erde in ökophysiologischer Betrachtung. Bd II: Die gemäßigten und arktischen Zonen. Jena Stuttgart

Walter H (1971/72) Ökologische Verhältnisse und Vegetationstypen in der Intermontanen Region des westlichen Nordamerikas. Verh Zool Bot Ges Wien 110/111:111–123

Walter H (1972) Der Wasserhaushalt der Pflanzen in kausaler und kybernetischer Betrachtung. Ber Dtsch Bot Ges 85:301–313

Walter H (1973) Die Vegetation der Erde in öko-physiologischer Betrachtung. Bd I: Die tropischen und subtropischen Zonen, 3rd edn. Jena Stuttgart

Walter H (1974) Die Vegetation Osteuropas, Nord- und Zentralasiens. (In Vegetationsmonogr d einzelnen Großräume, vol VII.) Stuttgart

Walter H (1975) Besonderheiten des Stoffkreislaufes einiger terrestrischer Ökosysteme. Flora 164:169–183

Walter H (1976) Die ökologischen Systeme der Kontinente (Biogeosphäre). Stuttgart

Walter H (1981) Über Höchstwerte der Produktion von natürlichen Pflanzenbeständen in NO-Asien. Vegetatio 44:37–41

Walter H, Bauer G (1937) Über das Einrollen der Blätter bei Farnen und Blütenpflanzen. Flora 131:387–399

Walter H, Box EO (1983) Overview of Eurasian deserts and semideserts. In: Ecosystems of the world, vol 5. Amsterdam, pp 3–269

Walter H, Haber W (1957) Über die Intensität der Bodenatmung mit Bemerkung zu den Lundegårdhschen Werten. Ber Dtsch Bot Ges 70:275–282

Walter H, Harnickell E, Mueller-Dombois D (1975) Klimadiagramm-Karten der einzelnen Kontinente und die ökologische Klimagliederung der Erde. Veget Monographien der einzelnen Großräume, vol 10. Stuttgart

Walter H, Kreeb K (1970) Die Hydratation und Hydratur des Protoplasmas der Pflanzen und ihre öko-physiologische Bedeutung. Protoplasmatologia II, C, 6. Wien

Walter H, Lieth H (1960–67) Klimadiagramm-Weltatlas. Jena

Walter H, Medina E (1971) Caracterizacion climatica de Venezuela sobre la base de climadiagramas de estaciones particulares. Bol Soc Venez Cienc Natur 29:211–240

Walter H, Stadelmann E (1968) The physiological prerequisites for the transition of autotrophic plants from water to terrestrial life. Bioscience 18:694–701

Walter H, Stadelman E (1974) A new approach to the water relations of desert plants. In: Brown Jr (ed) Desert biology, vol II. Academic Press, New York, pp 213–310

Walter H, Steiner M (1936) Die Ökologie der ostafrikanischen Mangroven. Z Bot 30:63–193

Walter H, Straka H (1970) Arealkunde. Floristisch-historische Geobotanik, 2nd edn. Stuttgart

Walter H, Volk H (1954) Die Grundlagen der Weidewirtschaft in SW-Afrika. Eugen Ulmer, Stuttgart

Walter H, Walter E (1953) Das Gesetz der relativen Standortskonstanz; das Wesen der Pflanzengemeinschaften. Ber Dtsch Bot Ges 66:227–235. Neudruck. In: Lauer W, Klink H-J (eds) Pflanzengeographie. Wiss Buchges, vol CXXX 1978. Darmstadt, pp 170–184

Walter H, Wiebe H (1966) Towards classification of the specific condition of water in protoplasm and in the cell wall of higher plants and its determination. Adv Front Plant Sci New Delhi 14:173–218

Walter H, Zimmermann W (1952) Ökologische CO_2-Absorptionsmessungen in verschiedenen Pflanzenbeständen. Z Bot 40:251–268

Weaver JE, Clements FE (1938) Plant ecology, 2nd edn

Wegener A (1936) Die Entstehung der Kontinente und Ozeane, 5th edn. Braunschweig

Weischet W (1977) Die ökologische Benachteiligung der Tropen. Teubner, Stuttgart

Weissenböck G (1969) Einfluß des Bodensalzgehaltes auf Morphologie und Ionenspeicherung von Halophyten. Flora B 158:369–389

Went FW, et al (1948, 1949, 1956) Ecology of desert plants I–IV. Ecology 29:242–353; 30:1–13, 26–38; 37:317–330

Went FW, Stark N (1968) Mycorrhiza. Bioscience 18:1035–1039

White LP (1970) Brousse tigrée patterns in Southern Niger. J Ecol 58:549–553

White LP (1971) Vegetation stripes on sheet wash surface. J Ecol 59:615–622

Whittaker RH (1970) The biochemical ecology of higher plants. In: Sandheimer E, Simeone JB (eds) Chemical ecology. Academic Press, New York, pp 43–70

Whyte RP (1976) Bioclimate and taxonomic consequences of tectonic movement and orogeny. Ann Arid Zone (Jodhpur) 15:247–269

Wilhelmy H (1954) Die klimamorphologische und pflanzengeographische Entwicklung des Trockengebiets am Nordrand Südamerikas seit dem Pleistozän. Die Erde (1954):244–273

Willert DJ von (1968) Tageswassermengen des Ionengehalts in *Salicornia europaea* in Abhängigkeit vom Standort und der Überflutung. Ber Dtsch Bot Ges 81:442–449

Wilmanns O (1978) Ökologische Pflanzensoziologie, 2nd edn. UTB 269, Heidelberg

Woodell SRJ, Mooney HA, Hill AJ (1968) The behaviour of *Larrea divaricata* (Creosote bush) in response to rainfall in California. J Ecol 57:37–44

Yunatov AA et al (1979) Vegetation Atlas of the Mongolian Peoples' Republic 1:1500000 (in Russian). Akad Nauk SSSR ai MNP, Moscow

Yurtsev BA (1972) Phytogeography of Northeastern Asia and the problem of transberingian floristic interrelations. In: Alan Grahaw (ed) Floristics and palaeofloristics of Asia and

Eastern North America. Amsterdam London New York, pp 24–30

Yurtsev BA (1974) Steppe communities of the Chukotka tundra and the "tundra steppe" of the Pleistocene (in Russian). Bot Z (Leningrad) 59:484–501

Yurtsev BA (1981) Relics of steppe complexes in northeast Asia (in Russian). Nauka, Novosibirsk

Zeller O (1951) Über Assimilation und Atmung bei tiefen Temperaturen. Planta (Berl) 39: 500–526

Zeller O (1973) Blührhythmik von Apfel und Birne im tropischen Hochland von Ceylon. Gartenbauwissenschaft 38:322–342

Zelniker JL (1968) Methods in ecophysiological investigations of plants in forest communities

(in Russian). Problems in Botany (Leningrad) X:5–18

Ziegler H, Vieweg H (1970) Poikilohydre Pteridophyten (Farngewächse). In: Walter H, Kreeb K (eds) Die Hydratation und Hydratur des Protoplasmas. Protoplasmatologia II, C, 6, Wien, pp 88–108

Zohary M (1952) Hydro-economic types in the vegetation of Near East deserts. Sympos biol and product of hot and cold deserts. London, pp 56–67

Zukrigl K, Eckhardt G, Nather J (1963) Standortskundliche und waldbauliche Untersuchungen in Urwaldresten der niederösterreichischen Kalkalpen. Mitt Forstl Bundesversuchsanst Mariabrunn, Heft 62. Wien

Subject Index

H. Walter

Vegetation of the Earth

and Ecological Systems of the Geo-biosphere

Translated from the 5th revised German edition by
O. Muise
Heidelberg Science Library
3rd revised and enlarged edition. 1985. 161 figures.
Approx. 275 pages. ISBN 3-540-13748-3

Contents: Introduction: The Classification of Ecological
Systems. – General Section. – Special Section. – Zonobiome
of the Equatorial Humid Diurnal Climate With Evergreen
Tropical Rain Forest. – Zonoecotone I/II – Semievergreen
Forest: Zonobiome of the Tropical Summer-Rain Region
With Deciduous Forests. – Zonoecotone II/III – Climatic
Savannas: Zonobiome of the Subtropical Arid Climate With
Deserts. – Zonoecotone III/IV – Semide sert: Zonobiome of
the Winter-Rain Region With an Arido-Humid Climate and
Sclerophyllic Woodlands. Zonobiome of the Warm-Temper-
ate Humid Climate. Zonobiome of the Temperate-Nemoral
Climate. – Zonoecotone VI/VII – Forest-Steppe: Zono-
biome of the Arid-Temperate Climate. – Zonoecotone VI/
VIII – Boreonemoral Zone: Zonobiome of the Cold-
Temperate Boreal Climate. – Zonoecotone VIII/IX: Forest
Tundra: Zonobiome of the Tundra Climate. – Summary. –
References. – Subject Index.

From the reviews of the 2nd edition:
"... His (the author's) world-wide experience and his origi-
nality speak to us on practically every page, with numerous
pointers for significant research that urgently needs attention. The book is therefore a powerful stimulant for all interested in ecological research. ... masterful, crisp style. ... a
concise, world-wide survey of the cause and effect chains of
ecological events in which vegetation necessarily plays a
dominant role. ... All ecologists will want to own this book;
certainly they ought to."   *Ecology*

"... This book provides an excellent insight into the world's
major vegetation types and the factors affecting vegetation.
It will prove invaluable to all those who in one way or
another are concerned with vegetation and its management."   *Commonwealth Forestry Review*

Springer-Verlag
Berlin
Heidelberg
New York
Tokyo

H. Remmert

Ecology

A Textbook
Translated from the German by M. A. Biederman-Thorson
1980. 189 figures, 12 tables. VIII, 289 pages. ISBN 3-540-10059-8

Contents: Ecology: the Basic Concept. – Autecology. – Population Ecology. – Ecosystems. – Outlook. – References. – General Books on Subjects Related to Ecology. – Subject Index.

The second edition of this outstanding textbook is now available in translation to English-speaking readers. Revised and expanded from the first edition, it brings into even greater focus the relationship between ecology and sensory physiology.

"We would expect an ecology textbook from one of Germany's best known ecologists to be highly original, and we certainly weren't disappointed. Professor Remmert has charged this overview of modern ecology with wit and expertise. Designed especially for students, it will prove equally valuabe for all those truly interested in the subject... The descriptions and choice of examples are extraordinarily original. This book will be a joy for the expert, a treasure-trove for the student, and food for the thought for environmental administrators."

translated from: *Anzeiger Ornithologische Gesellschaft*

H. Remmert

Arctic Animal Ecology

Translated from the German by J. Wieser
1980. 156 figures, 28 tables. VIII, 250 pages. ISBN 3-540-10169-1

Contents: Introduction: Delimitation of the Arctic. – Ecological Factors in the Arctic. – (Almost) Common Characteristics of Arctic Animals. – Peculiarities of the Systems. – Types of Arctic Climates. – Case Studies. – References. – Subject Index.

The heterogeneity of Arctic ecology has long fascinated researchers. At the same time, however, it has proved a hindrance to a synthetical approach to the study of northern polar regions. This book – the result of 15 years of ecological and physiological research in the Arctic – is the first to systematically describe factors typical to the area and their effect on the animal life there. Permanent light in summer, permanent darkness in winter, relatively constant temperatures and minimal differences in irridation values in the course of each 24-hour period are among the factors considered. An attempt is made to determine those characteristics common to all Arctic animals and the significance of these animals to Arctic ecosystems. Case studies exemplify the great variety of environments involved: from forests to Arctic desert, ice-free seas and permanent ice, from rivers that flow all year round to waters which appear only in summer. A brief review of conditions in the Antarctic is included to demonstrate the vast differences between the two polar regions.

Springer-Verlag
Berlin
Heidelberg
New York
Tokyo